CURRENT ORNITHOLOGY

VOLUME 1

Current Ornithology

Editorial Board

William R. Dawson, *Ann Arbor, Michigan*
Frances C. James, *Tallahassee, Florida*
Donald E. Kroodsma, *Amherst, Massachusetts*
Frank A. Pitelka, *Berkeley, California*
Robert J. Raikow, *Pittsburgh, Pennsylvania*
Robert K. Selander, *Rochester, New York*

A Continuation Order Plan is available for this series. A continuation order will bring delivery of each new volume immediately upon publication. Volumes are billed only upon actual shipment. For further information please contact the publisher.

CURRENT ORNITHOLOGY
VOLUME 1

Edited by
RICHARD F. JOHNSTON
University of Kansas
Lawrence, Kansas

PLENUM PRESS • NEW YORK AND LONDON

ISBN 0-306-41339-6

©1983 Plenum Press, New York
A Division of Plenum Publishing Corporation
233 Spring Street, New York, N.Y. 10013

All rights reserved

No part of this book may be reproduced, stored in a retrieval system, or transmitted
in any form or by any means, electronic, mechanical, photocopying, microfilming,
recording, or otherwise, without written permission from the Publisher

Printed in the United States of America

CONTRIBUTORS

JON E. AHLQUIST, Peabody Museum of Natural History and Department of Biology, Yale University, New Haven, Connecticut 06511

CYNTHIA CAREY, Department of EPO Biology, University of Colorado, Boulder, Colorado 80309

KENDALL W. CORBIN, Bell Museum of Natural History and Department of Ecology and Behavioral Biology, University of Minnesota, Minneapolis, Minnesota 55455

JOEL CRACRAFT, Department of Anatomy, University of Illinois, Chicago, Illinois 60680; and Field Museum of Natural History, Chicago, Illinois 60605

NORMAN L. FORD, Department of Biology, St. John's University, Collegeville, Minnesota 56321

M. GUMPEL-PINOT, Institut d'Embryologie, Nogent-sur Marne 94130, France

J. R. HINCHLIFFE, Zoology Department, University College of Wales, Penglais, Aberystwyth, Dyfed, SY23 3DA, Wales, United Kingdom

ELLEN D. KETTERSON, Department of Biology, Indiana University, Bloomington, Indiana 47405

LARRY D. MARTIN, Department of Systematics and Ecology and Museum of Natural History, University of Kansas, Lawrence, Kansas 66045

VAL NOLAN JR., Department of Biology, Indiana University, Bloomington, Indiana 47405

ROBERT E. RICKLEFS, Department of Biology, University of Pennsylvania, Philadelphia, Pennsylvania 19104

J. D. RISING, Department of Zoology, University of Toronto, Toronto, Ontario, Canada M5S 1A1

GERALD F. SHIELDS, Institute of Arctic Biology and Division of Life Sciences, University of Alaska, Fairbanks, Alaska 99701

CHARLES G. SIBLEY, Peabody Museum of Natural History and Department of Biology, Yale University, New Haven, Connecticut 06511

JEFFREY R. WALTERS, Department of Zoology, North Carolina State University, Raleigh, North Carolina 27607

DAVID W. WINKLER, Museum of Vertebrate Zoology and Department of Zoology, University of California, Berkeley, California 94720

PREFACE

The appearance of the first volume of a projected series is the occasion for comment on scope, aims, and genesis of the work. The scope of *Current Ornithology* is all of the biology of birds. Ornithology, as a whole-organism science, is concerned with birds at every level of biological organization, from the molecular to the community, at least from the Jurassic to the present time, and over every scholarly discipline in which bird biology is done; to say this is merely to expand a dictionary definition of "ornithology." The aim of the work, to be realized over several volumes, is to present reviews or position statements concerning the active fields of ornithological research. The reviews will be relatively short, and often will be done from the viewpoint of a readily-identified group or school.

Such a work could have come into being at any time within the past fifty years, but that *Current Ornithology* appears now is a result of events that are only seven to eight years old. One important event was the initiation in 1975–1976 of the Workshop on a National Plan for Ornithology, under the directorship of James R. King and Walter J. Bock, cosponsored by the American Ornithologists' Union and the National Science Foundation. Part of the Workshop's interests lay in publications resources, and certain kinds of information on publications were obtained by means of a questionnaire. Among other matters, most respondents thought that an annual or biennial review in ornithology would be a useful addition to the literature. At that time, five volumes of *Avian Biology*, edited by Donald S. Farner, James R. King, and Kenneth C. Parkes, had appeared, but the editors did not envision continuing the work on a regularly recurrent schedule. Some members of the Workshop consequently pursued prospects for initiating an annual volume and ultimately made an agreement with editors at Plenum Publishing Corporation for the production of the review.

Contents of Volume 1 concern aspects of ecology, systematics, behavior, physiology, and developmental biology. The interface of ecology and behavior is examined by Ford, looking at the degree to which conventional monogamous pair bonds are really monogamous, and Ketterson and Nolan, in an extended treatment of differential migration patterns. Ricklefs initiates the comparative study of avian demography, and shows that its development is in fact overdue. Winkler and Walters show that clutch size in precocial birds is at least as complex as that in altricial birds.

Martin presents support for the origin of birds from an ancestor in common with crocodilians, garnished with the hypothesis that flight originated from the behavior of arboreal jumpers. Carey treats the fundamentals of structure and function of eggs. Hinchliffe details recent developments in work on avian limb buds, with remarkable scanning electron micrographs. Shields summarizes recent work dependent on banding of bird chromosomes.

Sibley and Ahlquist discuss the techniques used in avian DNA hybridization studies, and demonstrate the use of such work to understanding avian phylogeny. Rising discusses another kind of hybridization, found in some of the birds living around the one-hundredth meridian of the North American Great Plains. Cracraft shows some of the consequences of entertaining alternative species concepts in the assessment of speciation in birds. Corbin presents the current picture of avian population genetics.

We are indebted to the members of the ornithological community who have suggested topics and authors for this and succeeding volumes of *Current Ornithology*. We hope suggestions continue, not only because any Editorial Board can use assistance, but also because the nature of the suggestions can provide us with a gauge of how well we serve our audience.

Richard F. Johnston

Lawrence, Kansas

CONTENTS

CHAPTER 1

COMPARATIVE AVIAN DEMOGRAPHY

ROBERT E. RICKLEFS

1. Introduction 1
2. Phenomenology and Hypotheses 2
3. Demographic Evolution 5
   3.1. Criterion for Fitness 6
   3.2. Parameterization 7
   3.3. Constraint and Optimization 8
   3.4. The Constraint Function 11
   3.5. The Optimization Criterion 14
   3.6. Interpreting Differences in Life-History Patterns 15
4. Density Dependence 18
5. Demographic Evolution in Varying Environments 22
   5.1. r- and K-Selection 23
   5.2. Bet Hedging 25
6. Correlated Environmental Factors 25
7. Discussion 26
   References 29

ix

CHAPTER 2

THE DETERMINATION OF CLUTCH SIZE IN PRECOCIAL BIRDS

DAVID W. WINKLER AND JEFFREY R. WALTERS

1. Introduction 33
2. Precocial Development 34
3. Clutch Size Theory and Terminology 36
4. Patterns in Clutch Size Variation 40
5. Factors Limiting Clutch Size 41
   5.1. Egg Formation Ability 42
   5.2. Parental Behavior 47
   5.3. Incubation Ability 50
   5.4. Nest Predation 52
   5.5. Explanations of Geographic and Seasonal Trends 53
6. Discussion 54
   6.1. The Web of Causation 54
   6.2. Levels of Explanation 55
   6.3. Complications 56
7. Conclusions 56
   References 61

CHAPTER 3

STRUCTURE AND FUNCTION OF AVIAN EGGS

CYNTHIA CAREY

1. Introduction 69
2. Mass and Contents of Eggs 71
   2.1. Mass and Contents of Fresh Eggs 71
   2.2. Contents of Pipped Eggs 74
3. Gas Exchange 75
   3.1. Exchange of Water 76
   3.2. Exchange of O_2 and CO_2 81
   3.3. Adaptation to Diverse Gaseous Environments 85
4. Gas Exchange and the Eggshell 87
   4.1. Eggshell 87
   4.2. Pores 92
   4.3. Modification of Eggshells in Various Environments 95
5. Summary 96
   References 98

CHAPTER 4

THE ORIGIN OF BIRDS AND OF AVIAN FLIGHT

LARRY D. MARTIN

1. Introduction 105
2. The Nature of *Archaeopteryx* 106
   2.1. Bipedalism 108
   2.2. Flight 108
3. The Origin of Birds 110
   3.1. The Pseudosuchian Origin 110
   3.2. Ornithischian Relationship 110
   3.3. Crocodilian Relationships 111
   3.4. The Coelurosaurian Origin 115
4. The Origin of Avian Flight 121
5. Conclusions 125
   References 126

CHAPTER 5

THE GREAT PLAINS HYBRID ZONES

J. D. RISING

1. Introduction 131
   1.1. Delimiting the Plains 133
   1.2. What is Hybridization? 133
2. Accounts of Hybridizing Taxa 134
   2.1. *Otus* 134
   2.2. *Colaptes* 134
   2.3. *Centurus* 137
   2.4. *Myiarchus* 137
   2.5. *Contopus* 138
   2.6. *Cyanocitta* 138
   2.7. *Parus* (Chickadees) 138
   2.8. *Parus* (Crested Titmice) 139
   2.9. *Sialia* 139
   2.10. *Sturnella* 140
   2.11. *Icterus* 140
   2.12. *Pheucticus* 145

2.13. *Passerina* 146
2.14. *Pipilo* 148
3. Discussion 149
   3.1. Stability of Zones 149
   3.2. Increased Variability in Zones 152
   3.3. Suture-Zones in the Great Plains 153
   3.4. Taxonomic Comments 155
   References 155

CHAPTER 6

SPECIES CONCEPTS AND SPECIATION ANALYSIS

JOEL CRACRAFT

1. Introduction 159
2. Species Concepts 161
   2.1. The "Biological Species" Concept 161
   2.2. Are "Biological Species" the Units of Evolution? 162
   2.3. A Proposed Species Concept for Ornithology 165
3. Speciation Analysis 174
   3.1. Introduction 174
   3.2. How Are Areas of Endemism Determined? 177
   3.3. How Is the History of Areas of Endemism Determined? ... 178
   3.4. How Might General Area-Cladograms Be Explained? 181
   3.5. Conclusions 184
   References 184

CHAPTER 7

BIRD CHROMOSOMES

GERALD F. SHIELDS

1. Introduction 189
2. The Diploid Number Problem 190
   2.1. Utility of Cell Culture Procedures 190
   2.2. Meiotic Procedures 193
3. Differential Banding Procedures 193
   3.1. C-Banding 193
   3.2. G-Bands 197
   3.3. R-Banding 205

3.4. NOR-Banding 206
3.5. Sequential Banding 206
4. Concluding Remarks 208
   References 208

CHAPTER 8

GENETIC STRUCTURE AND AVIAN SYSTEMATICS

KENDALL W. CORBIN

1. Introduction 211
2. Methods of Data Acquisition and Analysis 212
   2.1. Electrophoretic Studies 213
   2.2. Hardy–Weinberg Equilibria 214
   2.3. Genic Heterozygosity 214
   2.4. Genetic Distance between Taxa 220
   2.5. Analysis of Genetic Variance 221
3. Genetic Data and Their Role in Systematics 224
   3.1. Relationship between Genetic Structure and
        Higher Taxa 227
   3.2. Genetic Structure at the Species Level 237
   3.3. Genetic Structure within Species 239
4. Conclusions 241
   References 242

CHAPTER 9

PHYLOGENY AND CLASSIFICATION OF BIRDS BASED ON THE DATA OF
DNA–DNA HYBRIDIZATION

CHARLES G. SIBLEY AND JON E. AHLQUIST

1. Introduction 245
2. The DNA–DNA Hybridization Technique 246
   2.1. DNA Structure and Properties 246
   2.2. The Sequence Organization of the Genome in Relation to
        DNA–DNA Hybridization 247
   2.3. Factors That Determine the Rate and Extent of
        Reassociation, and the Thermal Stability of DNA–DNA
        Hybrids 251

2.4. The Hydroxyapatite (HAP) Column Chromatography
    Procedure 255
3. The Analysis of DNA–DNA Hybridization
   Data 256
    3.1. Definitions 256
    3.2. The Average Rate of Base Substitution 260
    3.3. Mathematical Analysis of DNA–DNA Data 260
    3.4. A "Robust" Clustering Method for DNA Hybridization
         Data 261
    3.5. Reliability, Sensitivity, and Reciprocity 264
4. Homology 267
5. The Evolution of DNA 270
    5.1. The Uniform Average Rate of DNA Evolution 270
    5.2. The New Zealand Wrens (Acanthisittidae) 272
    5.3. The Hawaiian Honeycreepers 274
    5.4. The Australo-Papuan Fairy-Wrens (Maluridae) 274
    5.5. Genetic Rate vs. Morphological Rates 275
    5.6. Congruence between Morphological and DNA–DNA
         Hybridization Data 278
    5.7. Generation Time 279
6. The Calibration Problem 281
7. Adaptive Radiation 282
8. Categorical Equivalence 285
9. Classification 286
   References 288

CHAPTER 10

EXPERIMENTAL ANALYSIS OF AVIAN LIMB MORPHOGENESIS

J. R. HINCHLIFFE AND M. GUMPEL-PINOT

1. Introduction 293
2. Mapping the Prospective Skeletal Areas 298
3. Regulation along Proximo-distal and Antero-posterior Axes ... 298
4. The ZPA and Experimental Analysis of Control of
   Antero-posterior Differentiation of the Wing 303
    4.1. The ZPA and Normal Development 309
    4.2. The Polar Coordinate Model and the Chick Wing Bud 310
5. Somite and Limb Bud Developmental Relations 313
    5.1. Developmental Dependence 313
    5.2. The Somitic Contribution of the Musculature 314

6. Innervation and Limb Development 315
    6.1. Role of Limb Pattern in Establishing Innervation 316
    6.2. Relationship between Neurons and Peripheral
         Conditions 318
    6.3. Relationship between Axons and Their Targets 318
    6.4. Innervation and the Pattern of Skeletal Differentiation 319
7. Conclusions 320
   References 321

CHAPTER 11

VARIATION IN MATE FIDELITY IN MONOGAMOUS BIRDS

NORMAN L. FORD

1. Introduction 329
2. Which Birds Are Monogamous? 330
3. Types of Variations in Mate Fidelity 332
    3.1. Extrapair Copulations 332
    3.2. Circumstantial Evidence for an EPC Strategy 334
4. Opportunistic and Facultative Polygyny 340
    4.1. North American Passerines 341
    4.2. European Passerines 344
    4.3. Non-Passerines 345
5. Polyandry 346
6. Conclusion 348
   References 351

CHAPTER 12

THE EVOLUTION OF DIFFERENTIAL BIRD MIGRATION

ELLEN D. KETTERSON AND VAL NOLAN JR.

1. Introduction 357
2. Winter Distribution of Eastern Migratory Juncos 360
3. Single-Factor Hypotheses for the Evolution of Differential
   Migration 362
    3.1. The Body-Size Hypothesis 362
    3.2. The Dominance Hypothesis 365
    3.3. The Arrival-Time Hypothesis 370

4. A Multifactor Hypothesis for the Evolution of Differential
   Migration 375
   4.1. The Migration-Threshold Hypothesis 375
5. Conclusions 388
   References 399

AUTHOR INDEX 403
BIRD NAME INDEX 411
SUBJECT INDEX 421

CHAPTER 1

COMPARATIVE AVIAN DEMOGRAPHY

ROBERT E. RICKLEFS

1. INTRODUCTION

Demography is the study of the life tables of populations which summarize probabilities of survival and rates of fecundity of individuals at each age. The particular values of these parameters for each population depend upon both the adaptations of individuals and attributes of the environments in which they live. Life-table parameters vary greatly within populations according to season, year, and individual differences, and they vary among populations according to locality, habitat, and ecological role. In some cases, the correlations of life-table parameters among themselves and with certain environmental factors have suggested mechanisms by which their values are determined.

Comparative demography makes use of such patterns to elucidate the ecological and evolutionary responses of populations to their environments (Bell, 1980). By means of suitable comparisons, one can isolate certain factors or groups of factors in the environment and estimate their effect on the life tables of populations. It is implicit in such comparisons that populations do not differ intrinsically in their manner of response to the environment; otherwise, attributes of organisms and attributes of environments would be uniquely related in each popu-

---

ROBERT E. RICKLEFS • Department of Biology, University of Pennsylvania, Philadelphia, Pennsylvania 19104.

lation and one could learn nothing by comparisons among them. Concern over this possible difficulty led Lack (1947, 1954) to urge that comparisons be made among close relatives, preferably among populations within species.

Because it is difficult to perform controlled experiments on natural populations, particularly when potential responses include evolutionary change, comparative studies have provided most of our insights into the relationships of populations to their environments. For birds, especially, comparative demographic studies have played a major role in the development of theories on the origin of life-history patterns.

David Lack must be considered the father of comparative avian demography, although he did not fully utilize available techniques of demographic analysis. Baker (1938), Moreau (1944), and others had contemplated patterns of variation in life-history traits, but Lack was the first to realize the potential contribution of the study of life-history phenomena to understanding the nature of evolutionary adaptation. Following Lack's (1947, 1948) papers on the significance of variation in clutch size, others have greatly expanded upon and elaborated his ideas. At present, we are confronted by many plausible mechanisms that relate life histories to the environment (see Stearns, 1976, 1977, 1980), but comparative studies have provided us with relatively little power to distinguish the roles of these mechanisms in generating observed patterns. As ecologists and evolutionists turn more frequently to experimental studies, there is growing need of a solid theoretical framework for the design of experiments and interpretation of results.

I have undertaken this essay to formalize the comparative approach to avian demography, particularly to place comparative demography in the context of evolutionary and population dynamics. Above all, I hope to provide a conceptual foundation for theoretical, comparative, and experimental efforts to understand the influence of factors in the environment on the diversification of avian life tables.

2. PHENOMENOLOGY AND HYPOTHESES

The life table lists those parameters, namely age-specific fecundity and survival, required to project population change through time. Fecundity and survival are abstractions that express all adaptations of form and function in their environmental setting. Fecundity is the product of clutch size, nesting success, and number of clutches attempted per season, which in turn depend upon parental behavior, development rate, and nest construction among the many other traits of the organism, and reflect food supply, weather, and the activities of predators among the many pertinent factors in the environment. The annual probability of survival by adults is influenced by patterns of seasonal migration, deposits of body fat, and the allocation of time and energy to reproduction, as well as by weather, predation, and food supply.

Patterns in life-history traits can be generalized by three statements. First, fecundity and survival vary widely among populations. Second, variation in life-history traits can be associated with variation in the environment. For example, the number of eggs laid tends to increase in direct relation to distance from the Equator, presumably in response to some climatological or biological factor or factors that vary consistently with latitude. Third, many life-history traits appear to vary together. The association of low fecundity, delayed reproduction, and long life is a well-documented example (Lack, 1968; Goodman, 1974; Nelson, 1977).

Beginning principally with Lack (1947), ecologists have attempted to interpret such patterns in ecological and evolutionary terms. Lack was concerned with the number of eggs laid by birds (clutch size), which he believed to be adjusted to the number of offspring parents can nourish. This hypothesis suggested to Lack that seasonal and geographical variation in clutch size is determined by the amount of food available to the parents.

In 1949, A. F. Skutch proposed, alternatively, that clutch size is adjusted in relation to the intensity of predation upon nests. Skutch suggested that predators might detect nests containing large broods more readily than they discover less noisy and less active nests having fewer offspring. Both Lack and Skutch sought to relate a single life-history trait to a factor in the environment. By their explanations for latitudinal gradients in clutch size, both argued that variation in the expression of a particular trait is caused overwhelmingly by variation in factors in the environment to which the trait responds directly, either by phenotypic adjustment or by evolutionary change, and that are independent of the activities of the population. Hence variation in clutch size could be explained by factors, food or predation, whose levels are intrinsic properties of the environment.

Ashmole (1963) recognized, however, that life-history traits can be interrelated ecologically through the effects of populations on their environments. In a population maintained at or near a constant level by density-dependent factors, adult mortality balances recruitment of young. A change in one necessarily results in a compensating response in the other. Hence, variation in clutch size might originate in factors that influence adult survival. Ashmole saw that differences in the reproductive rates of populations might be brought about by factors in the environment unrelated, or related only indirectly, to resource gathering. That is to say, variation in one trait (clutch size) might be coupled through density-dependent factors to variation in another trait (adult survival). A consideration not often recognized is that environmental factors influencing a population may be intrinsically correlated independently of the population, a point to which I shall return.

Cody (1966) introduced the principle of allocation to thinking about avian life-history traits. Every individual has a fixed amount of time at its disposal to distribute among its various activities. Favoring one activity, e.g., foraging, at the expense of another, e.g., preening, predator detection, may alter life-table parameters and result in a change in fitness if the life-table functions are linked. The principle of allocation dictates that individuals must operate within certain limits, that a decision concerning the particular activity of the moment may trade one life-history trait off against another, and that all such decisions are compromises.

Theoretical studies have argued that the optimum allocation (with respect to fitness) of time and resources among activities is influenced by the demography of the population as well as by factors in the environment that influence each life-history trait directly. Demographic properties include the rate of population growth (Cody, 1966; MacArthur and Wilson, 1967), population variability (Murphy, 1968), and expected future reproduction (Williams, 1966; Gadgil and Bossert, 1970; Cody, 1971; Charlesworth, 1973; Leon, 1976; Michod, 1979).

The complex ecological and evolutionary coupling of the population to its environment allows a great variety of explanations for variation among species in life-history traits. For example, latitudinal variation in clutch size has been considered in relation to (1) proximate environmental factors having direct effect upon clutch size, such as predation (Perrins, 1977; Skutch, 1949; Snow, 1970), abundance of food (Ricklefs, 1968), daylength (Lack, 1947, 1968), temperature (Royama, 1969), diversity of prey (Owen, 1977), and antipredator adaptations (Ricklefs, 1970), (2) factors acting indirectly through density-dependent feedback (Ashmole, 1963; Lack and Moreau, 1965; Ricklefs, 1980a) and overlap of molt and breeding (Foster, 1974), and (3) selection of the level of reproductive effort and allocation of resources in accordance with population density (r- and K-selection; MacArthur and Wilson, 1967), expectation of future reproduction (Williams, 1966; Gadgil and Bossert, 1970; Schaffer, 1974a; Goodman, 1974), and predictability of the environment (Goodman, 1979; Hastings and Caswell, 1979; Hirschfield and Tinkle, 1975; Murphy, 1968; Schaffer, 1974b). Many of these explanations have been discussed by Stearns (1976, 1977, 1980).

In the absence of experiments, we must rely on comparative studies to provide insights into such patterns as the relationship between clutch size and latitude. Tests of the hypothesis that such variation results from selection by factors that affect a trait directly rely upon correlations of the trait with environmental factors. Lack (1947) saw a correlation between clutch size and day length during the breeding season, and Skutch (1949) found clutch size related to nest predation. Tests of hypotheses based on density-dependence and demographic optimization rely on correlations among traits, such as between clutch size and population density (Ricklefs, 1980) or between clutch size and adult mortality (Cody, 1971; Goodman, 1974; Ricklefs, 1977b). The weakness of the comparative approach is that correlations are not uniquely predicted by one hypothesis, but may be equally well-explained by several alternative hypotheses. For example, hypotheses based upon density-dependent feedbacks among traits and upon evolutionary optimization of traits similarly predict an inverse relationship between fecundity and adult survival rate.

The challenge of comparative demography is to understand in general the causes of differences in the life tables or life-history traits of populations in terms of the environmental factors responsible, the manner in which these factors are influenced by the population, and the intrinsic constraints on phenotypic response imposed by the structure and function of the organism. Although the comparative approach is not fully adequate to solve these problems, it can yield useful information if its results are interpreted in the context of a complete theory of comparative demography. I shall begin with a discussion of demographic evolution as a basis for interpreting adaptations in terms of life-table parameters.

3. DEMOGRAPHIC EVOLUTION

The idea that life-history parameters are subject to evolutionary change presupposes that under particular conditions individuals having a particular value for a life-history trait are inherently more fit than others. By greater fitness we mean that the descendents of selected individuals are more numerous than those of other individuals having less fit traits. One approach to understanding the evolution of life-history patterns requires five steps (see Bell, 1980; Leon, 1976). First, we must adopt some measure of fitness having the property that individuals with higher values tend to leave more offspring than those with lower values. Second, we must describe life histories by objective and quantitative parameters that are measurable, at least in principle.

Third, we must derive expressions that relate fitness to life-history parameters. Unless the fitness criterion can be mapped onto the parameter space, it will not be possible to understand the evolution of patterns in life-history traits. Fourth, we must take into account that certain values and combinations of values of life-history parameters are not possible owing to mathematical or biological considerations. Probabilities of survival fall between zero and one. Infinite fecundity, although allowable in the parameter space, is not reasonable, nor is instantaneous development. More importantly, certain combinations of traits may be unreasonable if they cannot be attained owing to limited time or resources. The principle of allocation places certain restrictions on the permissible combinations of values within the total parameter space, and measuring these constraints is perhaps the greatest difficulty to understanding life-history patterns. Fifth, and last, we must adopt a criterion for choosing the most fit combination of life-history values under a particular set of conditions. Throughout this discussion I assume that phenotypic variation has a sufficiently large genetic component that selection results in adaptive change, and that the mechanism of heredity itself does not restrict phenotypic variation in any way.

When a life-history pattern is analyzed by these steps, it should be possible to evaluate how a change in the environment of a population, expressed either as a change in the value of some life-history parameter or as a shift in the boundaries of the permissible parameter space, influences the optimum (= most fit) life-history pattern. It is not my purpose here to fully characterize all aspects of the life histories of birds in this manner. But I shall illustrate this approach in some detail to demonstrate its application to questions concerning life-history patterns.

3.1. Criterion for Fitness

At the level of the gene, evolution by natural selection (Darwinian evolution) is the substitution of one allele by a more fit alternative with a different phenotypic expression. By definition, that allele which increases most rapidly in a population, by virtue of the fecundity and survival of its bearers, is the most fit. Population geneticists have derived expressions to define evolutionary fitness in terms of fecundity and viability in populations with discrete, nonoverlapping generations, such as those of annual plants and many insects with one generation per year (see Wilson and Bossert, 1971; Roughgarden, 1979; Charlesworth, 1980). In these models, evolutionary changes in the frequencies of alleles can be calculated directly from their fitness values.

Such simple models are not strictly applicable to birds because each generation overlaps prior and subsequent generations. That is, parents and offspring may occur together as reproducing adults in the same population. As each cohort of individuals ages, gene frequencies may change as less fit individuals are selectively removed from the population. Because of this complication, as well as the presence of two sexes whose life tables may differ, it has not been possible to write an explicit equation for changes in gene frequencies under selection, and thereby verify that any particular criterion for fitness is consistent with projected changes in allele frequencies. Using a model that is realistic for birds, Charlesworth (1970, 1973) has shown, however, that under weak selection the geometric rate of increase calculated from the life-table parameters of a group of individuals bearing a particular gene provides a good estimate of evolutionary fitness. Charlesworth further demonstrated that even under strong selection, relative values of fitnesses calculated in this manner provide a true rank ordering of the fitnesses of alleles and correctly predict the outcome of selection (see Charlesworth, 1980).

3.2. Parameterization

The geometric mean fitness of a population or a subset of a population may be calculated from age-specific schedules of fecundity and survival (see Wilson and Bossert, 1971). In the following discussion, the variable b_x is the expected number of female offspring of a female of age x. Males are left out of the calculations because of difficulties in ascertaining their fecundity, except in the most strictly monogamous situations. The variable l_x is the probability that an individual survives from birth to breed at age x. This survivorship is the product of the annual survival rates from the first year following birth (s_0) to the year leading up to age x (s_{x-1}), that is,

$$l_x = \prod_0^{x-1} s_i. \qquad (1)$$

The values of s_x, l_x, and b_x are conveniently set out in a life table, such as that of the Screech Owl (*Otus asio*) in Table I.

Life tables vary in response to the environment and with the genotype of the individual. When values of l_x and b_x remain constant over time, the population and each of its age classes assume a constant and identical geometric rate of growth λ, which is related to the life-table parameters according to the expression

TABLE I
Life Table of the Screech Owl (*Otus asio*) in Ohio[a]

Age (years)[b]	s_x	l_x	b_x[c]	$l_x b_x$	$x\lambda^{-x} l_x b_x$
0	0.305	1.000	0.00	0.000	0.000
1	0.594	0.305	1.04	0.317	0.317
2	0.632	0.181	1.30	0.235	0.470
3	0.667	0.115	1.30	0.150	0.450
4	0.750	0.086	1.30	0.112	0.448
5	0.750	0.064	1.30	0.083	0.415
6	0.750	0.048	1.30	0.062	0.372
7		0.036	1.30	0.047	0.329
Totals				1.006	2.801

[a]From VanCamp and Henny, 1975.
[b]Life table is arbitrarily truncated at 7 years, the age of the oldest individual recovered in the study.
[c]Fecundity (b_x) is the number of offspring reared divided by two because only females are considered in the life table; assumes that the sex ratio at fledging is 1:1.

$$1 = \sum_{0}^{\infty} \lambda^{-x} l_x b_x \qquad (2)$$

(see Lotka, 1956). The form of Eq. (2) is such that it cannot be solved analytically for λ, the value of which must be determined recursively. Nonetheless, Eq. (2) explicitly relates a reasonable measure of evolutionary fitness and the life-table parameters l_x and b_x, given certain assumptions about population structure and constancy of the environment. As we shall see below, when the life-table parameters are related to each other by explicit mathematical functions, Eq. (2) can be utilized to determine an optimized life-history pattern.

In the population described by Table I, births and deaths are nearly balanced and the value of λ that satisfies Eq. (2) is close to 1. Any change or combination of changes in values of l_x and b_x will cause a change in the value of λ for the population described by the life table.

3.3. Constraint and Optimization

Eq. (2) relates fitness λ to the life-table parameters. Selection acts to increase evolutionary fitness and therefore, in the absence of other influences and after the gene pool of the population has achieved an evolutionary equilibrium, the fitness of the selected genotype is the highest among the fitnesses of available alternatives. If life-table parameters s_x and b_x were biologically independent, selection would favor individuals with the largest values of each in the population and, so long as genetic variation were available, these parameters would increase. But it is more reasonable to suppose that the life-table parameters may be interrelated by biological constraints. These may dictate, for example, that an adaptation resulting in increased fecundity might also increase the risk of death of breeding adults. With such balances acting among the life-table parameters, selection seeks the particular combination of traits having the highest fitness. We shall call that combination the optimum life-history pattern. The mathematical steps to finding the optimum involve (1) writing an expression relating fitness to the life-history parameters, thereby making them variables in an explicit function, (2) differentiating the expression with respect to a particular trait or combination of traits, and (3) solving the equation at its maximum (derivative of λ with respect to the life-table variables = 0).

Formally, these steps begin with a function $f(\lambda, y_1, y_2, ...) = 0$ where the variables $y_1, y_2, ...$ are the life-history traits (l_xs, b_xs) for each age. Because many traits change simultaneously during the evolutionary adjustment of life-history patterns, the derivative of λ must be expressed either with respect to some dummy variable (E) which represents a set of simultaneous changes in traits (Schaffer, 1974a), or as a difference equation. These approaches differ primarily in notation. In the first case, the differential equation is

$$\frac{d\lambda}{dE} = \frac{dy_i}{dE}\frac{\partial \lambda}{\partial y_1} + \frac{dy_2}{dE}\frac{\partial \lambda}{\partial y_2} + \cdots \qquad (3)$$

in which the terms dy_i/dE relate changes in each life-table variable to a life-history trait E, and the terms $\partial \lambda/\partial y_i$ relate fitness to each of the life-table variables. One may envision the variable E as some trait, such as number of hours spent searching for food or size of nest, that influences one or more of the life-table variables l_x or b_x. The relationships described by the terms dy/dE relate changes in each variable to changes in life-history traits. A larger nest might provide increased thermal insulation and allow parents to feed more offspring with the same amount of food ($db_x/dE > 0$), but it might also make the nest more conspicuous to predators and increase the risk of death to the brood ($db_x/dE < 0$) or to the parent ($ds_x/dE < 0$).

I prefer to relate life-history parameters to each other directly in a difference equation, and will follow this approach here. The principal advantage is in deleting the dummy variable E in the mathematical analysis. We recognize, of course, that the life-table variables are not themselves adaptations, but rather they express the interaction between some adaptation (E) and the environment. We relate changes in λ to changes in the life-table variables by

$$\Delta\lambda = \Delta y_1 \frac{\partial\lambda}{\partial y_1} + \Delta y_2 \frac{\partial\lambda}{\partial y_2} + \cdots \qquad (4)$$

where Δ signifies "an arbitrarily small change in." We may express the biological constraints or restrictions interrelating the y's by

$$\Delta y_1 = \Delta y_2 \frac{\partial y_1}{\partial y_2} + \Delta y_3 \frac{\partial y_1}{\partial y_3} + \cdots . \qquad (5)$$

The terms $\partial y_i/\partial y_j$ represent the relationship between changes in the variables y_i and y_j determined by a function $f(y_1,y_2, ...) = 0$ that describes a surface bounding possible phenotypes in the population.

Starting from Eq. (2), $\Delta\lambda$ (Eq. 4) may be expressed as

$$\Delta\lambda = \frac{\sum_0^\infty \lambda^{-x}(l_x \Delta b_x + b_x \Delta l_x)}{\lambda^{-1}\sum_0^\infty x\lambda^{-x}l_x b_x} \qquad (6)$$

or, in terms of Δs_x's rather than Δl_x's, as

$$\Delta\lambda = \frac{\sum_0^\infty \lambda^{-x}l_x \Delta b_x + \sum_0^\infty s_x^{-1} \Delta s_x \sum_{x+1}^\infty \lambda^{-i}l_i b_i}{\lambda^{-1}\sum_0^\infty x\lambda^{-x}l_x b_x} \qquad (7)$$

(see Hamilton, 1966; Ricklefs, 1981b). The relationships between values of λ and values of s_x and b_x are illustrated for b_2 and s_2 of the Screech Owl in Fig. 1. Each diagonal line in the lefthand graph describes all the pairs of values of b_2 and s_2 that result in the same value of λ (0.8, 0.9, etc.). Note that for $b_2 = 1.30$ and $s_2 = 0.632$, the values observed in the population, λ is approximately 1.0. A change in one of the life-table parameters causes a predictable change in λ, as illustrated in the right-hand graph. If b_2 were to increase from 1.3 to 1.8, one would substitute $\Delta b_2 = 0.5$ into Eq. (7), and solve to obtain $\Delta\lambda = 0.032$ (all terms other than that containing Δb_2 are 0). A similar calculation shows that for $\Delta s_2 = 0.10$, for example, $\Delta\lambda = 0.026$.

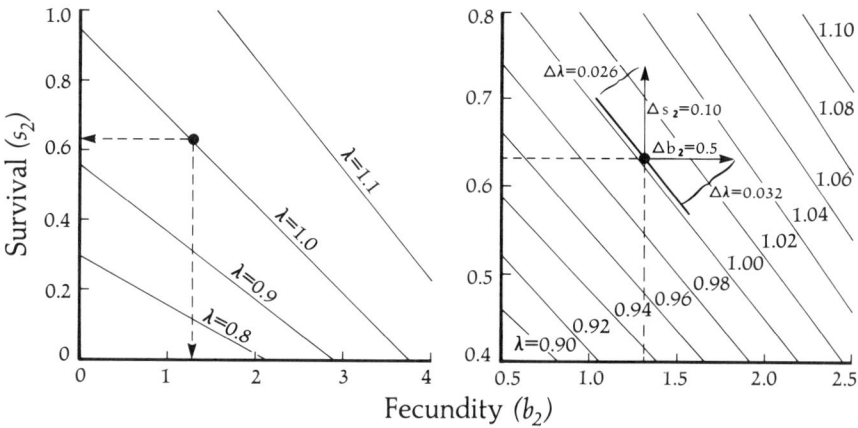

FIGURE 1. The relationship of geometric growth rate to combinations of values for survival rate s_2 and fecundity b_2 in the life table of the Screech Owl (Table I). The curves of equal fitness ($\lambda = 0.8, 0.9, \ldots$) are straight in comparisons between s's and b's or between b's at different ages, but are hyperbolas in comparisons between s's at different ages. At left, the solid circle and arrows indicate the observed life-history parameters for the Screech Owl. At right, the arrows indicate changes in geometric growth rate resulting from small increases in s_2 and b_2.

3.4. The Constraint Function

All other things being equal, an increase in either b_2 or s_2 results in an increase in λ. But biological considerations will likely cause b_2 and s_2 to be inversely related by some constraint function $f(b_2, s_2)$, such as that shown at the left of Fig. 2. The fitness of a combination of b_2 and s_2 on the constraint function is defined by $f(\lambda, b_2, s_2)$, which is the intersection of the constraint function and the family of lines each defined by a constant value of λ. These produce the relationship between λ and b_2 shown at the right of Fig. 2. The constraint function indicates a maximum of λ at a value of $b_2 = 2.8$ and $s_2 = 0.6$. Any other combination of values of b_2 and s_2 along the constraint function results in a lower value of λ.

The nature of the optimized life history depends in part on the shape and position of the constraint function. In the example in Fig. 2, the function $f(\lambda, b_2, s_2)$ has a single maximum. But one may conceive of other functions having more than one realistic maximum, a maximum at infinite values of each parameter, or an unrealistic maximum at which the values of one or more parameters are negative or imaginary. Multiple fitness peaks may result from discontinuities in phenotypic or environmental variables, or from phenotype–environment relation-

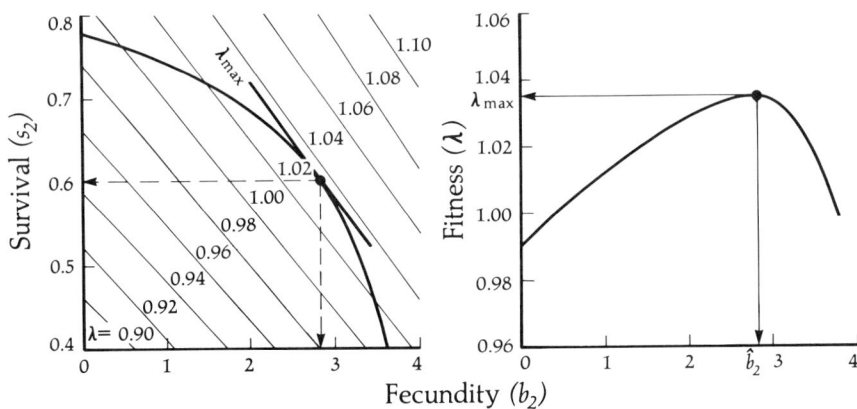

FIGURE 2. At right, the relationship between fitness (geometric growth rate) and b_2 when the relationship between b_2 and s_2 is determined by the constraint function shown by the curve at left. The line of equal geometric growth rate tangent to the constraint function determines the maximum fitness and the optimized values of b_2 and s_2.

ships that result in U-shaped fitness curves having maxima at extreme phenotypic values and a minimum in between.

The life-history pattern is molded primarily by negative relationships between life-table variables, for which $\partial y_j/\partial y_i < 0$; positive relationships ($\partial y_j/\partial y_i > 0$) require no evolutionary compromise. Negative relationships can take several forms, three of which are illustrated in Fig. 3. In all three, the variables are negatively related, i.e., $\partial y_j/\partial y_i < 0$. The curve at the left bulges outward from the origin of the graph and has a point of maximum fitness at some intermediate value of y_i and y_j. The shape of the curve is referred to as concave, meaning that the negative slope relating y_j and y_i at any point along the curve increases with increasing values of y_i, i.e., the second derivative of the function ($\partial^2 y_j/\partial y_i^2$) is less than 0.

The constraint curve at the center is convex ($\partial y_j/\partial y_i < 0$, $\partial^2 y_j/\partial y_i^2 > 0$). The points of maximum fitness on the constraint function can occur at either the extreme value of y_i or that of y_j; all intermediate phenotypes are less fit than one or the other extreme. Although one of the maxima may define a larger value of λ than the other, selection may not be able to reach it under some circumstances. If there were little phenotypic variation about the mean point (b_2, s_2) of the population, selection would cause the population to move away from the point of minimum fitness, always climbing toward higher values of λ regardless of the final value of λ achieved. If the population were polymorphic for b_2 and s_2, i.e., some individuals bred at age two and others did not, then the higher of the two maxima of λ might be achieved.

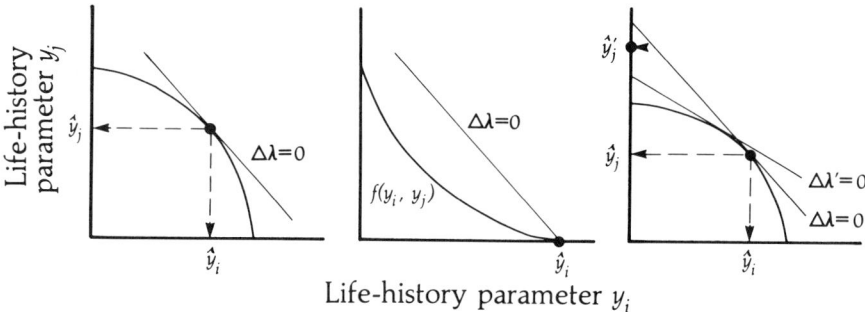

FIGURE 3. Three shapes of constraint functions relating life-history parameters y_i and y_j. The optimum is indicated in each case for the particular tangent lines drawn. Left, concave function with single maximum; center, convex function with a fitness peak at each intercept; right, complex curve with fitness peak at one intercept and a maximum determined by a concave function. See text for explanation.

In practice, few circumstances would normally result in a convex constraint function for life-history traits. Perhaps if the value of y_i were increased by extending a particular behavior over time, as by increasing the number of clutches attempted each season, then accumulated experience could reduce the impact of that behavior upon a second parameter y_j, say reproductive risk, and produce a convex function.

Concave functions express the "law of diminishing returns" whereby further investment nets progressively smaller benefits at increasing cost. For example, to increase clutch size from two to three may bring substantial increase in b_x at little risk (Δs_x), whereas the increase from, say six to seven eggs may so strain the female's physiology, because it is difficult to obtain the extra food needed to feed the nestlings, as to greatly increase her risk of death. In contrast, convex constraint functions (Fig. 3, center) imply that an increment in one variable has less effect on others as the value of the first increases.

Discontinuous constraint functions such as that at the right of Fig. 3, may result from discontinuities in the relationship between adaptations and life-table functions. In particular, the difference between not attempting to breed at a particular age and going through the behaviors prerequisite to any level of breeding, e.g., territory defense, nest building, no matter how small b_x, may cause a step change in s_x between $b_x = 0$ and $b_x > 0$. The function at the right of Fig. 3 has one point at $b_2 = 0$, at which s_2 represents the nonreproductive mortality; for all $b_2 > 0$, the minimum cost of breeding produces a step reduction in s_2. As b_2 increases, s_2 decreases further as a concave function representing the direct effect of increased effort to rear a larger brood (see Fig. 3, left).

In principle, the values of life-table functions at each age are influenced by those at the same or previous ages, although some constraints are more likely to be important than others. In a world of limited time and resources, reproduction (b_x) may be accomplished at the expense of self-maintenance ($\partial s_y/\partial b_x < 0$), or growth and storage ($\partial b_{y>x}/\partial b_x < 0$, perhaps $\partial s_y/\partial b_x < 0$), both of which could enhance fecundity and survival at later ages. Adaptations that enhance survival at a particular age (s_x) may be accomplished at the expense of reproduction at that age ($\partial s_s/\partial b_x < 0$), or growth and storage at a later age ($\partial b_{y>x}/\partial s_x < 0$, $\partial s_{y>x}/\partial s_x < 0$).

In birds, growth after maturity is negligible and stored materials, whether in the body or in food caches, are usually utilized on a short-term basis and rarely carried over from one age class to the next. Hence, tradeoffs between age classes may play a smaller role in birds than in groups in which growth continues after the onset of reproduction or nutrients are stored for long periods (see Hamilton, 1966).

For birds, the most important life-history constraint may be the relationship between fecundity and survival during a particular year, $f(b_x, s_x)$. Ricklefs (1977b) summarized evidence that mortality rates increase during the reproductive season. In some species, reserves of energy or nutrients are related to the initiation of breeding (Jones and Ward, 1976; Newton, 1977) and are depleted during the reproductive period (Breitenbach and Meyer, 1959; Johnson and West, 1973; Krapu, 1981; Ankney and MacInnes, 1978). Adult mortality has been related directly to breeding by Askenmo (1979) and Ainley and DeMaster (1980). Although constraints may be inferred from such observations, it may be impossible to characterize such relationships between life-history variables in detail. In order to understand the evolutionary optimization of life-history patterns, we may have to rely on indirect methods of analysis.

3.5. The Optimization Criterion

The optimum point on the constraint function $f(y_i, y_j)$ relating life-table variables y_j and y_i occurs at the tangent of the lines of equal fitness, as shown in Fig. 2. In practice, this optimum is difficult to analyze because the function $f(y_i, y_j)$ is unknown. But if we assume that the relationship between y_i and y_j is adjusted by evolution so as to maximize λ, then Eq. (7) and observed values for life-history parameters allow one to calculate at least the slope of the function (dy_j/dy_i) at its optimized point. Assuming λ is maximized at the optimum life-history pattern, $\Delta\lambda = 0$, and Eq. (7) may be written

$$\sum \lambda^{-x} l_x \Delta b_x + \sum s_x^{-1} \Delta s_x \sum_{x+1}^{\infty} \lambda^{-i} l_i b_i = 0. \tag{8}$$

To illustrate the calculation of dy_j/dy_i from Eq. (8), let us assume that the constraint of interest is between fecundity and adult survival and that the two are related only within age classes. Hence, $\Delta s_x = \Delta b_x(\partial s_x/\partial b_x)$. For simplicity, let us further assume that changes in b_x's at all ages occur in direct proportion to b_x and the same for all s_x's, so that $\Delta b_x = b_x \Delta B$ and $\Delta s_x = s_x \Delta S$. ΔB and ΔS are the fractional changes in b_x and s_x at all ages. Eq. (8) may now be written

$$\Delta B \sum_0^{\infty} \lambda^{-x} l_x b_x + \Delta S \sum_{x=0}^{\infty} \sum_{x+1}^{\infty} \lambda^{-i} l_i b_i = 0.$$

Substituting $\Delta S = \Delta B(\partial S/\partial B)$ we may solve to obtain

$$\begin{aligned}\frac{\partial S}{\partial B} &= -\frac{1}{\Sigma\Sigma \lambda^{-i} l_i b_i} \\ &= -\frac{1}{T - a}\end{aligned} \tag{9}$$

where T is the average age at which females give birth to their offspring weighted by population growth rate, i.e., $T = \Sigma x \lambda^{-x} l_x b_x$, and a is the age at first reproduction. For the Screech Owl, $T = 2.8$ yrs and $a = 1$ yr, hence $\partial S/\partial B = -1/1.80 = -0.56$. In general, the longer the reproductive lifespan, the smaller the negative value of $\partial S/\partial B$. So, for a given constraint function, the optimum point on the curve shifts towards lower fecundity and higher survival rates as other factors leading to longer reproductive life predominate.

3.6. Interpreting Differences in Life-History Patterns

The data of comparative demography are the life-history patterns of different populations. In the statistical identification of correlations or the testing of predictions of hypotheses, each population is a single data point; for statistical power, the number of populations in the sample must be large. From the standpoint of theory, however, we may ask how the life-histories of any two populations have come to differ. Suppose that population a is characterized by high adult survival and low fecundity, and population b by low survival and high fecundity. Such differences occur between populations of tropical passerines (a) and their temperate zone counterparts (b) (Ricklefs and Bloom, 1977). The differences between these patterns admit to two extreme interpretations and all intermediates between them. On one hand, the two populations might have identical constraint functions but different optimization criteria ($\partial S/\partial B$) (Fig. 4, left), as inferred from Eq. (9) above, for example. On the other hand, the different patterns might result from different constraint functions optimized by the same criterion or by criteria having identical slopes (Fig. 4, right). In the simplified model described above (Eq. 9), it is difficult to change the optimization criterion without altering the constraint function because the two are linked mathematically. In more complicated, and perhaps more realistic models, in which changes in s_x's and b_x's are less interdependent, the situation at the left of Fig. 4 might be obtained.

A complete description of the life-table parameters enables one to determine optimized (=observed) values of B and S, and, with certain assumptions about the importance of biological constraints, the slope of the constraint function, e.g., $\partial S/\partial B = -(T - a)^{-1}$. When values of $\partial y_j/\partial y_i$ in two populations are identical, differences in s_x and b_x must result from differences in constraint functions. When the optimization criteria differ, it is important to determine the sensitivity of optimized values of the life-history variables to changes in the criterion. In particular, one must determine how rapidly the slope of the constraint function changes in the region of optimization. When the change in

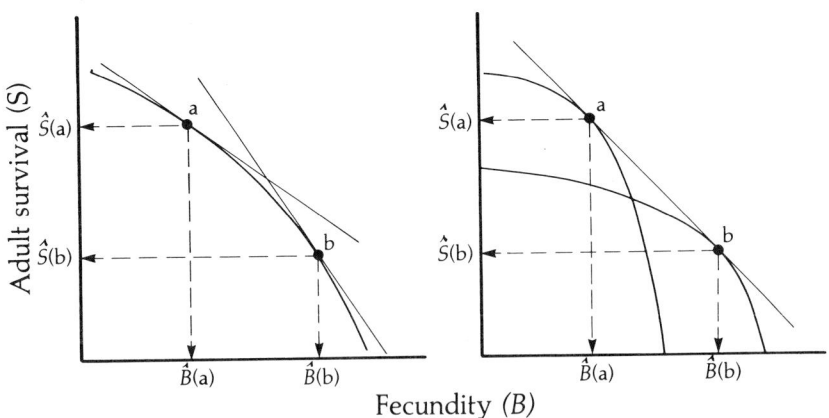

FIGURE 4. Alternative interpretations of observed differences between populations a and b in life-history parameters S and B. At left, the populations differ in their optimization criteria; at right, they differ in their constraint functions.

slope is gradual, as in the left-hand graph in Fig. 4, changes in the optimization criterion will produce correspondingly large responses in the optimized values of the life-history variables. When the constraint functions bend more sharply, as they do in the right-hand graph in Fig. 4, the optimized parameters are less sensitive to changes in the optimization criterion. In the extreme case of a rectangular constraint function, the optimized point is at the corner of the function regardless of the slope of the optimization criterion. Hence, it is crucial to be able to estimate the shape of the constraint function in order to interpret differences in observed life-history patterns. In principle, the shape of the function is difficult to characterize by observation or experiment, except under special circumstances (Ricklefs, 1977a), and must be estimated indirectly from the optimized life-history variables. An example of this approach, based on Ricklefs (1977a), is outlined immediately below.

Consider a concave constraint function of the form

$$Y_j = (1 - Y_i^Z) \tag{10}$$

where $Z > 1$. For an illustration of how Z in Eq. (10) can be estimated, let Y_i be fecundity and Y_j the component of adult survival related to reproduction. In this example, the annual adult survival S is the product of S_n, whose magnitude is independent of reproduction, and S_r, which is some function of fecundity $S_r(B)$. Adopting the general form of Eq. (10), we write

$$\begin{aligned} S &= S_n S_r(B) \\ &= S_n (1 - (cB)^Z) \end{aligned} \tag{11}$$

to describe the constraint function relating S to B; c is an arbitrary scaling constant. The first derivative of Eq. (11) is

$$\begin{aligned} \frac{dS}{dB} &= -S_n c^Z Z B^{Z-1} \\ &= -S_n Z (1 - S_r) B^{-1}. \end{aligned} \tag{12}$$

If we assume that changes in s_x and b_x at all ages are the same fraction of values of s_x and b_x, $\partial S/\partial B = -(T - a)^{-1}$, as we have shown above (Eq. 9). Therefore, at the optimized point,

$$-S_n Z (1 - S_r) B^{-1} = -(T - a)^{-1}$$

which may be rearranged to

$$Z = \frac{B}{(T - a)S} \frac{S_r}{(1 - S_r)}. \qquad (13)$$

For the Screech Owl, the term $B(T - a)^{-1}S^{-1}$ is $(1.3)(1.8^{-1})(0.67^{-1})$ = 1.08, and so Z is approximately the ratio of reproductive survival to reproductive mortality. In any long-lived bird, in which S is close to 1, Z is likely to be a large number. Even for temperate zone songbirds, in which S is closer to 0.5, only a fraction of the annual mortality is directly related to reproduction, hence S_r may be much higher than 0.5.

The value of S_r can be estimated by comparing survival rates in reproducing and nonreproducing segments of a population, e.g., female vs. male Brown-headed Cowbirds, *Molothrus ater* (Fankhauser, 1971), breeding vs. nonbreeding adult Adelie Penguins, *Pygoscelis adeliae* (Ainley and DeMaster, 1980), by experimentally manipulating B and observing the effect on adult survival (Askenmo, 1979; De Steven, 1980), or by comparing mortality rates during and outside the breeding season (Ricklefs, 1977b). Applied to data for a variety of species, the latter technique indicated that Z assumed values between 4–10 with an average of about 6, although true values may be much higher. Hence, the constraint function relating s_x to b_x not only is concave ($Z > 1$), but it is also likely to bend very sharply near the optimized point. As a result, differences between populations in values of s_x and b_x more likely reflect differences in the constraint functions, which result from the particular environments and adaptations of the species involved, than they reflect differences in the optimization criterion, which depends upon the demographic pattern of the population.

4. DENSITY DEPENDENCE

Up to this point we have considered density-independent models, which are characterized by constancy of life-table parameters regardless of whether the population increases, decreases, or remains constant in size. Such models realistically portray the behavior of rare phenotypes that affect the life table because such phenotypes have little effect on the dynamics of the population. But as a phenotype becomes common, its effect on the environment causes density-dependent adjustments in life-table parameters that influence the optimized life-history pattern. In general, survival, fecundity, or both, decrease with increasing local population size such that in a given environment $\lambda = 1$ at some finite and stably-regulated density (Lack, 1966; Klomp, 1980; Newton, 1980; O'Connor, 1980; Patterson, 1980).

Density dependence makes more difficult the interpretation of life-history adaptations even though optimization criteria are identical in models based upon density-independent processes, which maximize population growth rate, and those based on density-dependent processes, which maximize population size N (Charlesworth, 1971; Roughgarden, 1971; Ricklefs, 1981). Suppose that a population comprises individuals having a particular set of adaptations that results in growth rate λ. A mutant appears whose adaptations result in growth rate λ' such that $\lambda' > \lambda$. In a population unrestrained by density dependence, the number of the first kind of individual continues to grow at rate λ, but the mutant individuals leave more descendents and become a progressively larger fraction of the total population. With this change in genetic composition, the average population growth rate increases from λ eventually to λ'.

In populations regulated by density-dependent factors, the fitnesses of each genotype are decreasing functions of population density (Fig. 5). Ecological factors adjust each population to an equilibrium size $\hat{N}$, at which its growth rate λ is 1. Now suppose a mutant appears which at density $\hat{N}$ has growth rate $\lambda'(\hat{N}) > \lambda(\hat{N})$. The average growth rate of the population momentarily exceeds 1 and N increases. But as it does so, both $\lambda(N)$ and $\lambda'(N)$ decrease, tending to restore the average growth rate of the population to 1. These processes of increase and check continue simultaneously until the population comprises only the new

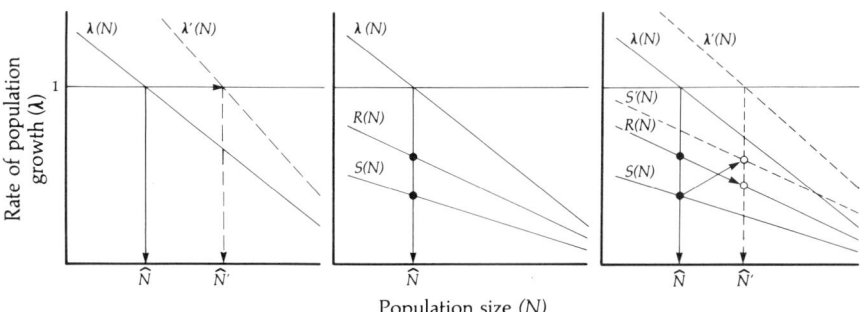

FIGURE 5. Rate of population growth and equilibrium population size when geometric growth rate of the population is a decreasing function of population size (N). Left, an increase in $\lambda(N)$ results in an increase in $\hat{N}$ under density dependence; center, the rate of geometric growth is subdivided into a survival function S(N) and a recruitment function R(N); right, an increase in survival at a given density results in higher equilibrium population density, but a lower recruitment rate.

genotype. Its density is then $\hat{N}'$, whereat $\lambda'(\hat{N}') = 1$. Because $\lambda'(N) > \lambda(N)$ for all densities between $\hat{N}$ and $\hat{N}'$, the second genotype replaces the first just as in the density-independent case. Because selecting increased λ and selecting increased N are identical in evolutionary result, the optimization criteria for density-dependent systems are identical to those for density-independent systems under the special condition of $\lambda = 1$ (Ricklefs, 1981).

Density-dependent responses of a population may cause life-history parameters to exhibit correlated patterns of variation even though they are not linked by biological constraints inherent to the phenotype. Population growth rate (λ) is the sum of the per capita rates of adult survival and recruitment of new individuals to the breeding population (Fig. 5, center). Recruitment is the product of fecundity and prereproductive survival. When either survival or recruitment is changed in a population regulated by density-dependent factors, the other inevitably responds. At the right in Fig. 5, a mutant appears with higher survival than the prevalent genotype at every density, but without effect on rate of recruitment. The mutant is strongly selected and the density of the population increases. One result is that the average survival rate of individuals in the population increases. But at the same time, recruitment rate decreases to balance population growth rate at the new equilibrium density.

Density-dependent adjustments of life-history parameters may cause the optimization criterion to shift and thereby select new values of life-history adaptations. But ecological effects of population density upon the constraint curves may influence life-history patterns more strongly, especially when the constraint function bends sharply. For example, in the curve relating adult survival to fecundity in Fig. 6 the intercepts indicate, on the vertical axis ($b_2 = 0$), the survival rate for individuals not reproducing, which depends upon adaptations for survival and the conditions of the environment. On the horizontal axis ($s_2 = 0$), the intercept expresses the ability of parents to rear offspring without regard to their own survival, which depends upon food availability and other factors in the environment. Populations whose adaptations place them at one or the other intercept, may be thought of as being specialized for either survival or producing offspring at that age. If the rate of survival were to increase owing to a new adaptation or a change in conditions affecting survival only, the population would increase to a new equilibrium value. The larger population might depress its resources because of the increased demand placed upon them, and perhaps lower an individual's ability to nourish its offspring. In this way, the constraint curve might shift, with the intercepts changing in inverse

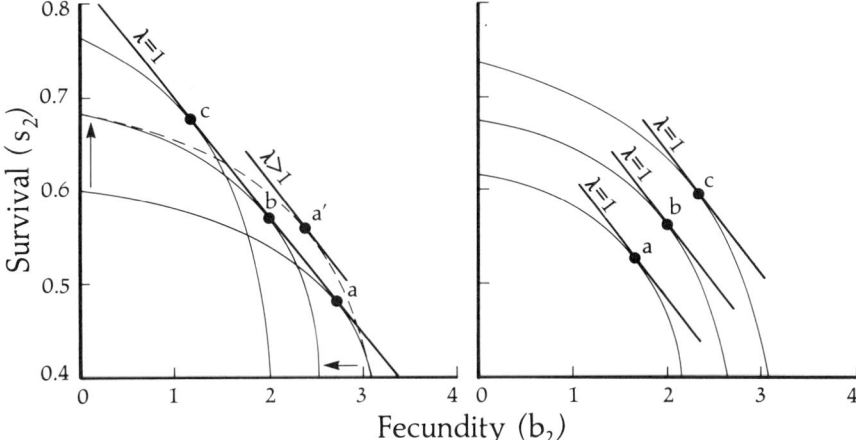

FIGURE 6. At left, change in optimized parameters s_2 and b_2 under density dependence when the optimization criterion does not shift. An initial increase in s_2 shifts the population from a to a', at which point λ is greater than 1. λ is restored to 1 by a decrease in b_2, which shifts the population from point a' to b. Further changes in either s_2 or b_2 might shift the population to point c or anywhere else along the curve of $\lambda = 1$. At right, simultaneous increases in b_2 and s_2 cause density-dependent responses in some other parameter, e.g., prereproductive survival s_0, so that the optimization criterion ($\lambda = 1$) shifts the optimized point from a to b to c.

relationship as shown in Fig. 6 (left). For example, if survival rate s_2 were to increase from 0.60 to 0.68 at the intercept ($b_2 = 0$) and by a similar proportion at other values of b_2 along the constraint curve, and if survival were to change by factors indepedent of those that determined fecundity, the rate of growth of the population (λ) would exceed 1. Under these conditions the population could be brought back into equilibrium by density-dependent responses that reduced fecundity, b_2. Changes in the values of b_x and s_x could also be balanced by density-dependent effects on other stages of the life cycle, perhaps prereproductive survival (s_0) or age at first reproduction. Under such conditions, positive correlations between values of fecundity and adult survival are conceivable (Fig. 6, right).

Density-dependent adjustment of population parameters has been invoked to explain both latitudinal gradients in clutch size exhibited by passerine birds and the uniformly one-egg clutch of pelagic seabirds (Ashmole, 1963; Ricklefs, 1980a). Both authors speculated that adult survival may be determined primarily by factors not directly related to the production of offspring, specifically by factors whose influence is expressed in survival rates during the nonbreeding season. According to this hypothesis, because populations of passerines are kept low by scarcity of winter resources at northern latitudes, each surviving adult has abundant resources available for reproduction during the more bountiful summer months. In the tropics, because most adults survive the nonbreeding season, population densities of breeders are great, and therefore fewer resources are available to each individual for reproduction. Ashmole (1963) argued similarly for seabirds. Populations of pelagic species are supported by the vast resources of the open ocean during the nonbreeding season, but by resources close to nesting sites when breeding. Ashmole supposed that populations increase, and reproductive rate decreases owing to density-dependent factors, until only one offspring can be reared, at which point further increase of the population is limited by recruitment. Populations of inshore-feeding species, which rear broods larger than one, may be more stringently limited by food supplies during the nonbreeding season because they do not disperse widely beyond their breeding ranges or are restricted to habitats of small extent during winter.

If we accept the role of density dependence in molding life-history patterns we must also accept that populations can deplete their food supplies and that intraspecific competition is intense. Both assertions are generally accepted by ecologists (Lack, 1954, 1966; MacArthur, 1972; Cody, 1974), although direct evidence for avian populations is rather limited (Holmes et al., 1979). If Ashmole and Ricklefs are correct about the role of density dependence in determining the life-history patterns of birds, we shall also have to accept several additional principles that find less agreement among avian biologists: populations of land birds are regulated during winter, territorial behavior does not primarily determine the level of resources for the adult but rather assures exclusive use, populations of seabirds depress their food resources greatly, breeding populations of pelagic seabirds are not limited by nesting sites where colonies do occur although inshore feeders may be, and so forth. Of course, these assertions have by no means been demonstrated, but they are strongly implied and must be tested along with other hypotheses concerning avian demography.

5. DEMOGRAPHIC EVOLUTION IN VARYING ENVIRONMENTS

Most theories about demographic evolution and population regulation, including those outlined above, assume that the environment is unvarying. In reality, however, measurements of climate and population parameters both indicate that conditions do change (Lack, 1966). A small body of theory has been developed recently to account for the effects of variation in life-table parameters. The theory suggests that adaptive adjustments of life histories should be sensitive to this variation. Two considerations appear to be important. First, evolution may proceed to a different optimum phenotype in a population that is growing than in one whose size is constant or decreasing. This is the basis for the different life-history patterns expected under r-selection (rapid growth) and K-selection (constant size) regimens.

Second, evolution may proceed to different optima depending on the degree of variation and uncertainty in life-history parameters. In general, theory predicts that selection should favor allocation of time and resources toward the stage of the life history at which survival or production is least variable, all other things being equal. When the survival of offspring is variable and unpredictable, selection weights adult survival more heavily than fecundity. When adult survival is less predictable, selection weights production of offspring more heavily than self-maintenance. These considerations are commonly referred to as "bet hedging" (Stearns, 1976).

5.1. r- and K-Selection

In an increasing population, young born today are more valuable, in terms of their contribution to fitness, than young born in the future. Each of today's offspring is a larger proportion of the total population than an individual born in a larger population in the future. Under conditions that favor rapid population growth, selection is generally thought to favor early reproduction and high fecundity over delayed reproduction and high adult survival (MacArthur and Wilson, 1967; Pianka, 1970). During phases of population decrease, the balance of selective forces shifts to favor delayed maturity and reduced fecundity more strongly.

Although these assertions are true for special situations leading to population increase or decrease, they may not be generally true. Consider the balance between fecundity and survival at a particular age x. By rearranging Eq. (8), we can show that at evolutionary equilibrium ($\Delta \lambda = 0$),

$$\frac{\partial s_x}{\partial b_x} = - \frac{1}{s_x^{-1} \sum_{x+1}^{\infty} \lambda^{x-i} b_i \prod_{x}^{i-1} s_j}. \tag{14}$$

When a population begins to grow following an increase in the survival of individuals younger than age x, only λ in Eq. (14) is affected. As λ increases, $\partial s_x/\partial b_x$ assumes a larger negative value and selection favors a shift towards higher fecundity at the expense of survival beyond age x. But when λ increases following an increase in either b_i or s_i at some age greater than x, the effect of the increase in λ on the denominator of Eq. (14) may be offset by the change in fecundity or survival. Selection could, under these conditions, favor higher survival rather than higher fecundity as λ increases. When the expected number of offspring at a particular age increases, selection favors adaptations that increase the probability of reaching that age, all other things being equal (Michod, 1979).

In general, an increase in population growth rate shifts the balance of selection to weight more heavily parameters expressed at younger ages. For example, by rearranging Eq. (8), we find that at evolutionary equilibirum ($\Delta\lambda = 0$)

$$\frac{\partial b_{x+y}}{\partial b_x} = -\frac{\lambda^y}{\prod_{x}^{x+y-1} s_i}. \tag{15}$$

Hence, as λ increases, $\partial b_{x+y}/\partial b_x$ assumes a larger negative value, indicating that selection will weight a change in fecundity at age x more strongly than a change y time units later, provided the increase in λ is not caused by an increase in survival between ages x and $x+y$.

The influence of population growth rate on evolutionary optimization could be treated in far more detail (see Charlesworth, 1980; Charlesworth and Leon, 1976; Michod, 1979), but the general pattern is clear. Increases in λ tend to select adaptive shifts favoring fecundity over adult survival and younger over older ages. But the expression of such shifts depends importantly on the shapes of the constraint functions, particularly their relationship to population growth rate and upon the counterbalancing influence of selection during phases of population increase and population decrease. If phases of population growth and decline were both caused by influences of the environment upon the same population parameters or set of parameters, adaptive shifts favored during one phase might be reversed during the other, and no net evolutionary change would result from variation in the life table. Such influences are difficult to evaluate because little is known about the demographic nature of change in avian populations. The current status of avian demography requires caution in ascribing life-history patterns to regimes of r- and K-selection.

5.2. Bet Hedging

When the environment varies unpredictably within the lifetime of the individual, selection may favor adaptations that either spread expected reproduction over many seasons or concentrate reproduction at younger ages, depending on the traits affected by environmental variation (Goodman, 1979; Hastings and Caswell, 1979; Murphy, 1968; Schaffer, 1974a; Stearns, 1976). In general, variation in adult survival is thought to favor early breeding and high reproductive effort. Variation in fecundity or prereproductive survival favors low reproductive effort and long adult lifespan. The effect of variation on fitness can be appreciated easily for the extreme case of a genotype that reproduces only once during its lifetime. Such a genotype cannot persist if realized reproduction in some years is occasionally and unpredictably 0.

The strength of selection for bet-hedging strategies is difficult to estimate analytically and depends upon both the amount of variation in demographic parameters and the distribution of that variation. In all but the simplest models (Bell, 1980; Hastings and Caswell, 1979), it is impossible to assess directly the effects of variation on optimization, and one must resort to computer simulations, such as that of Murphy (1968). As yet, however, so few data on variation in life-history traits are available, and so little experience with computer simulations has accumulated, that it is impossible to draw any general conclusion about the role of environmental variation in shaping the observed life histories of birds.

6. CORRELATED ENVIRONMENTAL FACTORS

Life-history parameters are determined in part by the direct influence of factors in the environment. As we have seen, correlations between parameters, such as that between fecundity and adult survival rate, may be caused by the effects of population density upon the environment, which indirectly result in correlations between environmental factors affecting different life-history parameters. But such correlations also may be intrinsic to the environment itself and independent of the populations affected, in which case we would not need to invoke demographic evolution and density dependence to explain correlation among life-history parameters. For example, general qualities of tropical habitats compared to those of temperate habitats might result both in fewer young reared and higher adult survival rates, regardless of density-dependent feedbacks. Nest predation is in large part responsible for low fecundity in tropical birds, but this factor almost certainly is independent of the causes of high adult survival. Because the two are not likely linked through density dependence, their balancing influence on avian demography, which is revealed in comparisons between regions, may derive, in part, from environmental correlations.

Although the effects of density dependence and environmental correlations might be similar, the two factors could operate in opposing directions and complicate the interpretation of life-history patterns. Certainly we should not attribute life-history correlations to either density-dependence or demographic response without first accounting for correlated trends in environmental factors. It is unlikely, however, that the direct effects of the environment upon life history parameters will precisely balance the life-table equation ($\lambda = 1$) as one goes from population to population. Therefore, density dependence would appear to be required at some level to adjust life-history patterns. But its influence cannot be evaluated in comparative studies without understanding patterns inherent in the environment.

7. DISCUSSION

I have not explicitly discussed many aspects of life histories, such as age-specificity, age at onset of reproduction, and parent-offspring conflict, nor have I treated explanations for patterns in life history in any detail. My purpose has been to provide a framework for interpreting life-history patterns and to suggest what we must know in order to fully understand differences among populations.

One source of demographic patterns is evolutionary optimization. The key to understanding this phenomenon is the constraint function interrelating the life-history variables. Through density dependence, the function itself responds to changes in life-history parameters, as does the criterion that determines the optimum life-history pattern. That is, the constraint curves and optimization criteria are adjusted simultaneously until equilibrium values of the life-history variables are reached, at which point, in general, $\lambda = 1$ and $\Delta\lambda = 0$.

In a practical sense, it is not possible to measure the entire constraint function, or even a sizeable portion of it, because phenotypic variation is small compared to differences between populations, heritabilities of traits are low, and genetic variation is difficult to measure. Experimental manipulations of fecundity are practical in some organisms, such as rotifers (Snell and King, 1977), insects (Murdoch, 1966), and fish (Reznick, 1980), especially in the laboratory. Similar experiments with birds have been relatively unsuccessful, first, because parents may not respond to larger numbers of eggs or young by increasing their effort, second, because parental care in birds is of long duration and so the outcome of the experiment may be difficult to measure, and third, because statsitical evaluation of differences in survival rates in control and experimental situations requires very large samples (Askenmo, 1979; De Steven, 1980). In most cases, the characterization of constraint functions will be limited to measuring observed optimized variables and to using models to estimate the slope of the optimization criterion from the life table. I have indicated one indirect approach to estimating the rate of change in the slope of the constraint function relating fecundity and adult survival at its optimized point. Often, it may be difficult to express such estimates in measurable terms.

In principle, it may be possible to construct models of certain types of constraints where life-history variables are interrelated by straightforward allocation of fixed resources. For birds, however, most tradeoffs between life-history parameters involve subtle behavioral and physiological factors that are beyond our ability to model in the absence of more detailed observation and experimentation. Such studies should be of high priority.

A second source of demographic pattern is density-dependent feedback. There is abundant evidence of competition in natural and confined populations of many kinds of animals and plants. But we know few details of the stage of the life cycle or time of year at which density-dependent factors act (Tanner, 1966), and it is precisely such details that dictate how density-dependent factors influence demography. A population limited primarily by adult survival in winter and one limited primarily by territorial behavior during the breeding season would likely exhibit different life histories in the same environment. Direct experimentation on factors responsible for population regulation (Watson, 1967, 1977; Perrins, 1965) and on the effect of birds upon their resources (Holmes et al., 1979) is feasible and will contribute substantially to understanding trends in life-history parameters.

A third source of demographic pattern is correlation among factors in the environment that is independent of the populations whose life-histories are being investigated. To understand these, it shall be necessary to relate variation among species and localities in factors that directly affect each life-history parameter to variation in the physical environment or to aspects of the biological environment that are not influenced by the population. One might investigate, for example, whether the availability of food is determined primarily by the productivity of the habitat or by the rate of harvesting by consumer populations.

Owing to the complexity of influences on avian life histories, it may never be possible to identify with confidence all the factors responsible for shaping the life table of a particular population. The study of life-history patterns will likely be most fruitful when three approaches are combined: comparison, experimentation, and modeling. Patterns revealed by comparative study establish the phenomenology of life histories, including correlations between life-history traits, and through correlations with environmental factors may suggest the major influences on life-history patterns. The life-table parameters themselves are not adaptations but are the expression of the interaction between the phenotype and the environment. Hence, they do not directly reveal evolutionary adjustment, but rather they contain information about both the environment and life-history adaptations.

Experimental studies are necessary to isolate the separate effects of the many environmental factors that influence life-history traits. Brood-size manipulation and both sterilization and hormone treatment to alter reproductive status in field situations are possible approaches in selected cases. Owing to the overwhelming importance of survival rates in evaluating life-history patterns, and to the technical and sampling problems involved in estimating survival rates, certain types of experiments may prove to be impractical. Nonetheless, carefully designed and analyzed experiments may reveal much about the genetic bases of life-history traits, their response to certain environmental factors, and the constraints that tie the life-history pattern together as a single system of integrated adaptations.

The role of modeling is principally to provide explicit, quantitative statements of our understanding of life-history patterns. The modeling process forces us to recognize all the components of fitness and the interactions between organisms and their environments, and it thereby reveals the areas in which we lack knowledge. The simple models presented in this paper underscore the central importance of constraint functions and density-dependent responses to understanding life-history patterns, thus emphasizing the difficulty of the problem. Models also may be used to evaluate the relative strengths of selective factors when certain quantities in the model can be guessed or measured with reasonable assurance. Although some hypotheses about life-history patterns are logical and plausible, modeling may reveal that they constitute trivial effects and may be bypassed in favor of more important effects. Occasionally, modeling may reveal unexpected relationships between variables that give rise to new hypotheses.

ACKNOWLEDGMENTS. I am grateful to A. E. Dunham, L. Freed, S. Stearns, and several anonymous reviewers for suggestions on a draft of the manuscript. This study was supported by National Science Foundation Grant DEB77-27071.

REFERENCES

Ainley, D. G., and DeMaster, D. P., 1980, Survival and mortality in a population of Adélie Penguins, Ecology **61**:522–530.

Ankney, C. D. and MacInnes, C. D., 1978, Nutrient reserves and reproductive performance of female Lesser Snow Geese, Auk **95**:459–471.

Ashmole, N. P., 1963, The regulation of numbers of tropical oceanic birds, Ibis **103b**:458–473.

Askenmo, C., 1979, Reproductive effort and return rate of male Pied Flycatchers, Am. Natural. **114**:748–753.

Baker, J. R., 1938, The evolution of breeding systems, in: Evolution: Essays on Aspects of Evolutionary Biology (G. R. de Beer, ed.), Oxford University Press, London.

Bell, G., 1980, The costs of reproduction and their consequences, Am. Natural. **116**:45–76.

Breitenback, R. P., and Meyer, R. K., 1959, Effect of incubation and brooding on fat, visceral weights, and body weight of the hen pheasant (Phasianus colchicus), Poultry Sci. **38**:1014–1026.

Charlesworth, B., 1970, Selection in populations with overlapping generations. I. The use of Malthusian parameters in population genetics, Theoret. Pop. Bio. **1**:352–370.

Charlesworth, B., 1971, Selection in density-regulated populations, Ecology **52**:469–474.

Charlesworth, B., 1973, Selection in populations with overlapping generations. V. Natural selection and life histories, Am. Natural. **107**:303–311.

Charlesworth, B., 1980, Evolution in Age-Structured Populations, Cambridge University Press, London.

Charlesworth, B., and Leon, J. A., 1976, The relations of reproductive effort to age, Am. Natural. **110**:449–459.

Cody, M. L., 1966, A general theory of clutch size, Evolution **20**:174–184.

Cody, M. L., 1971, Ecological aspects of reproduction, in: Avian Biology, Volume I (D. S. Farner and J. R. King, eds.), Academic Press, New York, pp. 461–512.

Cody, M. L., 1974, Competition and the Structure of Bird Communities, Princeton University Press, Princeton, New Jersey.

De Steven, D., 1980, Clutch size, breeding success, and parental survival in the Tree Swallow (Iridoprocne bicolor), Evolution **34**:278–291.

Fankhauser, D. P., 1971, Annual adult survival rates of blackbirds and starlings, Bird-Banding **42**:36–42.

Foster, M. S., 1974, A model to explain molt-breeding overlap and clutch size in some tropical birds, Evolution **28**:182–190.

Gadgil. M., and Bossert, W. H., 1970, Life historical consequences of natural selection, Am. Natural. **104**:1–24.

Goodman, D., 1974, Natural selection and a cost ceiling on reproductive effort, Am. Natural. **108**:247–268.

Goodman, D., 1979, Regulating reproductive effort in a changing environment, Am. Natural. **113**:735–748.

Hamilton, W. D., 1966, The moulding of senescence by natural selection, *J. Theoret. Biol.* **12**:12–45.

Hastings, A., and Caswell, H., 1979, Role of environmental variability in the evolution of life history strategies, *Proc. Natl. Acad. Sci. USA* **76**:4700–4703.

Hirshfield, M. F., and Tinkle, D. W., 1975, Natural selection and the evolution of reproductive effort, *Proc. Natl. Acad. Sci. USA* **72**:2227–2231.

Holmes, R. T., Schultz, J. C., and Nothnagle, P., 1979, Bird predation on forest insects: An exclosure experiment, *Science* **206**:462–463.

Johnson, S. R., and West, G. C., 1973, Fat content, fatty acid composition and estimates of energy metabolism of Adelie Penguins (*Pygoscelis adeliae*) during the early breeding season fast, *Comp. Biochem. Physiol.* **45**:709–719.

Jones, P. J., and Ward, P., 1976, The level of reserve protein as the proximate factor controlling the timing of breeding and clutch size in the Red-billed *Quelea quelea quelea*, *Ibis* **118**:547–574.

Klomp, H., 1980, Fluctuations and stability in Great Tit populations, *Ardea* **68**:205–224.

Krapu, G. L., 1981, The role of nutrient reserves in Mallard reproduction, *Auk* **98**:29–38.

Lack, D., 1947, The significance of clutch-size. Parts 1 and 2, *Ibis* **89**:302–352.

Lack, D., 1948, The significance of clutch-size. III. *Ibis* **90**:25–45

Lack, D., 1954, *The Natural Regulation of Animal Numbers*, Oxford University Press, London.

Lack, D., 1966, *Population Studies of Birds*, Clarendon Press, Oxford.

Lack, D., 1968, *Ecological Adaptations for Breeding in Birds*, Methuen, London.

Lack, D., and Moreau, R. E., 1965, Clutch-size in tropical passerine birds of forest and savanna, *Oiseau* **35**(suppl.):76–89.

Leon, J. A., 1976, Life histories as adaptive strategies, *J. Theoret. Biol.* **60**:301–335.

Lotka, A. J., 1956, *Elements of Mathematical Biology*, Dover, New York.

MacArthur, R. H., 1972, *Geographical Ecology*, Harper and Row, New York.

MacArthur, R. H., and Wilson, E. O., 1967, *The Theory of Island Biogeography*, Princeton University Press, New Jersey.

Michod, R. E., 1979, Evolution of life-histories in response to age-specific mortality factors, *Am. Natural.* **113**:531–550.

Moreau, R. E., 1944, Clutch-size: A comparative study, with special reference to African birds, *Ibis* **86**:286–347.

Murdoch, W. W., 1966, Population stability and life history phenomena, *Am. Natural.* **100**:5–11.

Murphy, G. I., 1968, Patterns in life history and the environment. *Am. Natural.* **102**:390–404.

Nelson, J. B., 1977, Some relationships between food and breeding in the marine Pelecaniformes, in: *Evolutionary Ecology* (B. Stonehouse and C. Perrins, eds.), Macmillan, London, pp. 77–87.

Newton, I., 1977, Timing and success of breeding in tundra-nesting geese, *Evolutionary Ecology* (B. Stonehouse and C. M. Perrins, eds.), Macmillan, London, pp. 113–126.

Newton, I., 1980, The role of food in limiting bird numbers, *Ardea* **68**:11–30.

O'Connor, R. J., 1980, Pattern and process in Great Tit (*Parus major*) populations in Britain, *Ardea* **68**:165–184.

Owen, D. F., 1977, Latitudinal gradients in clutch size: An extension of David Lack's theory, in: *Evolutionary Ecology* (B. Stonehouse and C. M. Perrins, eds.), Macmillan, London, pp. 171–179.

Patterson, I. J., 1980, Territorial behavior and the limitation of population density, *Ardea* **68**:53–62.

Perrins, C. M., 1965, Population fluctuations and clutch size in the Great Tit, *Parus major*, *J. Anim. Ecol.* **34**:601–647.
Perrins, C. M., 1977, The role of predation in the evolution of clutch size, in: *Evolutionary Ecology* (B. Stonehouse and C. M. Perrins, eds.), Macmillan, London, pp. 181–191.
Pianka, E. R., 1970, On r and K selection, *Am. Natural.* **104**:592–597.
Reznick, D. N., 1980, Life History Evolution in the Guppy (*Poecilia reticulata*), Ph.D. dissertation, University of Pennsylvania.
Ricklefs, R. E., 1968, On the limitation of brood size in passerine birds by the ability of adults to nourish their young, *Proc. Natl. Acad. Sci. USA* **61**:847–851.
Ricklefs, R. E., 1970, Clutch size in birds: Outcome of opposing predator and prey adaptations, *Science* **168**:599–600.
Ricklefs, R. E., 1977a, On the evolution of reproductive strategies in birds: Reproductive effort, *Am. Natural.* **111**:453–478.
Ricklefs, R. E., 1977b, A note on the evolution of clutch size in altricial birds, in: *Evolutionary Ecology* (B. Stonehouse and C. M. Perrins, eds.), Macmillan, London, pp. 193–214.
Ricklefs, R. E., 1980a, Geographical variation in clutch size among passerine birds: Ashmole's hypothesis, *Auk* **97**:38–49.
Ricklefs, R. E., 1980b, Fitness, reproductive value, age structure, and the optimization of life-history patterns, *Am. Natural.* **117**:819–825.
Ricklefs, R. E., 1981, The optimization of life-history patterns under density dependence, *Am. Natural.* **117**:403–408.
Ricklefs, R. E., and Bloom, G., 1977, Components of avian breeding productivity, *Auk* **94**:86–96.
Roughgarden, J., 1971, Density-dependent natural selection, *Ecology* **52**:453–467.
Roughgarden, J., 1979, *Theory of Population Genetics and Evolutionary Ecology: An Introduction*, Macmillan, New York.
Royama, T., 1969, A model for the global variation of clutch size in birds, *Oikos* **20**:562–567.
Schaffer, W. M., 1974a, Selection for optimal life histories: The effects of age structure, *Ecology* **55**:291–303.
Schaffer, W. M., 1974b, Optimal reproductive effort in fluctuating environments, *Am. Natural.* **108**:783–790.
Skutch, A. F., 1949, Do tropical birds rear as many young as they can nourish? *Ibis* **91**:430–455.
Snell, T. W., and King, C. E., 1977, Life span and fecundity patterns in rotifers: The cost of reproduction, *Evolution* **31**:882–890.
Snow, B. K., 1970, A field study of the Bearded Bellbird in Trinidad, *Ibis* **112**:299–329.
Stearns, S. C., 1976, Life-history tactics: A review of the ideas, *Quart. Rev. Biol.* **51**:3–47.
Stearns, S. C., 1977, The evolution of life history traits: A critique of the theory and a review of the data, *Annu. Rev. Ecol. Syst.* **8**:145–171.
Stearns, S. C., 1980, A new view of life-history evolution, *Oikos* **35**:266–281.
Tanner, J. T., 1966, Effects of population density on growth rates of animal populations, *Ecology* **47**:733–745.
VanCamp, L. F., and Henny, C. J., 1975, The Screech Owl: Its life history and population ecology in northern Ohio, *N. Am. Fauna* **71**:1–65.
Watson, A., 1967, Population control by territorial behavior in Red Grouse, *Nature* **215**:1274–1275.

Watson, A., 1977, Population limitation and the adaptive value of territorial behavior in Scottish Red Grouse *Lagopus l. scoticus*, in: *Evolutionary Ecology* (B. Stonehouse and C. M. Perrins, eds.), Macmillan, London, pp. 19–26.

Williams, G. C., 1966, Natural selection, the costs of reproduction, and a refinement of Lack's principle, *Am. Natural.* **100:**687–690.

Wilson, E. O., and Bossert, W. H., 1971, *A Primer of Population Biology*, Sinauer, Sunderland, Massachusetts.

CHAPTER 2

THE DETERMINATION OF CLUTCH SIZE IN PRECOCIAL BIRDS

DAVID W. WINKLER and JEFFREY R. WALTERS

1. INTRODUCTION

The evolution and regulation of clutch size has long been a central issue in ornithology. Early ornithologists realized that females of each species of bird lay a characteristic number of eggs, and we have been trying to determine ever since why this is so. In pursuit of the answer to this seemingly simple question, ornithologists have not only accumulated a wealth of egg data, but also have made important contributions to such diverse topics as life-history strategies, population regulation and group selection. Yet how clutch size is determined remains a controversial issue. The consensus that was once sought in the form of a central theory (Lack, 1968; Cody, 1966; Klomp, 1970; von Haartman, 1971) has disappeared in a sea of specific hypotheses. In this review we attempt to organize and summarize clutch size theories as they emerge in modified form from recent research and evaluate their ability to explain observed patterns in clutch size variation. We concentrate on the literature and concepts published since the review of Klomp (1970), but we incorporate earlier work when necessary.

The question of why birds lay the number of eggs they do may be

---

DAVID W. WINKLER • Museum of Vertebrate Zoology and Department of Zoology, University of California, Berkeley, California 94720.   JEFFREY R. WALTERS • Department of Zoology, North Carolina State University, Raleigh, North Carolina 27607.

answered at two levels (cf. Baker, 1938). On the proximate level, one seeks the physiological mechanisms that terminate egg production, thereby determining clutch size. On the ultimate level, one searches for the selective pressures that determine the optimal clutch size, thereby governing its regulation in evolutionary time. This distinction is often not maintained in clutch size studies. Our primary concern here is with the ultimate level of explanation, and we try to avoid purely mechanistic hypotheses (for an entry into the latter literature see Jones, 1978). Some of the hypotheses we consider, however, may function on both levels, forcing us to address the issue of levels of explanation in some detail.

We restrict our review to precocial species. Precocial and altricial species form natural ecological groups because chicks in the two groups differ fundamentally in the quality, timing, and duration of demands they make on their parents (Nice, 1962; Ricklefs, 1979a). Demands on parents are a crucial consideration in clutch size evolution (see Section 2), justifying a separate treatment of the two groups. Because the classification of birds by developmental mode is ambiguous, we first present a defense of the classification used to define the domain of this paper.

2. PRECOCIAL DEVELOPMENT

The classic precocial chick is downy, open-eyed, and capable of running about and feeding itself at hatching (Nice, 1962). This condition contrasts with that of the naked, blind, helpless young of typical altricial species. The trouble with the precocial–altricial dichotomy has always been that the traits used to define developmental modes do not covary perfectly, necessitating the use of intermediate categories such as semi-precocial and semi-altricial (Nice, 1962). Recent studies indicate that the development of down, locomotory ability, and foraging ability vary somewhat independently across species according to the particular needs of the young (Ricklefs, 1973, 1979a; Ricklefs et al., 1980; O'Connor, 1977; Blem, 1978).

We follow Ricklefs (1979a) in considering the physiological maturity of tissues, especially skeletal muscle, as being of primary importance in classifying modes of development. Larids, alcids, and the Procellariiformes, as well as classic precocial species, are thereby classified as precocial (Ricklefs, 1979a, 1979b; Ricklefs et al., 1980). Species might be further subdivided according to growth pattern, which depends on which tissues are mature at hatching and their relative importance in the species. Most importantly, current research suggests that mature tissues are not capable of rapid growth, and they may act

TABLE I
Classification of Avian Modes of Development

Class	Definitive characteristics	Member taxa
Precocial 1	Tissues generally mature, growth governed by leg muscles	Palaeognathiformes[a] Phoenicopteriformes Galliformes Gruiformes[b]
Precocial 2	Like Precocial 1, but small size of legs enables more rapid growth	Gaviiformes[c] Podicipediformes[c] Anseriformes Charadrii[d] Caprimulgidae[c]
Precocial 3	Like Precocial 2, but legs even smaller and growth more rapid	Lari Alcae[c]
Precocial 4	Growth slow, governed by pectoral muscles	Procellariiformes[e] Sphenisciformes[c]
Altricial 1	Rapid maturation of certain tissues, which then govern growth rate[f]	Ciconiiformes Opisthocomidae[c]
Altricial 2	Tissues generally immature	Pelecaniformes[g] Falconiformes[g] Columbiformes Psittaciformes Cuculiformes[g] Strigiformes Caprimulgiformes[g,h] Apodiformes Coliiformes Trogoniformes Coraciiformes Piciformes Passeriformes

[a]As defined by Cracraft (1974), includes ratites and tinamous.
[b]With possible exception of Heliornithidae.
[c]Placement speculative.
[d]With possible exception of Dromadidae.
[e]With possible exception of Diomedeidae.
[f]Tissues involved in escape behavior in Opisthocomidae, leg muscles in Ciconiiformes.
[g]There are some indications that these species may have special developmental patterns.
[h]Except Caprimulgidae.

as a brake on growth rate in other tissues of the developing chick. Thus, larids (and perhaps others such as Anseriformes, Charadrii, and Scolopaci), might be distinguished from the remaining precocial species because they grow faster. These species resemble other precocial birds in that growth rate is constrained by mature leg muscle, but this tissue is less constraining in at least the larids because the legs in adults are relatively smaller than those in the adults of other precocial species (Ricklefs, 1979a; Ricklefs et al., 1980).

A definitive classification of all birds is premature pending more detailed developmental studies like those of Ricklefs (1973; 1979b; Ricklefs et al., 1980) and Dunn (1976), but we follow a provisional classification based on the above considerations in determining our subject matter (Table I).

Ricklefs (1979a) relates mode of development to ecological factors, especially food supply. However, the phylogenetic conservatism of the precocial–altricial division in our classification is striking. For the most part, differences in developmental mode occur between orders, but not within (Table I), and thus appear to vary more with phylogeny than with ecology. The only apparent exception, the Heliornithidae, is a poorly known group of questionable affinities. Thus, mode of development appears to be of a very conservative character.

3. CLUTCH SIZE THEORY AND TERMINOLOGY

For the purpose of this review, reproductive effort is defined as all expenditure of time and energy that is necessary and sufficient for the production and rearing of young. Although this definition lacks the direct connection to demography and life-history theory implied by Trivers's (1972) concept of parental investment, it lacks the problems of measurement and applicability that plague the latter (e.g., Dawkins and Carlisle, 1976). The reproductive effort of birds consists of three components: mating, gametic, and parental efforts. Mating effort is all effort used to procure and retain a mate; gametic effort consists of the energy and materials devoted to the eggs and sperm; and parental effort is all effort devoted to incubation of eggs and protection and feeding of young up to, and sometimes after, fledging (cf. Low's, 1978, classification). Parents must make a series of "decisions" which involve not only the timing and magnitude of reproductive effort as a whole, but also the apportionment of effort to each of the three components and the categories within them (Fig. 1). Viewed from this perspective, clutch size is only a small part of the "bag of tricks" available to birds for tailoring their reproductive effort to the environment.

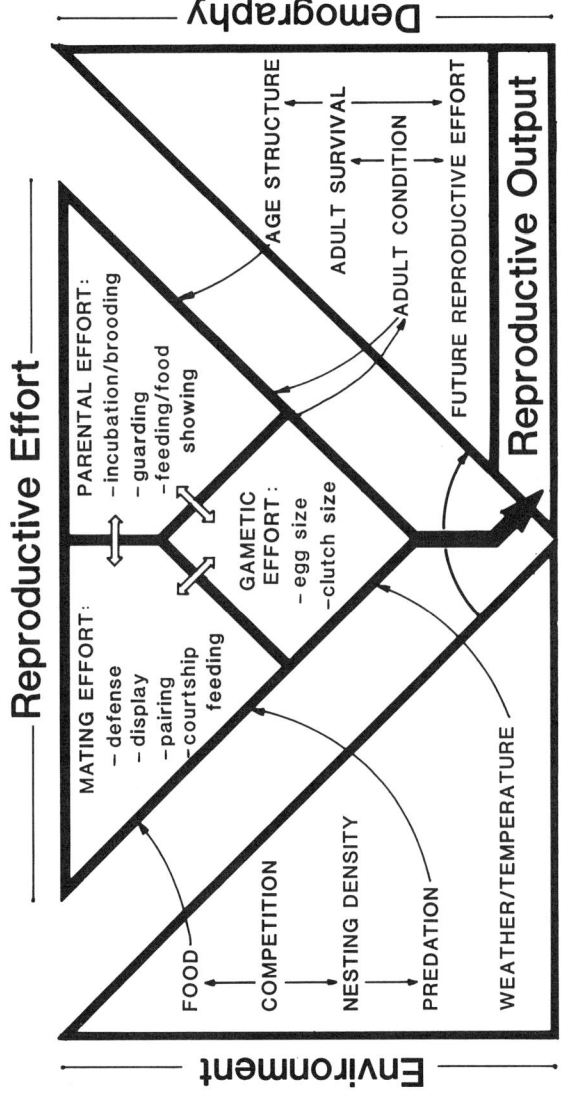

FIGURE 1. Interactions between environment, demography, and reproductive effort. Environmental effects can have both direct and indirect (via demography) effects on reproductive effort. Clutch size is only one of many subcomponents of reproductive effort. For further discussion and references see text.

This view implies a productive interaction between clutch size theory and life history theory. Clutch size is expected to respond both directly to characteristics of the environment and the population as well as indirectly to life history modifications which are themselves adjustments to the environment (see also MacArthur and Wilson, 1967; Foster, 1974a; Payne, 1974; Ricklefs, 1977a,b; Howe, 1978; Högstedt, 1980; Pugesek, 1981). Most clutch size research has focused on "direct environmental effects" rather than "indirect life history effects." Relations have been discerned (Fig. 1) between variation in clutch size and such environmental features as food and mineral supply (e.g., Ashmole, 1963; Lack, 1968; MacLean, 1975; Jones, 1976; Jones and Ward, 1976; Owen, 1977; Raveling, 1979; Ricklefs, 1980; Williams, 1981) and weather (Royama, 1966, 1969; Foster, 1974b; Yom-Tov and Hilborn, 1981); such interspecific interactions as competition (Cody, 1966) and predation (e.g., Ricklefs, 1970; Yom-Tov, 1974; Safriel, 1975; Perrins, 1977; Slagsvold, 1982); and such population features as age (e.g., Coulson, 1963; Coulson and White, 1958; Davis, 1975) and adult condition (e.g., Ankney and MacInnes, 1978; Raveling, 1979).

David Lack's (1947a, 1948, 1954, 1966, 1968) influential work focused on direct environmental effects on parental effort and its relation to clutch size. Building on previous work such as that of Heinroth (1922) and Moreau (1944), he championed the view that the modal clutch size corresponds to the maximum number of young that can be successfully fledged, and he identified the parents' ability to feed the young as the factor that limits fledging success in most birds. For species that do not feed young, Lack (1966, 1968) supported the view that clutch size is limited by the ability of females to procure sufficient food to produce eggs, although initially he was reluctant to do so (Lack, 1947a, 1954).

Lack applied his ideas to all birds (e.g., Lack, 1968), and he and his students conducted some of the first experiments in behavioral ecology (e.g., Lack and Lack, 1951) to support his views, which, as a consequence, came to dominate thinking on the evolution of clutch size in birds. Coincident with these developments in ornithology came some of the first explorations of life histories in the theoretical literature. Several of these early contributions (e.g., Williams, 1966; Charnov and Krebs, 1974; Goodman, 1974) investigated the relations between clutch size and other aspects of life history. These authors assumed: 1) that individuals maximize their reproductive output over their entire lifetimes rather than within a single breeding attempt and 2) that increasing reproductive effort reduces the chances of surviving to breed again. When these assumptions are met there will exist an optimal trade-off point between reproduction and survival that maximizes lifetime reproductive output. Reproductive effort tailored to this optimal tradeoff has been termed "optimal working capacity" (Royama, 1966). Lack's hypothesis may thus be restated to read that birds lay that number of eggs that results in the parents operating at the optimal working capacity. This modification of Lack's hypothesis may make it applicable not only to those species incapable of rearing a larger than normal brood (e.g., Crossner, 1977; Schifferli, 1978), but also to those species able to do so (e.g., von Haartman, 1971; Perrins and Moss, 1975; Haymes and Morris, 1977; Drent and Daan, 1980; Moss et al., 1981).

Another development in life-history theory relevant to clutch size theory has been the construction of hypotheses relating life-history to the constancy and predictability of the environment. Three "strategies" of covarying life-history traits (r-strategy, K-strategy, bet-hedger) have been proposed for various regimes of the time-scale (relative to the organism's lifespan) and predictability of environmental variation (Table II; Murphy, 1968; Schaffer, 1974; Pianka, 1970, 1976; Stearns, 1976, 1977). These hypotheses make predictions for the total reproductive effort to be expended, but how the effort should best be apportioned among the various components of reproduction is just beginning to be explored (e.g., Smith and Fretwell, 1974; Brockelman, 1975).

In sum, any given clutch size differences between two populations might be the result of environmental factors having a direct effect on the laying female or the result of evolutionary adjustments of the populations' life histories. Evaluating the relative importance of these two sorts of factors is a major research challenge, but most results discussed

TABLE II
Covarying Life-History Traits Expected under Various Environmental Conditions[a]

Trait	Strategy		
	r-strategy	K-strategy	Bet-hedger
Environmental predictability			
For adults	Low	High	High
For juveniles	Low	High	Low
Length of life	Short	Long	Long
Reproductive effort per episode	Large	Large	Small
Leads to:	High number of young	High quality of young	Minimizing risks of failure

[a]After Murphy, 1968; Schaffer, 1974; Pianka, 1970; Stearns, 1976.

below relate to direct environmental effects on clutch size. We will often refer to indirect life history effects as well, but such effects on clutch size will be poorly understood until life-history theory can make clear predictions for the apportionment among components within the overall reproductive effort.

4. PATTERNS IN CLUTCH SIZE VARIATION

Birds that do not feed their young (henceforth "non-feeders") form a taxonomically well-defined group consisting of the ratites, Tinamiformes, Anseriformes, Galliformes, and most Charadrii and Scolopaci. Many of these species have large clutches; clutches of ten or more eggs are not unusual among the Galliformes and Anseriformes. However, some taxa such as rheas, guans, and shorebirds have modal clutches of only three to five eggs, like those of precocial species that feed their young (henceforth "feeders"). Of the latter, only the Rallidae have clutch sizes that approach those of non-feeders (Appendix I). There is thus a general inverse relation between feeding of young and clutch size.

A major source of variation in clutch sizes is the characteristic differences in clutch sizes between taxa. The greatest differences occur between representatives of different orders (Appendix I), but marked and consistent differences also exist between lower taxa within the orders. For example, jacanas and lapwings have larger clutches than plovers (Walters, in press), and geese and stiff-tailed ducks have smaller clutches than tree ducks or diving ducks (Appendix II), independent of geography (see below).

One striking aspect of intraspecific variation in clutch size of precocial birds is the dichotomy in their clutch size distributions. In some species there is a sharp upper limit to clutch size, whereas in the remainder the distribution is normal or nearly so (Fig. 2). The existence of a sharp truncation, characteristic especially of the Charadriiformes, has important implications for the limitation of clutch size in these groups (see below).

The most pervasive and consistent intraspecific trend is the tendency for mean clutch size to decline steadily throughout the nesting season. Such declines occur in virtually every species of feeder and non-feeder studied (for reviews see Klomp, 1970; von Haartman, 1971; see also Niethammer, 1966; Glutz von Blotzheim, 1973, 1975, 1977; Palmer, 1962, 1976a,b; Cramp, 1977; Batt and Prince, 1979), and while some species show a very weak or nonexistent seasonal trend, counter-examples in which mean clutch size increases with time of laying are extremely rare (e.g., Mills and Shaw, 1980). Much of this variation may be due to smaller clutches being laid in replacement clutches, but in those species which have been studied carefully, the same seasonal trend obtains even when replacement clutches are ignored.

Perhaps the most well-known type of clutch size variation is geographic. Several taxa do not exhibit such variation, e.g., Sphenisciformes and Procellariiformes, but among those that do, clutches tend to be larger in Arctic and northern temperate regions than in the tropics. This sort of variation is present in pheasants (Lack, 1947b), waterfowl (Cody, 1966; Appendix II), shorebirds (Walters, in press), and gulls and terns (Fig. 3). Such variation is also likely in Podicipediformes (Sealy, 1978) and Rallidae (Ripley, 1977; but see McFarlane, 1975). Among waterfowl and shorebirds, clutches in the southern temperate zone are relatively small, trending more toward those of tropical regions than those of northern temperate and Arctic regions (Appendix II; Walters in press) contrary to the pattern in altricial birds (Moreau, 1944), gulls and terns (Fig. 3). In Anseriformes and Laridae, clutches tend to be smaller in Arctic regions than in northern temperate regions (Johnsgard, 1973; Appendix II; Fig. 3). Most of these geographic patterns are found primarily in interspecific comparisons presumably because most species do not occupy a large enough latitudinal range to manifest the trend.

Clutch size also varies on finer spatial scales. For example, clutch size in the European Oystercatcher (*Haematopus ostralegus*) varies latitudinally across Europe, and it is larger along the coast than in inland localities (Glutz von Blotzheim, 1975). Clutch size differences within the same general geographic area can be quite large in larids (e.g., Lemmetyinen, 1973; Nisbet, 1978) and can also vary strikingly between habitats and colony areas (e.g., Poslavskii and Krivonosov, 1976; Veen, 1977; Pierotti, 1982; see also Högstedt, 1980). In terns, clutch size is also markedly variable from season to season (e.g., Langham, 1974; Coulson and Horobin, 1976).

5. FACTORS LIMITING CLUTCH SIZE

We will now summarize the various leading hypotheses for explaining clutch size variation in precocial birds, discuss their position in the framework outlined in Fig. 1, and evaluate their ability to explain the aforementioned trends in clutch size.

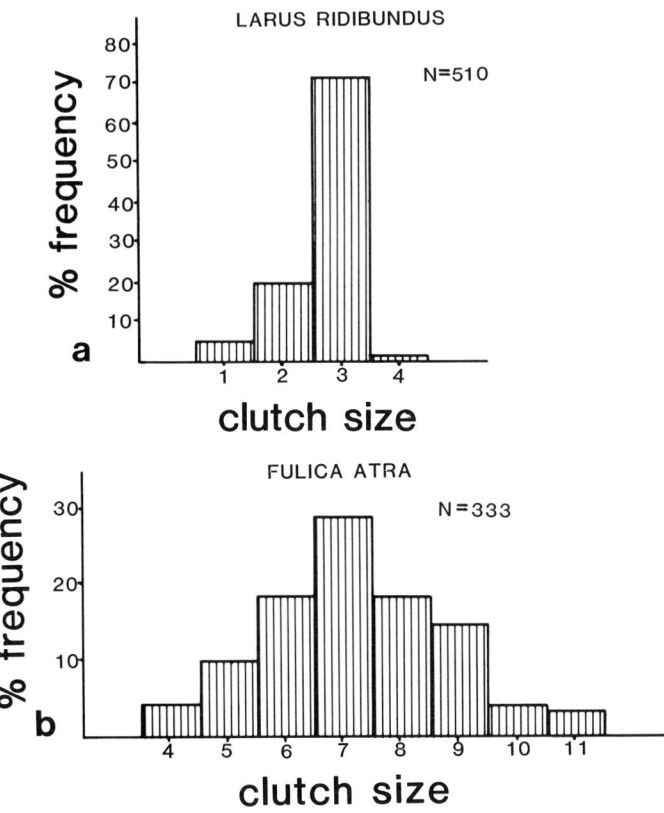

FIGURE 2. Typical truncated (a) and normal clutch size (b) distributions. Data from Glutz von Blotzheim (1964) and Fiala (1978), respectively, represent patterns found throughout precocial bird species and were chosen only for their large sample sizes.

5.1. Egg Formation Ability

5.1.1. General Discussion of the Hypothesis

Because precocial young can often feed themselves and independently regulate their body temperatures, they are presumably less expensive to care for than are altricial chicks, especially in non-feeders. Taken to its logical end, this notion has led to the widespread belief that parental effort cannot limit clutch size in precocial non-feeders and that clutch size in this group is limited by gametic effort, specifically by the ability of females to form eggs. The egg formation ability hypothesis thus refers to a direct environmental effect, that of food, on gametic effort.

The large clutch sizes of non-feeders tend to support the hypothesis, and consistent relationships between egg size and body weight and between clutch weight and body weight (Rahn et al., 1975; Ross, 1979) can be invoked to account for species with small clutches.

This does not constitute conclusive evidence for the egg formation ability hypothesis, however, and the best evidence for a limitation on egg-formation ability comes from studies of Arctic-nesting geese. In such geese, females feed little, if at all, after their arrival on the breeding grounds prior to egg-laying. They depend on nutritional reserves accumulated and stored in their bodies just before and during their northward migrations to the breeding grounds for the materials and energy to form eggs. The female alone incubates, and incubation constancy is high, so the female must also depend on her reserves during incubation (Ryder, 1970; Inglis, 1977). If energy reserves are exhausted during incubation the female may die on the nest or be forced to abandon the clutch (Inglis, 1977; Ankney and MacInnes, 1978). It is obviously disadvantageous for females to deplete their reserves below a certain level during egg-laying, and egg production ceases when reserves fall to a threshold level, regardless of the number of eggs produced (Ankney and MacInnes, 1978). Clutch size thus appears to be limited by the resources of the female, and it is strongly correlated with the amount of reserve present when egg-laying commences (Fig. 4; see also Raveling, 1979).

It is difficult to generalize from Arctic-nesting geese to other precocial species because the geese are extreme in their reliance on prebreeding reserves. Females of other precocial species feed (or are fed) to a variable degree during egg laying and incubation, and thus can at least partially adjust for deficits incurred before or during laying. Evidence from several duck species (e.g., Korschgen, 1977; Drobney, 1980; Krapu, 1981) indicates, however, that even in such species the size of female reserves has an important effect on clutch size

Further evidence for the egg formation ability hypothesis consists of a correlation between intraspecific variation in clutch size and food supply, which has been demonstrated in waterfowl (Bengtson, 1971) and the Galliformes (Siivonen in Lack, 1966, pp. 6–7; Moss et al., 1981; see also Klomp, 1970; von Haartman, 1971; Savory, 1975). However, at least some species appear to respond to enormous reductions in food supply with only very small reductions in clutch size (Bengtson, 1971), and the clutch sizes of others are insensitive to food supply (e.g., Högstedt, 1974). In the most thoroughly studied galliform, the Red Grouse (*Lagopus lagopus scoticus*), clutch size is only slightly sensitive to feeding conditions, but other demographic factors, especially the survival of young immediately after hatching, are quite sensitive to food supply (Moss et al., 1981). This suggests that food could limit clutch

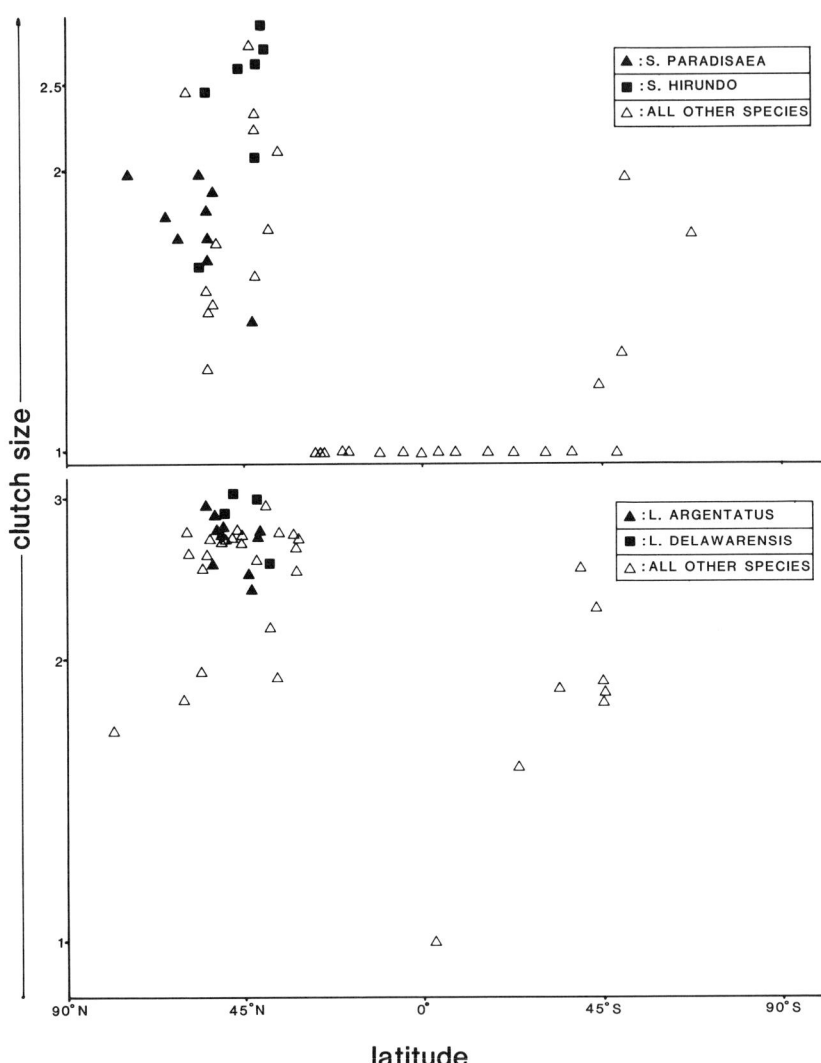

FIGURE 3. Clutch size vs. latitude in Sterninae (above) and Larinae. Data sources are available from Winkler on request.

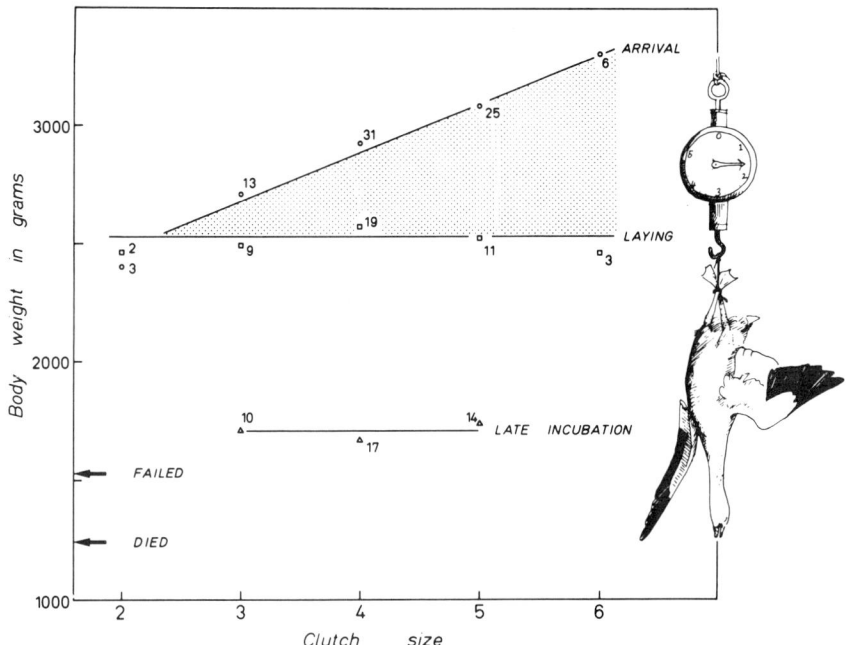

FIGURE 4. Change in body weight in relation to clutch size of female Lesser Snow Geese, *Chen caerulescens*, in the course of breeding in Arctic Canada. Numbers indicate sample sizes of birds collected at the various stages of reproduction. For comparison, mean weights of failed breeders, i.e., those abandoning eggs, and of females dying from starvation on the nest are shown (assembled from data in Ankney and MacInnes, 1978; reprinted from Drent and Daan, 1980; courtesy of *Ardea*).

size through egg formation ability via effects on egg quality as well as egg number. The egg formation ability hypothesis predicts a correlation between feeding conditions and either clutch size or chick viability, or both. Such correlations appear to be widespread among precocial birds that do not feed young, but they are not universal.

One might also expect some indication that food is at a premium during egg-formation. Yet in at least some shorebirds (*Vanellus* spp.), time budgets do not appear to be constrained during egg formation (Walters, in press). Large amounts of time are devoted to loafing during this period, time that could be devoted to foraging if females had difficulty procuring sufficient food. Loafing is indeed reduced in later demanding stages of the breeding cycle (Walters, in press). Food quantity therefore does not appear limiting, although food quality might be.

The egg formation ability hypothesis has most often been invoked for precocial non-feeders, but it is actually better supported for some precocial feeders than for most non-feeders. Relations between food supply and clutch size have long been suggested in the Laridae (e.g., Bateson and Plowright, 1959; Lack, 1968; Lemmetyinen, 1973; Coulson and Horobin, 1976). More recently, Nisbet (1977) has shown clutch size in the Common Tern (*Sterna hirundo*) to be influenced positively by the food intake of the female during the egg formation period. The female's food intake depends on courtship feeding by the male, and the amount and quality of food she is fed affects virtually all aspects of her (and her mate's) reproductive output (Nisbet, 1973, 1978). In a very similar situation, Winkler (1982) has shown differences in clutch size and egg size in two populations of California Gulls (*Larus californicus*) to be related to differences in the spring food supplies for laying females. As in the terns, these differences in food supplies are mediated through differences in the relative frequency of courtship feeding (see also Tasker and Mills, 1981). This evidence is as good as exists for Arctic-nesting geese except that the consequences of laying an excessive number of eggs have not been demonstrated.

It is interesting that the aforementioned species bring a larger number of follicles to near ovulation size than they will lay (e.g., Davis, 1944; Paludan, 1952; Barry, 1962), thus casting further doubt on an absolute limitation of egg-laying ability and lending more importance to optimal working capacity considerations.

5.1.2. The Determinant–Indeterminant Dichotomy

Determinant layers are those species that will not replace fresh eggs removed from the nest, while indeterminant species are those which will lay replacement eggs for those removed early in incubation. At first sight, the distinction between determinant and indeterminant might appear to be relevant to the egg-formation ability hypothesis, as it suggests that some species are much less inclined to lay supernumerary eggs than others. The "determinancy" dichotomy, however, is plagued by problems both in its definition and in its application to clutch size theory. First, determinancy has been measured with two distinctly different methods. The Galliformes and ducks have been shown to be indeterminant because they will lay a full complement of replacement eggs if all but one egg of the original clutch is removed as soon as laying is complete. Other species, especially the Charadriiformes, will not lay replacement eggs in this experimental regime (e.g., Davis, 1942) and will be misclassified as determinant layers (e.g., Lack, 1947a; VanTyne and Berger, 1976, p. 483). If, instead, eggs are removed from the nest as they are laid, gulls and at least one shorebird and one grebe can be induced to lay supernumerary eggs (e.g., Rinkel, 1940, Klomp, 1951; Paludan, 1952; Weidmann, 1956; Parsons, 1976; Fugle and Rothstein, 1977). No clear classification and comparison of species' determinancies can be made until all species to be compared are sujected to the same experimental regime.

Even a clear categorization based on consistent experimental regime would be of limited use, however, because classification of a species as an indeterminant layer does not preclude the possibility that supernumerary laying constitutes an effort beyond the optimal working capacity. Furthermore, determinant species need not be limited by female egg resource levels. If, however, supernumerary laying could be shown not to reduce a species lifetime reproductive output, then this would comprise good evidence against the egg-formation ability hypothesis for that species. Given our present imperfect understanding of the mechanisms causing replacement of removed eggs, it seems wisest to regard species such as Arctic-nesting geese and the Galliformes as opposite ends of a spectrum of propensities to lay replacement eggs, with the Charadriiformes somewhere in between. A rigid dichotomy may obscure fundamental similarities in the physiology and ecology of egg production, even though some species may differ in the mechanisms of laying termination. For example, gulls probably respond to tactile stimuli from the eggs in terminating egg production (e.g., Paludan, 1952; Weidmann, 1956; Parsons, 1976), while the Galliformes appear to respond to visual cues from the clutch (Steen and Parker, 1981). Although the physiology of clutch size determination is accessible to experimentation, it remains a poorly understood aspect of the reproductive biology of most birds.

5.2. Parental Behavior

The traditional alternative to the egg-formation hypothesis, especially for feeders, is that clutch size is limited by the costs and benefits of parental behavior. These costs and benefits are determined by the environment, and they in turn constrain clutch size as described by Lack (see above).

The best evidence for this hypothesis in precocial species comes from twinning studies on Procellariiformes and other sea birds (e.g., Huntington in Palmer, 1962; Rice and Kenyon, 1962; Harris, 1966). In these single-egg species, it is quite apparent that the parents would be unable to feed two chicks. The Gruidae (Miller, 1973), and at least the *Eudyptes* penguins (Williams, 1980) are other precocial groups with small clutches that seem incapable of feeding or protecting a larger number of chicks.

The single-egg species have long been recognized as special in that their food is sparse or expensive to obtain and in that they lack the potential to reduce clutch size (Ricklefs, 1968, 1979a). They are hence especially likely to be limited in their ability to feed young. The evidence that precocial species with larger clutches are limited in this way is less convincing. For example, gulls of various species are able to raise larger numbers of young than their normal brood size, and young in these artificially-enlarged broods fledge in as good a condition and at as great a weight as chicks from broods of normal or smaller sizes (e.g., Coulson and White, 1958; Harris and Plumb, 1965; Pearson, 1968; Ward, 1973; Haymes and Morris, 1977). Lack (1968) and others have attributed this ability to raise extra young to artificially favorable feeding conditions due to use of refuse dumps and fish cleaning stations. The only study in which this possibility was carefully investigated, however, indicated that chicks in both normal and super-normal broods had poorer growth and survival in pairs that capitalized on such food sources than in pairs feeding on natural food (Ward, 1973). Whether the increased parental effort required to feed chicks in enlarged broods has any effect on the parents' mortality and future reproductive effort has not been determined.

It has generally been assumed that young of non-feeders are not demanding in parental time or energy (e.g., Kendeigh, 1952; Parmelee and Payne, 1973; Emlen and Oring, 1977; Welty, 1982, p. 293). A corollary of this assumption is that parental behavior is not sufficiently demanding to limit clutch size in such species. There certainly are species, such as a brood-parasitic duck (*Heteronetta*) and megapodes (Frith, 1956, 1962; Clark, 1964; Weller, 1968) in which the cost of parental care is negligible, since the young receive no care, but generally the assumption has no empirical basis. In fact, there is a growing body of data indicating considerable time and energy costs of parental care in at least some non-feeders.

Parental behavior in precocial non-feeders consists primarily of vigilant behavior, anti-predatory behavior, incubation, brooding, and leading and following young. Time budget studies of shorebirds (Gibson, 1978; Ashkenazie and Safriel, 1979; Maxson and Oring, 1980; Walters, 1982, in press) suggest that these behaviors require considerable time in some species, although not in others. Walters's study (1980, 1982, in press) of four species of lapwing is a case in point. In two species, adults tend their young actively, moving with them constantly, and these species devote considerable time to parental behavior (Table III). This commitment detracts from foraging time, suggesting that the time required to tend young could limit the number of young for which adults can care and still meet their maintenance energy requirements (Table III). In the remaining two species tending is inactive, that is, adults monitor their young from fixed positions, moving to new positions relatively infrequently. The time cost of parental behavior is much less in these species (Table III).

The assumption that parental behavior has a uniformly low cost

TABLE III
Differences between Actively- and Inactively-Tending Lapwings

Characteristic	Active tender	Inactive tender
Time devoted to parental behavior during brood-tending period	55% (S)[a]	26% (B)
Foraging time for pre-breeding period vs. brood-tending period[b]	67% vs. 23% (S)*	70% vs. 60% (B) 35% vs. 17% (L)*
Time devoted to parental behavior when tending 1 chick vs. when tending >1 chick	50% vs. 62% (S)*	17% vs. 18% (L)[c]
Foraging time when tending 1 chick vs. tending >1 chick	26% vs. 17% (S)*	22% vs. 15% (L)
Mean distance between chick and nearest adult	8.5 m (S)	13.5 m (L) 27.5 m (B)
Mean adult–chick distance when chick inactive vs. when chick foraging[d]	5.4 vs. 6.6 m (S,d) 10.2 vs. 11.1 m (S,sj)	2.4 vs. 8.8 m (B,d)* 16.2 vs. 25.8 m (B,sj)*
Mean adult–chick distance when brood contains 1 chick vs. when brood contains >1 chick[e]	7.4 vs. 10.2 m (S)	9.6 vs. 13.9 m (L) 7.9 vs. 15.5 m (B)

[a]Data from Walters (1980, 1982, in press). Time values are percent of daylight hours. Unless otherwise noted, data are based on the period from hatching until the young are able to fly. Species are indicated by the capital letter in parentheses: S = Southern Lapwing (*Vanellus chilensis*), B = Blacksmith Plover (*V. armatus*), and L = Long-toed Lapwing (*V. crassirostris*).
[b]Significant differences between time budgets ($p<0.05$) are indicated by *.
[c]Time devoted to mildly vigilant behavior was not measured in Long-toed Lapwing, thus estimate represents minimum time devoted to parental behavior.
[d]A chick that was resting, sitting, or preening was considered inactive. For this comparison the brood-rearing period is divided into two segments, that during which the chicks had downy plumage (d = downy) and that during which they had juvenile plumage but were not yet able to fly (sj = small juvenile). Significant differences ($p<0.05$) between the distances compared are indicated by *.
[e]For foraging chicks not yet able to fly, except for Blacksmith Plover, for which data from chicks engaged in other activities and chicks able to fly were included due to small sample sizes.

among precocial non-feeders is likely incorrect. In any case, that parental behavior is costly is neither a necessary nor a sufficient condition for it to limit clutch size. The crucial condition is that there be a relationship between brood size and either the costs to parents or the benefits to young of parental behavior. In lapwings, the time cost of parental behavior increases with brood size in active tenders, but the benefits of parental care probably remain fairly constant with increasing brood size if the benefits depend on adult–young distances (Table III, see below). By contrast, the costs of inactive tending probably increase little, if at all, with increasing brood size, but the benefits to young dependent on adult–young distances may decrease with increasing brood size (Table III).

Safriel (1975) has proposed that the farther a chick is from a tending parent, the more likely that it will be preyed upon. He further proposes that this limits clutch size in shorebirds because predators tend to take whole broods once a chick is detected. He found that artificially enlarged broods of Semipalmated Sandpipers (*Calidris pusilla*) were more likely to disappear, presumably due to predation, than were broods of normal size, so that reproductive success decreased with increased brood size. Walters (1982; in press) suggests that Safriel's model be restricted to inactive tenders, and that prevention of chicks becoming lost is another benefit dependent on adult–young distance. Thus, increasing disparities between costs and benefits with increasing brood size could lead to a limitation of clutch size in both actively tending and inactively tending shorebirds.

Evidence that parental behavior could limit clutch size in other non-feeders is sparse. Data from a species of Arctic-nesting goose (*Anser brachyrhynchus*) (Lazarus and Inglis, 1978) indicate that parental behavior is only moderately demanding and that its cost is not very sensitive to brood size (Fig. 5). By contrast, a study (Pellis and Pellis, 1982) of the Australian Cape Barren Goose (*Cereopsis novaehollandiae*) suggests that the costs of active tending may be similar to those in lapwings (Fig. 5). The relationship between brood size and benefits of parental behavior was not measured in either of these studies. Until more is known of the dispersion of duck, ratite, and galliform broods and their costs and benefits of tending, the applicability of the parental behavior hypothesis to these non-feeders cannot be evaluated.

5.3. Incubation Ability

Support for the hypothesis that precocial bird clutch sizes are limited by the ability to incubate eggs comes from experiments with Char-

CLUTCH SIZE IN PRECOCIAL BIRDS 51

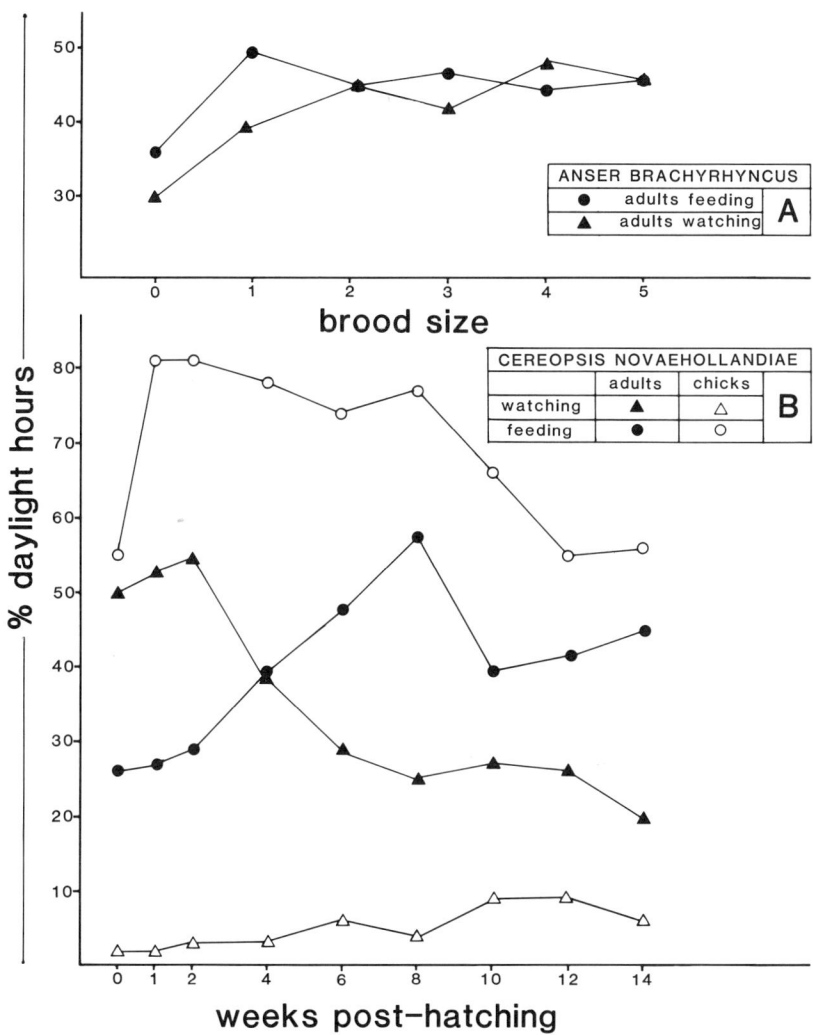

FIGURE 5. Parental care in geese in relation to brood size (A) in *Anser brachyrhynchus* (Lazarus and Inglis, 1978) and in relation to brood age (B) in *Cereopsis novaehollandiae* (Pellis and Pellis, 1982). Although increasing brood size in *Anser* is associated with slightly increased time watching for predators, this activity results in reduced maintenance activities (not shown) rather than reduced feeding. By contrast, as broods increase in age in *Cereopsis*, adult watching decreases and adult feeding increases.

adriiformes that indicate that some species have difficulty incubating unusually large clutches (Beer, 1961; Baerends et al., 1970; Gibson, 1971; Andersson, 1976; Hills, 1980). The incubation ability hypothesis might also apply to at least the Eudyptes penguins (Williams, 1980) and could well apply to the Procellariiformes.

Andersson (1978) suggested that shorebird clutch sizes are limited by three factors: 1) the need for large and advanced precocial young, 2) the need for good heat economy in the clutch (see also Miller, 1979), and 3) the ability of birds to cover eggs (determined by brood patch area). This hypothesis can explain the sharp truncation of shorebird clutch sizes at the upper end of their distributions (Fig. 2); however, it may beg the question of adaptation. Although incubation ability may constrain maximal clutch size, it determines clutch size only if the other two premises of Andersson's argument are satisfied. Selective factors driving brood patch area per se must also be sought. Brood patch number (and area) varies among gull species (e.g., Johnston, 1956; Beer, 1965; Harris, 1970; Howell, et al., 1974), suggesting variability in this trait over evolutionary time. Thus, incubation capacity and clutch size may coevolve in response to whatever factors determine clutch size (Andersson, 1976). Regardless, the hypothesis that more than the normal number of eggs cannot be incubated successfully is easily tested, and it should be tested in any situation in which it appears at all likely.

In any event, the incubation ability hypothesis apparently applies only to those taxa with truncated clutch size distributions. In other groups such as rallids, anatids, and phasianids the number of eggs hatched continues to increase with egg number far beyond the modal clutch size, although the percentage hatched declines (Lack, 1947b; Leopold, 1951; Hilden, 1964; Fredrickson, 1969; Heusmann, 1972). These species can clearly incubate a number of eggs greater than the usual clutch size. Finally, in those ratites and tinamous in which several females lay eggs in a single nest, the tending male obviously can incubate more eggs than an individual female lays.

5.4. Nest Predation

Because females lay at best one egg per day, clutch size is positively correlated with the length of time during which a nest is exposed to predation. Nest predators generally consume entire clutches, so prolonging laying increases the risk that eggs will be lost. Perrins (1977) showed that if clutch size is large and predation pressure is sufficiently high, then birds will be selected to limit their clutch sizes. However, his calculations indicate that the clutch size and predation rates required for this to occur preclude this hypothesis from applying to most precocial birds. The predation rates required in his model would result in less than 1% of nests being successful for ratites and shorebirds, and about 5% being successful for most waterfowl and galliforms, which is clearly unrealistic. Only for a few waterfowl and some galliforms is the nest predation required conceivable, and no field tests have been conducted to determine if the required predation rates occur in these species.

This hypothesis entails a direct environmental effect of predation. No other mechanism by which predation pressure could determine clutch size in precocial birds through a direct effect has yet been specified (but see Yom-Tov, 1974; Slagsvold, 1982). Predation pressure may, of course, be one of the factors affecting indirect life-history adjustments in clutch size, regardless of its role as a direct limiting factor.

5.5. Explanations of Geographic and Seasonal Trends

As our understanding of the determination of clutch size increases, we should grow better able to explain geographic and seasonal trends. Geographic patterns may be interpreted in at least two ways. First, they may reflect regional differences in factors affecting life-history strategies, as, for example, between the tropics and temperate zone, or between the maritime habitats used by shorebirds and waterfowl in the southern temperate zone and the inland habitats they frequent in the northern temperate zone (Walters, in press). Second, geographic variation may be related to regional differences in directly-acting environmental factors. Hesse (1923) and Lack (1947a; et seq.) suggested that birds at higher latitudes have more time available to forage, and therefore can raise more young. The data present several problems for this interpretation, perhaps the most serious being that nocturnal birds display the same latitudinal trends as do diurnal birds (e.g., von Haartman, 1971). As an alternative to the day-length hypothesis, Ricklefs (1980) resurrected Ashmole's (1963) hypothesis that more food is available per individual at higher latitudes due to the seasonality of resource production. This hypothesis appears to explain latitudinal trends better than any other food-related hypothesis, and it may also help explain the lower clutch sizes in the generally less seasonal, more maritime environments of the southern hemisphere. Ashmole's hypothesis entails a strong emphasis on food supply as a limiting factor on clutch size, either directly through limitation of egg formation ability or indirectly through limitation of chick-rearing ability. Other parental behavior hypotheses and the incubation ability hypothesis cannot account for geographic variation except through indirect life-history effects on reproductive effort. It will be difficult to determine to what extent geographic variation is due to direct environmental effects vs. indirect life history effects, but this is an essential task. Studies of species that are exceptions to the usual trends should be especially instructive if one can identify potential limiting factors or aspects of life history in which exceptional taxa experience an environmental cline different from typical taxa.

The nearly ubiquitous seasonal declines in clutch size are not easily explained by any single direct environmental effect because it is difficult to imagine any environmental factor that changes in a similar direction with season in all the diverse habitats and different times of year in which a seasonal decline in clutch size has been observed. Young seabirds are known to breed later and lay smaller clutches (e.g., Mills, 1973; Ryder, 1980), suggesting that the seasonal decline may be an age-related life history optimization. In at least kittiwakes (*Rissa tridactyla*), however, there is a negative effect of time of breeding on clutch size, even when the effects of age are eliminated (Coulson and White, 1961). Thus, direct environmental effects may play a role in seasonal declines in specific cases. None of these interpretations are rejectable at present, and a better understanding of the covariation of age, time of breeding and clutch size is an important area for further research (see Perrins, 1970; Hannon et al., 1982).

6. DISCUSSION

6.1. The Web of Causation

Given the enormous variety of life-styles among birds, it would be surprising indeed if clutch size did not vary as widely as it does. The great ecological diversity of birds also makes it unlikely that all their clutch sizes should be controlled by the same factors, or that the clutch size of a given species should everywhere and always be determined in the same way. Instead, a web of demographic and ecological factors interact to determine clutch size (Fig. 1), and ecologists must beware of "getting nicked by Occam's razor" (Hilborn and Stearns, 1982). This is not to suggest that the determination of clutch size is a mystical, unpredictable process. Specific hypotheses may be generated from Fig. 1 and tested for individual species. This single-species approach is a limited one, but no more so than the single-factor approach that has produced the current disjointed literature on clutch size, and it has the potential to lead to a general theory by defining which of the proposed pathways of Fig. 1 are most important for groups with various characteristics.

6.2. Levels of Explanation

The distinction between proximate and ultimate causation which has often been blurred in the literature on clutch size is important because it concerns the nature of the causal link between the organism's environment and its adaptations in breeding biology. Proximate causation lends itself to mechanistic "how" explanations, while ultimate causation lends itself to evolutionary "why" explanations. Proximate and ultimate causes can act side-by-side: the variation in clutch size in Arctic-nesting geese (see above) is due to the proximate modifying effect of food supply, with the limit to clutch size being determined by the lower threshold of resources below which the female will not lay. The lower resource threshold ultimately depends on such indirect life history effects as the energy requirements of parental care, intensity of predation and food availability.

Although food supply probably has a proximate effect on gametic effort in these geese, this same relationship between food supply during egg laying and clutch size may reflect an ultimate effect in species in which clutch size is limited by parental effort. That is, food supply during egg-laying may be a good predictor of food supply during brood rearing, and it may thus be used as a cue to adjust clutch size to future conditions.

Bengtson's (1971) study provides some interesting results in this regard. He found clutch sizes to be reduced in four species of waterfowl in years in which densities of invertebrate prey eaten by both adults and young were low. Clutch size was not reduced in an additional species in which neither adults nor young used these prey. These results are consistent with either a proximate or an ultimate effect of food supply during egg-laying. However, in two additional species only young would be expected to use these prey. One of these species displayed a reduced clutch, and the other displayed a strong but statistically insignificant trend in that direction. This implies that food is involved in an ultimate chain of causation, acting as a cue for future conditions.

The relative importance of proximate and ultimate factors in clutch size determination probably depends on how well the female can anticipate future conditions. If she must rely on long-term averages for a prediction of conditions such as the future food supply for the young, one would expect clutch size to be relatively insensitive to conditions during the laying period.

In sum, levels of explanation pose two problems in clutch size research. First, hypothesized ultimate causation must be compatible with known proximate mechanisms. Second, the possibility that a directly acting environmental factor can act as either the proximate or ultimate causes of clutch size variation must be considered. These issues are especially critical in the case of the direct environmental effect of food on egg-formation ability.

6.3. Complications

There are numerous considerations specific to particular taxa that further complicate the determination of clutch size. For example, in ostriches, rheas, and many tinamous, several females lay in a single nest, so that brood size is to some extent independent of the clutch size of a single female (Sauer and Sauer, 1966; Jenni, 1974; Bruning, 1974; Davies, 1976). Hence, an individual female's egg production may be limited by social factors. For example, dominant female ostriches lay more eggs than other females and toss eggs laid by others out of the nest (Bertram, 1979). Subordinate females may be able to lay more eggs than they do, but social circumstances limit their opportunity to produce viable eggs.

Peculiar social systems can present additional complications. "Supernormal clutches" in colonial precocial birds have been ascribed, at least since Call (1891), to intraspecific nest parasitism. Intraspecific nest parasitism is widespread in waterfowl (e.g., Weller, 1959; Andersson and Eriksson, 1982), and the incidence of two females laying in a single nest is high in rails. Supernormal clutches are associated with polygynous matings and female–female pairs in several gull species (Hunt and Hunt, 1977; Conover et al., 1979; Ryder and Somppi, 1979; Hunt, 1980). These unusual departures from single female laying provide rich research opportunities for the evolutionary biologist to test hypotheses relating mating systems to clutch size regulation (see also Vehrencamp, 1977).

7. CONCLUSIONS

In the years since the reviews of Lack (1968) and Klomp (1970) there has been a growing awareness of the need to build life-history

CLUTCH SIZE IN PRECOCIAL BIRDS   57

considerations into explanations of avian clutch sizes. Life-history effects might account for much of the observed geographic variation, differences between major taxa, and differences between species within taxa in clutch size. But such applications of life-history theory will not be possible until ornithologists can gather more sophisticated longitudinal data on the reproductive efforts and demography of wild birds, particularly on the link between reproductive effort and future reproductive success. This assumed relationship is the basis of all postulated life-history effects, and it is often crucial for the identification of direct environmental limitation of clutch size. Studies in which enlarged clutches are successful can disprove direct environmental determination only if the resulting change in reproductive effort has no detrimental effects on future reproduction.

Most empirical work has attempted to identify the environmental factors that limit clutch size. The evidence demonstrates variation among precocial birds with respect to the mechanism for clutch size limitation. Lack's (1968) proposal for non-feeders, that females are limited by their ability to form eggs, appears to apply to Arctic-nesting geese and probably megapodes when allowance is made for optimal working capacity considerations, and it may apply to additional taxa. However, other hypotheses are equally viable for these groups.

The most promising alternative to the egg-formation hypothesis for non-feeders is that clutch sizes are limited by parental behavior, at least in shorebirds and one species of goose. Data on the relationship between brood size and costs to adults and benefits to young of parental behaviors are needed for ratites, ducks, and galliforms to evaluate the parental behavior hypothesis for these groups.

Among feeders, Lack's parental behavior hypothesis remains viable, but growing evidence from the Laridae suggests that the egg-formation hypothesis also needs to be considered in this group. Further progress requires elucidation of the relationship between brood size, reproductive effort, and future reproduction, as well as the effects on clutch size of food supply during egg-laying. Only for seabirds with single-egg clutches is there unequivocal evidence for the parental behavior hypothesis.

Clutch size determination by incubation ability may have limited application, but it is easy to test by experiment. Nest predation appears to have little effect on the determination of precocial bird clutch sizes. Both these factors, however, require additional theoretical and empirical exploration.

The general theory of clutch size in precocial birds has made significant advances since Lack's (1968) work, yet in some ways little has changed. His ideas about limiting factors have been built into a complex theory, but have not been discredited in their now restricted role. In fact, they are still leading hypotheses. This is due in part to the lack of sufficiently detailed empirical studies of the reproductive behavior of precocial species, although promising starts have been made on geese, grouse, shorebirds, and larids. A lesson from these studies seems to be that few if any birds have only one main mechanism of clutch size determination. Analyses of clutch size should therefore benefit from comprehensive study of all components of reproductive effort and their effects on future breeding and survival.

APPENDIX I
Clutch Sizes of Precocial Birds

	Taxa	Modal clutch size of typical species[a]
Species that do not feed young		
Ratites	Struthionidae	3–8
	Rheidae	4–5
	Casuariidae	3–5
	Dromiceidae	8–9
	Apterygidae	1
Tinamiformes	Tinamidae	3–6
Anseriformes	Anhimidae	2–5
	Anatidae	4–11
Galliformes	Megapodiidae	[b]
	Cracidae	2–3
	Tetraonidae	7–9
	Phasianidae	6–14
	Numididae	8–10
	Meleagrididae	10–11
Charadriiformes	Jacanidae	4
	Ibidorhynchidae	4
	Recurvirostridae	4
	Thinocoridae	4
	Charadriidae	2–4
	Scolopacidae	3–4
Species that feed young		
Sphenisciformes	Spheniscidae	1–2
Gaviiformes	Gaviidae	2

APPENDIX I (Continued)

Taxa		Modal clutch size of typical species[a]
Species that feed young (cont.)		
Podicipediformes	Podicipedidae	3–5
Procellariiformes	Diomedeidae	1
	Procellariidae	1
	Hydrobatidae	1
	Pelecanoididae	1
Phoenicopteriformes	Phoenicopteridae	1
Gruiformes	Mesitornithidae	1–3
	Turnicidae	3–4
	Pedionomidae	4
	Gruidae	2
	Aramidae	4–8
	Psophiidae	7?
	Rallidae	4–8
	Heliornithidae	2
	Rhynochetidae	1
	Eurypigidae	2–5
	Cariamidae	2
	Otidae	2–4
Charadriiformes	Rostratulidae	2–4
	Haematopodidae	2–3
	Burhinidae	2
	Glareolidae	2–4
	Dromadidae	1
	Chionididae	2–3
	Pluvianellus	2
	Stercorariidae	2
	Laridae	1–3
	Rynchopidae	2–4
	Alcidae	1–2
Caprimulgiformes	Caprimulgidae	2–3

[a]References are Mackworth-Praed and Grant, 1957; Thomson, 1964; Johnson, 1965; Sauer and Sauer, 1966; Lack, 1968; Maclean, 1972; Bruning, 1974; Davies, 1976; Cramp, 1977; Pizzy, 1980; Johnsgard, 1981.
[b]Megapodes do not lay discrete clutches.

APPENDIX II
Geographic Variation in Clutch Sizes of Waterfowl[a]

Taxa[b]	Region[c]	Typical modal clutch size			
		<5	5–7	8–10	>10
Anserini (geese)	Arctic	5	5	0	0
	N temperate	0	3	0	0
	Tropics	1	0	0	0
Anserini (swans)	Arctic	1	0	0	0
	N temperate	0	2	0	0
	S temperate	0	3	0	0
Dendrocygnini (tree ducks)	Tropics	0	0	4	3
	S temperate	0	0	0	1
Tadornini and Tachyerini (shelducks)	N temperate	0	0	2	0
	Tropics	0	2	2	0
	S temperate	0	9	2	1
Anatini and Merganettini (dabbling ducks)	Arctic	0	1	2	0
	N temperate	0	0	10	2
	Tropics	0	3	7	0
	S temperate	3	6	4	0
Aythyini (diving ducks)	Arctic	0	1	2	0
	N temperate	0	0	8	0
	Tropics	0	1	1	0
	S temperate	0	1	2	0
Mergini (eiders and mergansers)	Arctic	2	5	5	0
	N temperate	0	0	3	1
Cairinini (perching ducks)	N temperate	0	0	0	2
	Tropics	0	1	6	2
	S temperate	0	0	1	0
Oxyurini (stiff-tailed ducks)	N temperate	0	1	1	0
	Tropics	0	3	0	0
	S temperate	2	1	0	0

[a] Data sources as in Appendix I. Values are the number of species with the indicated model clutch size.
[b] Classification follows Johnsgard (1978).
[c] Determined by location of majority of breeding range: Arctic = above 60°N; N temperate = 23.5°N–60°N; Tropics = 23.5°S–23.5°N; S temperate = below 23.5°S.

ACKNOWLEDGMENTS. Paul B. Allard provided valuable assistance with the preparation of the figures. Malte Andersson, Gilbert Grant, Susan Hannon, Scott Hatch, Frank A. Pitelka, Peter Pressley, and Samuel Zeveloff provided important suggestions on earlier versions of the manuscript. This is paper No. 8748 of the Journal Series of the North Carolina Agricultural Research Service, Raleigh, N. C., 27650.

REFERENCES

Andersson, M., 1976, Clutch size in Long-tailed Skua *Stercorarius longicaudus*—some field experiments, *Ibis* **118**:586–588.
Andersson, M., 1978, Optimal egg shape in waders, *Ornis Fennica* **55**:105–109.
Andersson, M., and Eriksson, M. O. G., 1982, Nest parasitism in goldeneyes (*Bucephala clangula*): Some evolutionary aspects, *Am. Natural.* **120**:1–16.
Ankney, C. D., and MacInnes, C. D., 1978, Nutrient reserves and reproductive performance of female Lesser Snow Geese, *Auk* **95**:459–471.
Ashkenazie, S., and Safriel, U. N., 1979, Time-energy budget of the Semipalmated Sandpiper (*Calidris pusilla*) at Barrow, Alaska, *Ecology* **60**:783–799.
Ashmole, N. P., 1963, The regulation of numbers of tropical oceanic birds, *Ibis* **103**:458–473.
Baerends, G. P., Drent, R. H., Glas, P., and Groenewold, H., 1970, An ethological analysis of incubation behaviour in the Herring Gull, *Behaviour Suppl.* **17**:134–235.
Baker, J. R., 1938, The evolution of breeding seasons, in: *Evolution: Essays in Aspects of Evolutionary Biology* (G. R. DeBeer, ed.), Clarendon Press, Oxford, pp. 161–177.
Barry, W. T., 1962, Effect of late seasons on Atlantic Brant reproduction, *J. Wildl. Mgmt.* **26**:19–26.
Bateson, P. P. G., and Plowright, R. C., 1959, The breeding biology of the Ivory Gull in Spitzbergen, *Br. Birds* **1959**:105–114.
Batt, B. D. J., and Prince, H. H., 1979, Laying dates, clutch size and egg weight of captive Mallards, *Condor* **81**:35–41.
Beer, C. G., 1961, Incubation and nest-building behaviour of Black-headed Gulls. I. Incubation behaviour in the incubation period, *Behaviour* **18**:62–106.
Beer, C. G., 1965, Clutch size and incubation behavior in Black-billed Gulls (*Larus bulleri*), *Auk* **82**:1–18.
Bengtson, S. A., 1971, Variations in clutch-size in ducks in relation to the food supply, *Ibis* **113**:523–526.
Bertram, B. C. R., 1979, Ostriches recognize their own eggs and discard others, *Nature* **279**:233–234.
Blem, C. R., 1978, The energetics of young Japanese Quail, *Coturnix coturnix japonica*, *Comp. Biochem. Physiol.* **59**(2A):219–223.
Brockelman, W. K., 1975, Competition, the fitness of offspring, and optimal clutch size, *Am. Natural.* **109**:677–699.
Bruning, D. F., 1974, Social structure and reproductive behavior in the Greater Rhea, *Living Bird* **13**:251–294.
Call, A. B., 1891, Notes at random, *Oologist* **8**:198.
Charnov, E. L., and Krebs, J. R., 1974, On clutch-size and fitness, *Ibis* **116**:217–219.
Clark, G. A., Jr., 1964, Life histories and the evolution of megapodes, *Living Bird* **3**:149–168.

Cody, M. L., 1966, A general theory of clutch size, *Evolution* **20**:174–184.
Conover, M., Miller, D. E., and Hunt, G. L., 1979, Female–female pairs and other unusual reproductive associations in Ring-billed Gulls and California Gulls, *Auk* **96**:6–9.
Coulson, J. C., 1963, Egg size and shape in the Kittiwake (*Rissa tridactyla*) and their use in estimating the age composition of populations, *Proc. Zool. Soc. Lond.* **140**:211–227.
Coulson, J. C., and Horobin, J., 1976, The influence of age on the breeding biology and survival of the Arctic Tern *Sterna paradisaea*, *J. of Zool., Lond.* **178**:247–260.
Coulson, J. C., and White, E., 1958, The effect of age on the breeding biology of the Kittiwake, *Rissa tridactyla*, *Ibis* **100**:40–51.
Coulson, J. C., and White, E., 1961, An analysis of the factors influencing the clutch size of the Kittiwake, *Proc. Zool. Soc. Lond.* **B6**:207–217.
Cracraft, J., 1974, Phylogeny and evolution of the ratite birds, *Ibis* **116**:494–521.
Cramp, S., ed., 1977, *Handbook of the Birds of Europe the Middle East and North Africa: The Birds of the Western Palearctic, Volume 1, Ostrich to Ducks*, Oxford University Press, Oxford.
Crossner, K. A., 1977, Natural selection and clutch size in the European Starling, *Ecology* **58**:885–892.
Davies, S. J. F., 1976, The natural history of the Emu in comparison with that of other ratities, *Proc. XVI Int. Ornithol. Cong.* 109–120.
Davis, D. E., 1942, Number of eggs laid by Herring Gulls, *Auk* **59**:549–554.
Davis, D. E., 1944, The occurrence of burst atretic follicles in birds, *Anat. Rec.* **90**:307–309.
Davis, J. W. F., 1975, Age, egg-size, and breeding success in the Herring Gull *Larus argentatus*, *Ibis* **117**:460–473.
Dawkins, R., and Carlisle, T. R., 1976, Parental investment and mate desertion: A fallacy, *Nature* **262**:131–133.
Drent, R. H., and Daan, S., 1980, The prudent parent: Energetic adjustments in avian breeding, *Ardea* **68**:225–252.
Drobney, R. D., 1980, Reproductive bioenergetics of wood ducks, *Auk* **97**:480–490.
Dunn, E. H., 1976, Development of endothermy and existence energy expenditure of nestling Double-crested Cormorants, *Condor* **78**:350–356.
Emlen, S. T., and Oring, L. W., 1977, Ecology, sexual selection, and the evolution of mating systems, *Science* **197**:215–233.
Fiala, V., 1978, Beitrag zur Populationsdynamik und Brutbiologie des Blasshuhns (*Fulica atra*), *Folia Zool.* **27**:349–369.
Foster, M. S., 1974a, A model to explain molt-breeding overlap and clutch size in some tropical birds, *Evolution* **28**:182–190.
Foster, M. S., 1974b, Rain, feeding behavior, and clutch size in tropical birds, *Auk* **91**:722–726.
Fredrickson, L. H., 1969, An experimental study of clutch size of the American Coot, *Auk* **86**:541–550.
Frith, H. J., 1956, Breeding habits in the family Megapodiidae, *Ibis* **98**:620–640.
Frith, H. J., 1962, *The Mallee-Fowl*, Angus and Robertson, Sydney.
Fugle, G. N., and Rothstein, S. I., 1977, Clutch size determination, egg size, and eggshell thickness in the Pie-billed Grebe, *Auk* **94**:371–373.
Gibson, F., 1971, The breeding biology of the American Avocet (*Recurvirostra americana*) in central Oregon, *Condor* **73**:444–454.
Gibson, F., 1978, Ecological aspects of the time budget of the American Avocet, *Am. Midl. Natural.* **99**:65–82.

Glutz von Blotzheim, U. N., 1964, *Die Brutvogel der Schweiz*, Verlag Aargauer Tagblatt A G, Aarau.
Glutz von Blotzheim, U. N., ed., 1973, *Handbuch der Vögel Mitteleuropas, Band 5,Galliformes and Gruiformes*, Akademische Verlagsgesellschaft, Wiesbaden.
Glutz von Blotzheim, U. N., ed., 1975, *Handbuch der Vögel Mitteleuropas, Band 6, Charadriiformes (1. Teil)*, Akademische Verlagsgesellschaft, Wiesbaden.
Glutz von Blotzheim, U. N., ed., 1977, *Handbuch der Vögel Mitteleuropas, Band 7, Charadriiformes (2. Teil)*, Akademische Verlagsgesellschaft, Wiesbaden.
Goodman, D., 1974, Natural selection and a cost ceiling on reproductive effort, *Am. Natural.* **108**:247–268.
von Haartman, L., 1971, Population dynamics, in: *Avian Biology, Volume I* (D. S. Farner and J. R. King, eds.), Academic Press, New York, pp. 392–461.
Hannon, S. J., Lennart, G. S., and Zwickel, F. C., 1982, Spring movements of female Blue Grouse: Evidence for socially induced delayed breeding in yearlings, *Auk* **99**:687–694.
Harris, M. P., 1966, Breeding biology of the Manx Shearwater *Puffinus puffinus*, *Ibis* **108**:17–33.
Harris, M. P., 1970, Breeding ecology of the Swallow-tailed Gull, *Creagrus furcatus*, *Auk* **87**:215–243.
Harris, M. P., and Plumb, W. J., 1965, Experiments on the ability of Herring Gulls and Lesser Black-backed Gulls to raise larger than normal broods, *Ibis* **107**:256–257.
Haymes, G. T., and Morris, R. D., 1977, Brood size manipulations in Herring Gulls, *Can. J. Zool.* **55**:1762–1766.
Heinroth, O., 1922, Die Beziehungen zwischen Vogelgewicht, Eigewicht, Gelegegewicht und Brutdauer, *J. Ornithol.* **70**:172–285.
Hesse, R., 1923, Die Bedeutung der Tagesdauer für die Vogel, *Sizungsber. Naturh. Ver. Preuss. Rheinlande Westfalens* **1922A**:13–17.
Heusmann, H. W., 1972, Survival of Wood Duck broods from dump nests, *J. Wildl. Mgmt.* **36**:620–624.
Hilborn, R., and Stearns, S. C., 1982, On inference in ecology and evolutionary biology: the problem of multiple causes, *Acta Biotheor.* **31**:145–164.
Hilden, O., 1964, Ecology of duck populations in the island group of Valassaaret, Gulf of Bothnia, *Ann. Zool. Fenn.* **1**:153–279.
Hills, S., 1980, Incubation capacity as a limiting factor of shorebird clutch size, *Am. Zool.* **20**:774 (abst).
Högstedt, G., 1974, Length of the prelaying period in the lapwing *Vanellus vanellus* L., in relation to its food resources, *Ornis. Scand.* **5**:1–4.
Högstedt, G., 1980, Evolution of clutch size in birds—adaptive variation in relation to territory quality, *Science* **210**:1148–1150.
Howe, H. F., 1978, Initial investment, clutch size, and brood reduction in the Common Grackle (*Quiscalus quiscula* L.) *Ecology* **59**:1109–1122.
Howell, T. R., Araya, B., and Millie, W. R., 1974, Breeding biology of the Gray Gull, *Larus modestus*, *Univ. Calif. Publ. Zool.* **104**:1–57.
Hunt, G. L., Jr., 1980, Mate selection and mating systems in seabirds, in: *Behavior of Marine Animals, Volume 4, Marine Birds* (J. Burger, B. L. Olla, and H. E. Winn, eds.), Plenum Press, New York, pp. 113–151.
Hunt, G. L., and Hunt, M. W., 1977, Female–female pairing in Western Gulls (*Larus occidentalis*) in southern California, *Science* **196**:1466–1467.
Inglis, I. R., 1977, The breeding behavior of the Pink-footed Goose: Behavioural correlates of nesting success, *Anim. Behav.* **25**:747–764.
Jenni, D. A., 1974, Evolution of polyandry in birds, *Am. Zool.* **14**:129–144.

Johnsgard, P. A., 1973, Proximate and ultimate determinants of clutch size in Anatidae, *Wildfowl* **24**:144–149.

Johnsgard, P. A., 1978, *Ducks, Geese, and Swans of the World*, University of Nebraska Press, Lincoln.

Johnsgard, P. A., 1981, *The Plovers, Sandpipers, and Snipes of the World*, University of Nebraska Press, Lincoln.

Johnson, A. W., 1965, *The Birds of Chile and Adjacent Regions of Argentina, Bolivia, and Peru, Volume 1*, Platt establicimientos graficos S.A., Buenos Aires.

Johnston, D. W., 1956, The annual reproductive cycle of the California Gull, II. Histology and female reproductive system, *Condor* **58**:206–221.

Jones, P. J., 1976, The utilization of calcareous grit by laying *Quelea quelea*, *Ibis* **118**:575–576.

Jones, P. J., and Ward, P., 1976, The level of reserve protein as the proximate factor controlling the timing of breeding and clutch size in the Red-billed Quelea, *Quelea quelea*, *Ibis* **118**:547–573.

Jones, R. E., ed., 1978, *The Vertebrate Ovary: Comparative Biology and Evolution*, Plenum Press, New York.

Kendeigh, S. C., 1952, Parental care and its evolution in birds, *Illinois Biol. Monogr.* **22**:1–356.

Klomp, H., 1951, Over de achteruitgang van de Kievit, *Vanellus vanellus* (L.), in Nederland en gegevens over het legmechanisme en het eiproductivermogen, *Ardea* **39**:143–182.

Klomp, H., 1970, The determination of clutch-size in birds, a review, *Ardea* **58**:1–124.

Korschgen, C. E., 1977, Breeding stress of female eiders in Maine, *J. Wildl. Mgmt.* **41**:360–373.

Krapu, G. L., 1981, The role of nutrient reserves in mallard reproduction, *Auk* **98**:29–38.

Lack, D., 1947a, The significance of clutch-size, *Ibis* **89**:302–352.

Lack, D., 1947b, The significance of clutch-size in the Partridge (*Perdix perdix*), *J. Anim. Ecol.* **16**:19–25.

Lack, D., 1948, The significance of clutch-size, III., *Ibis* **90**:25–45.

Lack, D., 1954, *The Natural Regulation of Animal Numbers*, Clarendon Press, Oxford.

Lack, D., 1966, *Population Studies of Birds*, Clarendon Press, Oxford.

Lack, D., 1968, *Ecological Adaptations for Breeding in Birds*, Methuen, London.

Lack, D., and Lack, E., 1951, The breeding biology of the swift *Apus apus*, *Ibis* **93**:501–546.

Langham, N. P. E., 1974, Comparative breeding biology of the Sandwich Tern, *Auk* **91**:255–277.

Lazarus, J., and Inglis, I. R., 1978, The breeding behaviour of the Pink-footed Goose: Parental care and vigilant behaviour during the fledgling period, *Behaviour* **65**:62–88.

Lemmetyinen, R., 1973, Clutch size and timing of breeding in the Arctic Tern in the Finnish archipelago, *Ornis Fenn.* **50**:19–28.

Leopold, F., 1951, A study of nesting Wood Ducks in Iowa, *Condor* **53**:209–220.

Low, B. S., 1978, Environmental uncertainty and the parental strategies of marsupials and placentals, *Am. Natural.* **112**:197–213.

MacArthur, R. H., and Wilson, E. O., 1967, The theory of island biogeography, *Princeton Monogr. Pop. Biol.* No. 1.

McFarlane, R. W., 1975, Notes on the Giant Coot (*Fulica gigantea*), *Condor* **77**:324–327.

Mackworth-Praed, C. W., and Grant, C. H. B., 1957, *Birds of Eastern and Northeastern Africa*, Volumes 1-2, Longmans, London.

Maclean, G. L., 1972, Clutch size and evolution in Charadrii, *Auk* **89**:299–324.

MacLean, S. F., Jr., 1975, Lemming bones as a source of calcium for Arctic sandpipers (*Calidris* spp.), *Ibis* **116**:552–557.
Maxson, S. J., and Oring, L. W., 1980, Breeding season time and energy budgets of the polyandrous Spotted Sandpiper, *Behaviour* **74**:200–263.
Miller, E. H., 1979, Egg size in the Least Sandpiper *Calidris minutilla* on Sable Island, Nova Scotia, Canada, *Ornis. Scand.* **10**:10–16.
Miller, R. S., 1973, The brood size of cranes, *Wilson Bull.* **85**:436–441.
Mills, J. A., 1973, The influence of age and pair-bond on the breeding biology of the Red-billed Gull *Larus novaehollandiae scopulinus*, *J. Anim. Ecol.* **42**:147–162.
Mills, J. A., and Shaw, P. W., 1980, The influence of age on laying date, clutch size, and egg size of the White-fronted Tern, *Sterna striata*, *N. Z. J. Zool.* **7**:147–153.
Moreau, R. E., 1944, Clutch-size: A comparative study, with special reference to African birds, *Ibis* **86**:286–347.
Moss, R., Watson, A., Rothery, P., and Glennie, W. W., 1981, Clutch size, egg size, hatch weight and laying date in relation to early mortality in Red Grouse *Lagopus lagopus scoticus* chicks, *Ibis* **123**:450–462.
Murphy, G. I., 1968, Patterns in life history, *Am. Natural.* **102**:391–403.
Nice, M. M., 1962, Development of behavior in precocial birds, *Trans. Linn. Soc. N. Y.* **8**:1–211.
Niethammer, G., ed., 1966, *Handbuch der Vögel Mitteleuropas, Band I, Gaviiformes–Phoenicopteriformes*, Akademische Verlagsgesellschaft, Frankfurt.
Nisbet, I. C. T., 1973, Courtship-feeding, egg-size and breeding success in Common Terns, *Nature* **241**:141–142.
Nisbet, I. C. T., 1977, Courtship-feeding and clutch size in Common Terns, *Sterna hirundo*, in: *Evolutionary Ecology* (B. M. Stonehouse and C. M. Perrins, eds.), University Park Press, Baltimore, pp. 101–109.
Nisbet, I. C. T., 1978, Dependence of fledging success on egg-size, parental performance and egg-composition among Common and Roseate terns, *Sterna hirundo* and *S. dougallii*, *Ibis* **120**:207–215.
O'Connor, R. J., 1977, Growth strategies in nestling passerines, *Living Bird* **16**:209–238.
Owen, D. F., 1977, Latitudinal gradients in clutch size: an extension of David Lack's theory, in: *Evolutionary Ecology* (B. M. Stonehouse and C. M. Perrins, eds.), MacMillan, London, pp. 171–180.
Palmer, R. S., 1962, *Handbook of North American Birds, Volume I, Loons through Flamingos*, Yale University Press, New Haven.
Palmer. R. S., 1976a, *Handbook of North American Birds, Volume II, Waterfowl Part 1*, Yale University Press, New Haven.
Palmer, R. S., 1976b, *Handbook of North American Birds, Volume III, Waterfowl Part 2*, Yale University Press, New Haven.
Paludan, K., 1952, Contributions to the breeding biology of *Larus argentatus* and *Larus fuscus*, *Vidensk. Medd. Dansk Natur. For.* **114**:1–128.
Parmelee, D. F., and Payne, R. B., 1973, On multiple broods and the breeding strategy of arctic Sanderlings, *Ibis* **115**:218–226.
Parsons, J., 1976, Factors determining the number and size of eggs laid by the Herring Gull, *Condor* **78**:481–492.
Payne, R. B., 1974, The evolution of clutch size and reproductive rates in parasitic cuckoos, *Evolution* **28**:169–181.
Pearson, T. H., 1968, The feeding biology of sea-bird species breeding on the Farne Islands, Northumberland, *J. Anim. Ecol.* **37**:521–553.
Pellis, S. M., and Pellis, V. C., 1982, Do post-hatching factors limit clutch size in the Cape Barren Goose, *Cereopsis novaehollandiae* Latham? *Austral. Wildl. Res.* **9**:145–149.

Perrins, C. M., 1970, The timing of birds' breeding seasons, Ibis **112**:242–255.
Perrins, C. M., 1977, The role of predation in the evolution of clutch size, in: *Evolutionary Ecology* (B. M. Stonehouse and C. M. Perrins, eds.), University Park Press, Baltimore, pp. 181–191.
Perrins, C. M., and Moss, D., 1975, Reproductive rates in the Great Tit, *J. Anim. Ecol.* **44**:695–706.
Pianka, E. R., 1970, On r and K selection, *Am. Natural.* **104**:592–597.
Pianka, E. R., 1976, Natural selection of optimal reproductive tactics, *Am. Zool.* **16**:775–784.
Pierotti, R., 1982, Habitat selection and its effect on reproductive output in the Herring Gull in Newfoundland, *Ecology* **63**:854–868.
Pizzy, G., 1980, *A Field Guide to the Birds of Australia*, Princeton University Press, Princeton.
Poslavskii, A. N., and Krivonosov, G. A., 1976, Ecology of the Sandwich Tern (*Thalasseus sandvicensis* Lath.) at the boundary of the distribution range, *Soviet J. Ecol.* **7**:232–236.
Pugesek, B. H., 1981, Increased reproductive effort with age in the California Gull (*Larus californicus*), *Science* **212**:822–823.
Rahn, H., Paganelli, C. V., and Ar, A., 1975, Relation of avian egg weight to body weight, *Auk* **92**:750–765.
Raveling, D. G., 1979, The annual cycle of body composition of Canada Geese with special reference to control of reproduction, *Auk* **96**:234–252.
Rice, D. W., and Kenyon, K. W., 1962, Breeding cycles and behaviour of Laysan and Black-footed albatrosses, *Auk* **79**:517–567.
Ricklefs, R. E., 1968, Patterns of growth in birds, *Ibis* **110**:419–451.
Ricklefs, R. E., 1970, Clutch-size in birds: Outcome of opposing predator and prey adaptations, *Science* **168**:599–600.
Ricklefs, R. E., 1973, Patterns of growth in birds. II. Growth rate and mode of development, *Ibis* **115**:177–201.
Ricklefs, R. E., 1977a, On the evolution of reproductive strategies in birds:Reproductive effort, *Am. Natural.* **111**:453–478.
Ricklefs, R. E. 1977b, A note on the evolution of clutch-size in altricial birds, in: *Evolutionary Ecology* (B. M. Stonehouse and C. M. Perrins, eds.), University Park Press, Baltimore, pp. 193–214.
Ricklefs, R. E., 1979a, Patterns of growth in birds. V. A comparative study of development in the Starling, Common Tern, and Japanese Quail, *Auk* **96**:10–30.
Ricklefs, R. E., 1979b, Adaptation, constraint, and compromise in avian postnatal development, *Biol. Rev.* **54**:269–290.
Ricklefs, R. E., 1980, Geographical variation in clutch size among passerine birds: Ashmole's hypothesis, *Auk* **97**:38–49.
Ricklefs, R. E., White, S., and Cullen, J., 1980, Postnatal development of Leach's Storm-Petrel, *Auk* **97**:768–781.
Rinkel, G. L., 1940, Waarnemingen over het gedrag van de Kievit (*Vanellus vanellus* (L.)) dedurende de broedtijd, *Ardea* **29**:108–147. (Dutch with English summary).
Ripley, S. D., 1977, *Rails of the World*, Godine, Boston.
Ross, H. A., 1979, Multiple clutches and shorebird egg and body weight, *Am. Natural.* **113**:618–622.
Royama, T., 1966, Factors governing feeding rate, food requirements, and brood size of nestling Great Tits, *Parus major*, *Ibis* **108**:313–347.

Royama, T., 1969, A model for the global variation of clutch size in birds, *Oikos* **20**:562–567.
Ryder, J. P., 1970, A possible factor in the evolution of clutch size in Ross' Goose, *Wilson Bull.* **82**:5–13.
Ryder, J. P., 1980, The influence of age on the breeding biology of colonial nesting seabirds, in: *Behavior of Marine Animals, Volume 4, Marine Birds* (J. Burger, B. L. Olla, and H. E. Winn, eds.), Plenum Press, New York, pp. 153–168.
Ryder, J. P., and Somppi, P. L., 1979, Female–female pairing in Ring-billed Gulls, *Auk* **96**:1–5.
Safriel, U. N., 1975, On the significance of clutch size in nidifugous birds, *Ecology* **56**:703–708.
Sauer, E. G. F., and Sauer, E. M., 1966, The behavior and ecology of the South African Ostrich, *Living Bird* **5**:45–75.
Savory, C. J., 1975, Seasonal variation in the food intake of captive Red Grouse, *Br. Poult. Sci.* **16**:471–479.
Schaffer, W. M., 1974, Optimal reproductive effort in fluctuating environments, *Am. Natural.* **108**:783–790.
Schifferli, L., 1978, Experimental modification of brood size among House Sparrows, *Passer domesticus*, *Ibis* **120**:365–369.
Sealy, S. G., 1978, Clutch size and nest placement in the Pied-billed Grebe in Manitoba, *Wilson Bull.* **90**:301–302.
Slagsvold, T., 1982, Clutch size variation in passerine birds: The nest predation hypothesis, *Oecologia* **54**:159–169.
Smith, C. C., and Fretwell, S. D., 1974, The optimal balance between size and number of offspring, *Am. Natural.* **108**:499–506.
Stearns, S. C., 1976, Life-history tactics: A review of the ideas, *Quart. Rev. Biol.* **51**:3–47.
Stearns, S. C., 1977, The evolution of life history traits: A critique of the theory and a review of the data, *Annu. Rev. Ecol. Syst.* **8**:145–171.
Steen, J. B., and Parker, H., 1981, The egg-numerostat—A new concept in the regulation of clutch size, *Ornis Scand.* **12**:109–110.
Tasker, C. R., and Mills, J. A., 1981, A functional analysis of courtship feeding in the Red-billed Gull, *Larus novaehollandiae scopulinus*, *Behaviour* **77**:221–241.
Thomson, A. L., ed., 1964, *A New Dictionary of Birds*, Nelson, London.
Trivers, R. L., 1972, Parental investment and sexual selection, in: *Sexual Selection and the Descent of Man* (B. Campbell, ed.), Aldine, Chicago, pp. 136–179.
Van Tyne, J., and Berger, A. J., 1976, *Fundamentals of Ornithology*, 2nd ed., Wiley-Interscience, New York.
Veen, J., 1977, Functional and causal aspects of nest distribution in colonies of the Sandwich Tern (*Sterna sandvicensis* Lath.), *Behaviour Suppl.* **20**:1–193.
Vehrencamp, S. L., 1977, Relative fecundity and parental effort in communally nesting anis, *Crotophaga sulcirostris*, *Science* **197**:403–405.
Walters, J. R., 1980, The Evolution of Parental Behavior in Lapwings, Ph.D. thesis, University of Chicago.
Walters, J. R., 1982, Parental behavior in lapwings (Charadriidae) and its relationship with clutch sizes and mating systems, *Evolution* **36**:1030–1040.
Walters, J. R., in press, The evolution of parental behavior and clutch size in shorebirds, in: *Behavior of Marine Animals, Volume 5, Shorebirds* (J. Burger and B. Olla, eds.), Plenum Press, New York.
Ward, J. G., 1973, Reproductive Success, Food Supply, and the Evolution of Clutch-Size in the Glaucous-winged Gull, Ph.D. thesis, University of British Columbia, Vancouver.

Weidmann, U., 1956, Obervations and experiments on egg-laying in the Black-headed Gull (*Larus ridibundus* L.), *Anim. Behav.* **4**:150–161.

Weller, M. W., 1959, Parasitic egg laying in the Redhead (*Aythya americana*) and other North American Anatidae, *Ecol. Monogr.* **29**:333–365.

Weller, M. W., 1968, The breeding biology of the parasitic Black-headed Duck, *Living Bird* **7**:169–208.

Welty, J. C., 1982, *The Life of Birds*, 3rd ed., W. B. Saunders Co., Philadelphia, p. 293.

Williams, A. J., 1980, Offspring reduction in Macaroni and Rockhopper penguins, *Auk* **97:754–759.**

Williams, A. J., 1981, Why do penguins have long laying intervals? *Ibis* **123**:202–204.

Williams, G. C., 1966, Natural selection, the costs of reproduction, and a refinement of Lack's principle, *Am. Natural.* **100**:687–690.

Winkler, D. W., 1982, Clutch size and its relations to food supply, courtship feeding and egg size in the California Gull (*Larus californicus*), (abst.) 1982 Meeting, Am. Ornithol. Union, Chicago.

Yom-Tov, Y., 1974, Effect of food and predation on breeding density and success, clutch size and laying date of crow (*Corvus corone* L.) *J. Anim. Ecol.* **43**:479–497.

Yom-Tov, Y., and Hilborn, R., 1981, Energetic constraints on clutch size and time of breeding in temperate zone birds, *Oecologia* **48**:234–243.

CHAPTER 3

STRUCTURE AND FUNCTION OF AVIAN EGGS

CYNTHIA CAREY

1. INTRODUCTION

It is widely accepted that birds must breed successfully in order to invade and to colonize a new habitat. Although avian eggs have traditionally received credit for their important role in producing the next generation of birds and in serving as the objects of adult incubation behavior, the role of structure and function of eggs in avian reproductive biology and in adaptation of birds to extreme environments has been largely ignored.

Some of this neglect has been due to lack of information about physiology and morphology of avian eggs. Of course, eggs of domesticated birds, particularly chickens (*Gallus domesticus*), have received intensive study for decades from embryologists and poultry scientists (Romanoff and Romanoff, 1949). But, other than the massive descriptive work on the physical properties of eggs of wild birds by Schönwetter (1960–1980), relatively little has been known about these eggs until recently. Beginning in 1970, a group of physiologists asked the question, "How do eggs breathe?" (Wangensteen et al., 1970/71; Wangensteen and Rahn, 1970/71; Wangensteen, 1972; Paganelli et al., 1975). These studies led to detailed evaluations of functional and morphological

---

CYNTHIA CAREY • Department of EPO Biology, University of Colorado, Boulder, Colorado 80309.

features of eggs that contribute to embryonic survival. These findings have broad implications for research in many areas of ornithology.

The ability of an embryo to contribute to the reproductive success of its parents depends importantly on whether its needs are met within its tolerance limits. Most of its requirements, particularly heat, defense from predators, and periodical turning are provided by the adults (Drent, 1975). In fact, much of the successful breeding of birds in stressful environments, particularly in very hot or cold habitats, depends almost entirely on adult behavior (Carey, 1980a). However, this review concentrates on certain other requirements of the embryo and features of the egg that meet these needs.

Some embryonic necessities, such as energy-rich substances, nutrients, water, and minerals, are already packaged in the egg at laying. Although contents of domesticated bird eggs have been thoroughly analyzed (Romanoff, 1967), eggs of wild species differ significantly from these eggs in mass, duration of incubation period, proportion of various materials inside the egg, and the developmental maturity of the hatchling. The interrelations among these factors are just beginning to be analyzed.

Embryos also require exchange of O_2, CO_2, and water vapor with the environment. The rate of gaseous diffusion is governed by the characteristics of the shell and the gas tensions surrounding the egg. Shell structure has apparently evolved in conjunction with the incubation period and the typical nest microclimate, with the result that almost all embryos develop in similar gaseous environments inside the shell, despite considerable diversity in external conditions, duration of incubation periods, and egg mass. However, embryos apparently have regulatory capacities for dealing with variation in gaseous exchange that also contribute importantly to survival.

Finally, requirements of embryos for successful hatching are intimately involved with structural features of the eggshell. The eggshell serves as an interesting study in evolutionary compromise to multiple, if not antagonistic, demands. It must be strong enough to support the masses of the egg contents and the incubating adults yet brittle enough for the embryo to puncture it and hatch. It must be sufficiently porous to afford diffusion of enough O_2 for normal growth and maintenance, but it must restrict excessive losses of CO_2 and water vapor and prevent invasion of microorganisms. The eggshells laid in diverse gaseous environments are perhaps the most interesting of all, since selection appears not to have provided optimal compromises but has been sensitive to particular priorities.

The range in sizes of eggs laid by birds is one final feature that may prove useful to ornithologists. This range is large enough that evaluation of the properties of eggs as a function of mass is possible. Allometric equations not only allow predictions of numerous properties of eggs based on mass, but also afford identification of those eggs that do not fit normal trends. A selection of the allometric equations that have been formulated over the last decade is presented in the Appendix. Equations from this list that are discussed in the text will be identified by the letter which designates them in the Appendix.

2. MASS AND CONTENTS OF EGGS

Each avian species produces an average clutch size containing eggs of a particular mass that require incubation for a specific duration of time and produce hatchlings at a characteristic stage of developmental maturity. The reasons why natural selection has favored a particular set of features for the reproduction of each species have been the subject of considerable debate. Although the data that are beginning to accumulate on the interrelations of egg mass, egg content, incubation period, and developmental patterns do not yet provide substantive explanations, they serve to provide ideas for future discussion.

2.1. Mass and Contents of Fresh Eggs

Masses of avian eggs vary from the 0.2 g egg of the smallest hummingbirds to the 9 kg egg of extinct *Aepyornis* (Schönwetter, 1960–1980). Egg mass is directly related to body mass of the adult female according to the relation

$$W = 0.277 \, B^{0.770} \tag{1}$$

where both egg mass (W) and body mass (B) are in g (Rahn et al., 1975). As these authors point out, this generalized equation is far less useful than specific ones calculated for each family or subfamily, since considerable taxonomic diversity exists. For instance, females of 12 orders weighing 100 g lay eggs varying in size by a factor of 4.7. Generally, however, a ten-fold increase in body mass of the adult is associated with a 4.73-fold increase in egg mass and a 1.47-fold increase in duration of the incubation period (Rahn et al., 1975). Investigators should not use the equation presented by Rahn et al. (1975) to estimate egg mass for the Fringillidae, since it is not based on actual egg masses but on linear measurements.

Egg mass generally constitutes a larger proportion of adult body mass of smaller birds than of larger ones. Of course, there are interesting exceptions. On the basis of body size, Common Kiwis (*Apteryx australis*) should lay eggs representing about 3% of body mass, but in fact they lay eggs averaging about 18% (Calder, 1978). When egg mass is multiplied by average clutch size, the mass of the clutch may constitute as much as 100% of the body mass of Charadriiformes and the smallest birds in the Anatidae, but only 14–16% of body mass of the larger members of the Anatidae and Phasianidae (Rahn et al., 1975).

Larger eggs provide more total supplies and calories than smaller ones for the growing embryos. For example, a 600-fold increase in egg mass of Zebra Finches (*Poephila guttata*) and Australian Cassowaries (*Casuarius casuarius*) is associated with an 830-fold increase in total calories (Carey et al., 1980). Since large eggs produce large chicks, part of the additional contents are used to synthesize hatchling body tissue. But, since the incubation period is directly related to egg mass [Eq. (A), Appendix], larger eggs require more calories to sustain the embryo for prolonged periods of time. Eggs with incubation periods longer than predicted on the basis of egg mass, such as those of kiwis and procellariform birds, have larger yolks and higher caloric contents than eggs of similar size with normal incubation periods (Carey et al., 1980).

The general principles governing egg size and caloric content that have emerged thus far are based on averages of a few eggs per species. Only a few data permit assessment of the significance of populational variation in these features. Larger eggs often produce more viable hatchlings than smaller eggs of the same species or population (Parsons, 1970; Nisbet, 1978). Larger eggs are associated with greater yolk content in certain precocial species (Romanoff and Romanoff, 1949; Ricklefs et al., 1978), but not in altricial species (Ricklefs and Montevecchi, 1979; Ricklefs, 1983). Enhanced survivability of hatchlings of large tern (*Sterna*) eggs is attributed to an increased amount of albumen (Nisbet, 1978).

The reasons for variation in mass and relative proportions of yolk and albumen in eggs of a clutch laid by a single female are far from being understood, but differences in these characteristics are directly or indirectly related to fledging success in Common Grackles (*Quiscalus quiscula*) that lay large clutches (Howe, 1976, 1978). Since females begin incubation before the clutch is complete, embryos hatch first from the eggs laid earliest and have a competitive advantage for food. But eggs laid later in the sequence are larger and produce heavier chicks that can attempt to compete successfully for food with the older siblings. These offsetting strategies produce successful fledging by all hatchlings in years when food is plentiful, but the younger hatchlings are allowed to starve in years during food scarcity (Howe, 1976, 1978).

Although observations that birds hatch at different developmental stages were made centuries ago, we are just beginning to understand how such diversity is associated with the contents and physiological and morphological features of eggs. Nice (1962) was the first to document the relation between modes of development and yolk content of fresh eggs. She categorized eggs into eight groups based on the ability of hatchlings to acquire food, the amount of down, locomotory ability, and the development of vision. The average amount of yolk as a proportion of fresh egg contents varies from 24–65% in altrical and the most precocial eggs, respectively (Carey et al., 1980). Since yolk and albumen differ in the relative quantities of water, protein, and lipid, the proportions of these three components vary as a function of yolk content (Fig. 1).

Nice's eight categories are lumped into four groups for a preliminary analysis of egg contents of precocial, semi-precocial, semi-altrical,

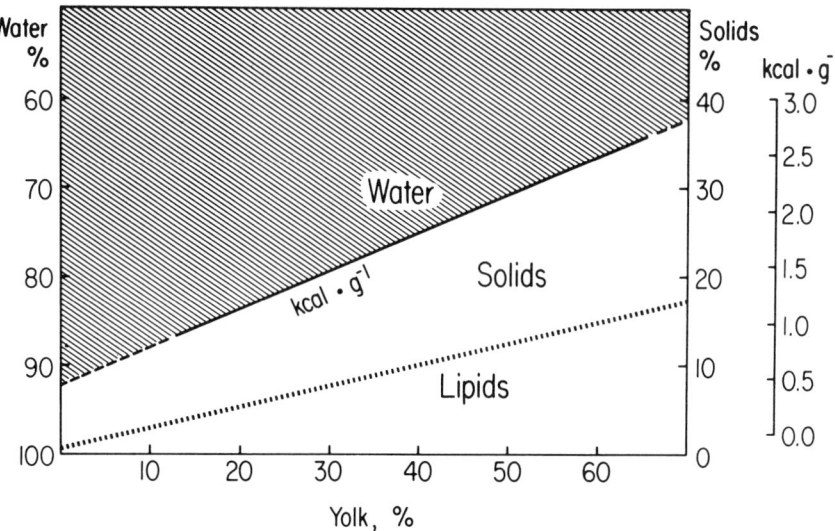

FIGURE 1. Relations among solids, water, and lipid constituents of avian eggs with varying yolk contents. All values are expressed as percent of egg content. Solid line represents the regression between dry mass (D, % egg content) and yolk content (Y, % egg content): $D = 8.02 + 0.420\ Y$. The dotted line is the regression between lipid content (L, % of dry mass) and yolk content (Y, % egg content): $L = 0.749 + 0.232\ Y$. The solid line may also be used to predict an approximate caloric value (Kcal · g^{-1} wet mass) using the ordinate on the right. Reprinted from Carey et al. (1980) by permission of *The Condor*.

TABLE I

Mean ± SD Values for Contents of Fresh and Pipped Avian Eggs in Four Developmental Modes[a]

Egg property		Precocial	Semi-precocial	Semi-altricial	Altricial
		F = 9,22 P = 2,8	F = 7,24 P = 3,7	F = 6,15 P = 5,15	F = 17,48 P = 3,16
Water % contents	F P	74.7 ± 0.03 73.2 ± 0.02	76.5 ± 0.02 77.4 ± 0.02	81.2 ± 0.02 *83.8 ± 0.03	84.4 ± 0.02 84.5 ± 0.01
Lipid % contents	F P	10.3 ± 0.16 9.5 ± 0.16	9.6 ± 0.34 * 5.8 ± 0.16	6.6 ± 0.16 * 4.5 ± 0.16	5.9 ± 0.15 * 3.8 ± 0.15
Lipid % dry mass	F P	41.0 ± 0.05 *35.6 ± 0.07	40.5 ± 0.15 *25.3 ± 0.05	35.2 ± 0.08 *28.0 ± 0.11	37.1 ± 0.07 *24.1 ± 0.02
Kcal·g^{-1} wet content	F P	1.83 ± 0.05 1.50 ± 0.15	1.66 ± 0.17 *1.50 ± 0.14	1.36 ± 0.22 *1.09 ± 0.15	1.11 ± 0.16 1.03 ± 0.09
Kcal·g^{-1} dry content	F P	7.24 ± 0.23 *6.19 ± 0.15	7.05 ± 0.24 *6.64 ± 0.19	7.10 ± 0.29 *6.78 ± 0.40	7.06 ± 0.33 *6.65 ± 0.08

[a]Data for fresh eggs are from Carey et al. (1980) and those for pipped eggs are preliminary unpublished data of Carey. F denotes fresh and P signifies pipped eggs. The first digit after F and P in each column signifies the number of species and the second equals the number of eggs. Asterisks indicate significant differences between fresh and pipped eggs within the same developmental mode as judged by T-tests.

and altricial eggs (Table I). Water constitutes 74.7% and 84.4% of the contents of precocial and altricial eggs, respectively. Values for intermediate groups fall between these averages. The variation is attributable to an increase in the proportion of solids in the egg, due to replacement of a volume of albumen (85–90% water) with yolk (43–66% water; Ricklefs, 1977). The larger proportion of yolk content in semi-precocial and precocial eggs is associated with a larger lipid and caloric (kcal·g^{-1} wet mass) content, but the proportion of lipid in dry matter and the kcal·g^{-1} dry mass do not vary significantly among groups.

2.2. Contents of Pipped Eggs

A few data are available for a comparison of the contents of fresh and pipped eggs of the same species (Table I). These results speak to a frequent assertion in the literature that altricial chicks have exhausted their yolk supplies by hatching (Drent, 1970). Neutral lipids average 24.1% of the dry mass of pipped altricial embryos and 3.8% of the total contents; the proportion of lipid in the dry contents of pipped precocial chicks is 36.5%. These data support Schmekel's (1960) findings that yolk in altricial hatchlings varies from 6.75% of body mass of piciform birds to 14% of columbiform species. Yolk contents of precocial hatchlings of galliform and struthioniform birds average 15% and 25%, respectively. Therefore, it seems certain that a portion of the initial provision of yolk in altricial eggs is designated to support existence after hatching.

The final relative water content of fresh and pipped eggs of the same developmental mode match closely, as previously noted by Ar and Rahn (1980). The precision with which these fractions coincide in each developmental category implies that maintenance of the same relative level of hydration throughout incubation is important for embryonic survival (Ar and Rahn, 1980; Section 3.1.3). Ar and Yom-Tov (1978) have concluded that the additional water in altricial and semialtricial eggs represents an extra water resource for hatchlings that are apt to be fed dry food. However, water contents of eggs of columbiform birds, which produce young initially fed a liquid diet, are indistinguishable from those of other altricial eggs (Carey et al., 1980).

Therefore, an evolutionary derivation of altricial from precocial birds has apparently been associated with a decrease in the body mass of the adult, and a reduction in the size of the egg and in the proportion of solids to liquids in the egg. Birds producing altricial eggs appear to have traded a high energetic commitment to manufacture of eggs for increased responsibilities of parental care after hatching (Ar and Yom-Tov, 1978). The evolution of the spectrum of developmental modes from precocial to altricial also may have been associated with physiological capacity of cells to function in varying hydric environments inside the egg.

3. GAS EXCHANGE

The development of a hatchling requires combustion of energy-rich substrates inside the egg. Since this is essentially an aerobic process, O_2 is required from the ambient atmosphere. Additionally, certain amounts of the waste product of such combustion, CO_2, must be removed from the interior of the egg since its buildup would disrupt the acid–base status of the embryo. These gases travel through small pores in the shell of the egg by the process of diffusion (Wangensteen and Rahn, 1970/71; Wangensteen et al., 1970/71; Wangensteen, 1972). Further, the interior of the egg is fully saturated with water vapor (Ar et al., 1974). Since the nest microclimate is usually less humid than the interior of the egg, water vapor continuously diffuses out of the pores. Experimental evidence confirms that O_2, CO_2, and water vapor share common diffusive pathways through the shell (Paganelli et al., 1978).

The diffusion of each gas through the shell can be described by a modification of Fick's first law of diffusion (Wangensteen and Rahn, 1970/71):

$$\dot{M} = (D/RT) \cdot (Ap/L) \cdot \Delta P \qquad (2)$$

where $\dot{M}$ = gas flux ($cm^3 STPD \cdot sec^{-1}$), D = binary diffusion coefficient ($cm^2 \cdot sec^{-1}$), RT = gas constant and absolute temperature (cm^3 STPD $\cdot cm^{-3} \cdot torr^{-1}$), Ap = functional pore area of shell (cm^2), L = length of diffusion path, or shell thickness (cm), and ΔP = partial pressure difference of gas across shell (torr). If $\dot{M}$ for any gas is compared among eggs of different species under similar conditions of temperature, pressure, and ambient gas tensions, Eq. (2) indicates that any variation found in $\dot{M}$ will solely be a function of variation in shell thickness (L) and/or the sizes and numbers of pores (Ap) among the eggs.

The factors $(D/RT) \cdot (Ap/L)$ are often combined into the term "G" ($mg \cdot day^{-1} \cdot torr^{-1}$) which describes the conductance of the eggshell to gases (Ar et al., 1974). Therefore, Eq. (2) becomes:

$$\dot{M} = G \cdot \Delta P. \qquad (3)$$

G is a convenient term to use since it is considerably easier to measure accurately than Ap and L (Section 4.2.1). Although conductance of the egg to O_2, CO_2, an water vapor can all be calculated, traditionally G_{H_2O} is measured because it is the easiest method. Then, if necessary, G_{O_2} and G_{CO_2} can be estimated from G_{H_2O} using the appropriate corrections for differences in D of each gas (Paganelli et al., 1978; Hoyt et al., 1979). Details of the method for estimating G_{H_2O} are given by Ar et al. (1974).

3.1. Exchange of Water

3.1.1. Patterns of Water Loss

With few exceptions, almost all eggs are incubated in a microclimate that is not fully saturated with water vapor. The ΔP_{H_2O} across the eggshell provides the driving force for continuous loss of water from the egg during incubation. It was initially assumed that G_{H_2O} and daily water loss, $\dot{M}_{H_2O}$, are constant throughout incubation (Drent, 1970), but both these features increase in the early stages of incubation in at least six species of passerines (Carey, 1979; Hanka et al., 1979; Sotherland et al., 1980; Taigen et al., 1980; Birchard and Kilgore, 1980). This phenomenon has now also been observed in eggs of a parrot (A. Ar, personal communication) and several ducks (H. Rahn, personal communication). Therefore, it is important to compare G_{H_2O} of eggs of birds in the latter stages of incubation when G_{H_2O} is independent of age.

As might be anticipated, $\dot{M}_{H_2O}$ is directly proportional to egg mass [Eq. (B), Appendix]. However, the utility of this equation is limited as predictive device, since the 95% confidence interval indicates considerable variation. For instance, a 50 g egg is predicted to lose between 138–465 mg · day^{-1} (Ar and Rahn, 1980). The variation is decreased substantially and the meaning of water loss for embryonic existence becomes more apparent when $\dot{M}_{H_2O}$ is inserted into this equation:

$$F = \frac{\dot{M}_{H_2O} \cdot I}{W} \quad (4)$$

where F = percent of the initial egg mass (W, in g) lost as water vapor during the incubation period (I, in days; Rahn and Ar, 1974). When all the available values of $\dot{M}_{H_2O}$ are analyzed in this manner, eggs lose between 10–23% of their initial mass as water vapor during incubation; the overall average for 81 species is 15% (Ar and Rahn, 1980).

The general similarity in the proportion of water lost during incubation despite very large ranges in mass and incubation periods of avian eggs appears to result largely from the shell structure of the egg. Rahn and Ar (1974) have formulated the relation:

$$G_{H_2O} = \frac{5.2\ W}{I} \quad (5)$$

which indicates that G_{H_2O} and I are inversely proportional for a given egg mass. Although exceptions have been found and others certainly will be described in the future, the generalization exists that the slope of the regression line approaches 1.0 if $\dot{M}_{H_2O}$ and G_{H_2O} are plotted as a function of W/I (Rahn and Ar, 1980; Fig. 2). A practical example is afforded by a comparison of eggs of chickens and Wedge-tailed Shearwaters (*Puffinus pacificus chlororhynchus*). Eggs of both species weigh about 60 g and lose about 15% of their initial mass as water vapor, but the incubation period of the shearwater eggs is over twice that of the chicken eggs. The similarity in fractional water loss is due to a reduction in G_{H_2O} of the shearwater eggs to less than half that of the chicken eggs (Ackerman et al., 1980).

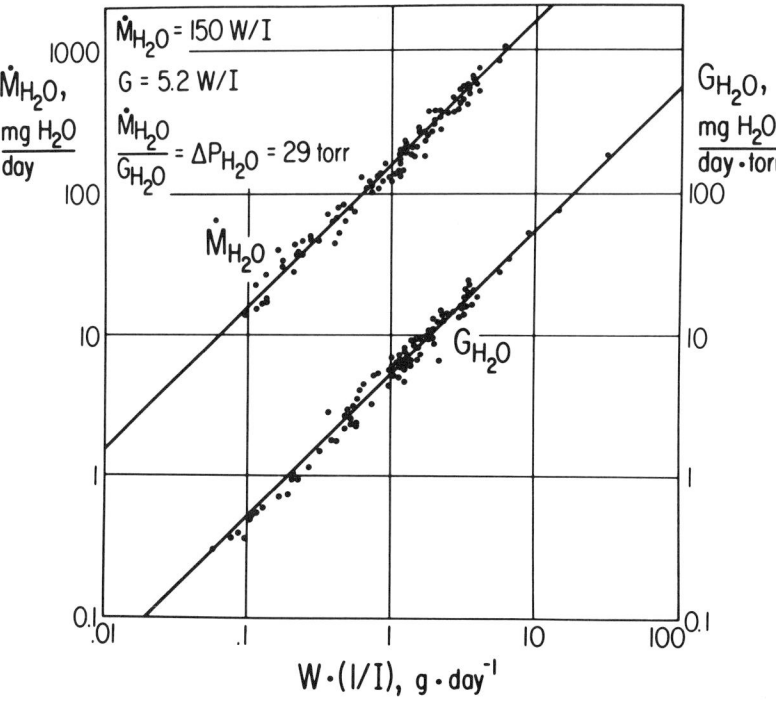

FIGURE 2. Daily water loss ($\dot{M}_{H_2O}$) and water vapor conductance (G_{H_2O}) plotted as a function of W/I. The equations for the regression lines are given in the upper left corner. The ratio of $\dot{M}_{H_2O}/G_{H_2O}$ yields the average ΔP_{H_2O} across the eggshell. Reprinted by permission of *American Zoologist* from Rahn and Ar (1980).

3.1.2. Importance of Water Loss

Solids are catabolized and metabolic water is produced continuously throughout incubation. If water were not lost from the egg, the solid content would become progressively more diluted. Since the proportions of water in contents of fresh and pipped eggs are constant (Table I), the amount of water lost compensates both for the reduction in solids and for the production of metabolic water.

The loss of water from the egg may be important for embryonic survival for at least two reasons. First, the volume of water lost is replaced by an equivalent volume of air in the blunt end of the egg, the air cell (Sections 3.2.1 and 4.1.1). The chick pips into the air cell in the final stages of incubation and uses this volume to inflate its lungs (Rahn et al., 1974). This procedure begins the transition between diffusive respiration via the chorioallantois to convective breathing via the lungs and establishes the final acid–base status of the blood (Tazawa, 1980). While poultry breeders are aware that embryonic mortality increases if the "optimal" air cell volume is not established by sufficient water loss, the cause of mortality is not known (Lundy, 1969).

Water loss also serves to maintain the same relative level of hydration of the fresh egg and pipped embryo (Section 2.2). Although it is unknown why such levels are important, the precision with which relative hydration is maintained has prompted Ar and Rahn (1980) to propose that water loss must be precisely regulated to achieve optimal hatchability.

3.1.3. Tolerance of Variation in Water Loss

The assumption that a characteristic G_{H_2O}, as a product of Ap and L, has resulted from natural selection to achieve a prescribed level of water loss for each species is based both on circumstantial and experimental evidence. The similarity of values for F, as well as the precision of relative hydration of egg contents, serve as circumstantial evidence that water loss performs some vital function during incubation (Ar et al., 1974; Ar and Rahn, 1980). The experimental evidence to date suggests that mortality of chicken embryos rises sharply if the rate of daily water loss is either increased or decreased by variation of incubator humidity (Lundy, 1969; Ar and Rahn, 1980; Snyder and Birchard, 1982).

Acceptance of the importance of selection for G_{H_2O} for a given I and W in the evolutin of each species involves two assumptions: (1) that differential mortality will result if G_{H_2O} is either increased or decreased substantially, and (2) that the embryo is a passive occupant of the shell without regulatory abilities to deal with variation in $\dot{M}_{H_2O}$. Recent evidence suggests that the implied importance of G_{H_2O} needs reevaluation.

Chicken embryos that were stressed during the last $\frac{1}{3}$ of incubation by withdrawal of 6 cm^3 of water from the allantoic fluid, or by increasing shell porosity by 3–4 times, hatched normally, though they lost approximately 25% of their body water and hatched with significantly reduced body masses (Simkiss, 1980). This indicates that embryos have regulatory capacities for dealing with variation in rates of water loss and that the importance of G_{H_2O} in the evolution of each species has been exaggerated (Simkiss, 1980).

Eggshell conductances of naturally incubated eggs of Red-winged Blackbirds (Agelaius phoeniceus) were modified on day 1 of incubation, either by poking holes in the shell over the air cell or by covering the same area with wax. These treatments produced variation in $\dot{M}_{H_2O}$ between 0.01 and 0.25 g·day^{-1}. Embryos hatched successfully from eggs losing between 0.03–0.17 g · day^{-1} and lived at least 7 days after hatching. This 3.5-fold range in water loss was associated with variation between 77–89% in the final water content of pipped embryos in similarly treated eggs; the control value was 86%. Therefore, water loss and final water content can vary substantially without increasing embryonic mortality (Carey, unpublished data).

These results are supported by observations that variabilitiy in G_{H_2O} of wild populations and in eggshell characteristics of domesticated birds is very large (Tyler and Geake, 1960; Carey et al., 1983). By using average values for each species in the formulation of general concepts, we may have ignored the importance of variability in features of the eggshell that can influence embryonic survival.

These data do not refute the "relative hydration hypothesis" (Ar and Rahn, 1980) nor invalidate the circumstantial evidence that fractional water loss and relative levels of hydration fall within restricted limits for most species. However, they suggest that G_{H_2O} has evolved in conjunction with the general nesting conditions of each species and that the regulatory properties of the embryo may contribute importantly to survival by "fine-tuning" the internal environment in which the embryo is developing.

3.1.4. Microclimate of the Nest

According to Eq. (3), water loss is both a function of G_{H_2O} and ΔP_{H_2O}. Since the water vapor pressure in the nest microclimate (P_N, in torr) comprises the lower limit of ΔP_{H_2O}, it is relevant to consider how it might affect embryonic survival.

The term P_N can be estimated if the average temperature of the egg, which sets the water vapor pressure inside the egg (P_A, in torr) is known (Ar et al., 1974):

$$P_A - P_N = \dot{M}_{H_2O}/G_{H_2O}. \qquad (6)$$

Published values of P_N vary from 3–36 torr and average 20.8 torr (Walsberg, 1980). These calculated values are, of course, averaged over one or more days during incubation and do not describe the rapid fluctuations in P_N that must occur when the incubating adult leaves or returns to the nest. However, the accuracy of these calculations has been checked directly by radiotelemetry and the average P_N obtained by both methods agree well (Howey et al., 1977).

Water vapor travels from the interior of the egg to the external environment by a two-step process: (1) by diffusion out of the egg into the microclimate of the nest, and (2) by convection from the nest into the atmosphere (Rahn et al., 1976, 1977). The importance of the central role of P_N in this process has prompted the suggestion that P_N must be regulated by the incubating adult (Rahn et al., 1976, 1977). In order to perform such regulation, an adult would require sensory capability to detect P_N, a knowledge of what P_N should be to match the G_{H_2O} of the eggs, and the ability to adjust P_N without compromising the needs of the embryo for heat or the foraging requirements of the adult (Walsberg, 1980).

This possibility has been tested by varying the water content of air recirculated through nest cups of House Finches (Carpodacus mexicanus) and Phainopeplas (Phainopepla nitens). Exceedingly dry or fully-saturated air circulated through the nests produced a 2–3-fold change in P_N and a 1.8–2.4-fold range in $\dot{M}_{H_2O}$, but caused no change in adult behavior (Walsberg, 1983). Further, direct observations of adults incubating eggs under identical ambient conditions indicate no correlation between nest attentiveness and microclimate. One pair of Snowy Plovers (Charadrius alexandrinus) spent 90% of a time interval shading their eggs while another pair 25 m away spent 100% of the same interval sitting loosely on their eggs (Grant, 1982).

A computer model developed by Walsberg (1980) suggests why regulation of P_N may not prove critical for survival of embryos. The results show that P_N must be varied to extreme limits before $\dot{M}_{H_2O}$ approaches possibly lethal limits. While more information is clearly needed to understand the complex interrelations among $\dot{M}_{H_2O}$, G_{H_2O}, P_N, and adult attentiveness, this model suggests that regulation of $\dot{M}_{H_2O}$ of each species is primarily achieved by production of an egg with the appropriate structural features for the average conditions of the nest environment.

3.2. Exchange of O_2 and CO_2

3.2.1. Patterns of Exchange of O_2 and CO_2

Although O_2, CO_2, and water vapor share common paths for diffusion through the shell (Paganelli et al., 1978), the patterns of exchange of O_2 and CO_2 differ considerably from that of water vapor. With some exceptions, $\dot{M}_{H_2O}$ is relatively constant throughout incubation (Section 3.1.1), but $\dot{M}_{O_2}$ and $\dot{M}_{CO_2}$ vary substantially as embryonic metabolism increases during development.

Oxygen consumption has now been measured in about 40 species. At least two developmental patterns have emerged. Oxygen consumption in precocial embryos increases exponentially for the first 80% of the incubation period, then plateaus or even drops briefly, before increasing at the onset of internal pipping (Hoyt et al., 1978; Vleck et al., 1979; Vleck et al., 1980; Hoyt and Rahn, 1980). Another rise in $\dot{M}_{O_2}$ is evident during hatching, and, within a few days after hatching, $\dot{M}_{O_2}$ attains 80–90% of the value of adult birds of equivalent size (Hoyt and Rahn, 1980). The plateau period during incubation is associated with a cessation or decrease in growth (Vleck et al., 1980). Oxygen consumption of semi-altricial and altricial embryos increases exponentially throughout the incubation period. This increase is associated with continuous growth in these embryos (Vleck et al., 1979; Vleck et al., 1980).

The yolk sac serves as the respiratory organ during the early phases of development until the chorioallantoic membrane forms. It functions until "internal pipping," when the embryo pips the inner shell membrane and initiates lung ventilation. The documentation of the dynamics of blood perfusion and gas exchange in the chorioallantois of chicken embryos (Tazawa, 1978, 1980) represents one of the finest technical achievements in research on eggs. However, we know practically nothing about these features in eggs of wild birds.

Oxygen diffuses from the external environment through pores in the shell and shell membranes (Sections 4.1.1 and 4.2.1), and finally through the epithelium of the chorioallantois to reach the blood. The reverse route is taken by CO_2. The wet shell membranes are highly resistant to O_2 diffusion in the first few days of incubation (Kutchai and Steen 1971; Tullett and Board, 1976; Lomholt, 1976a) but after they dry, the resistance of the shell and outer shell membranes to gaseous diffusion remain fixed for the duration of incubation (Wangensteen et al., 1970/71).

Since the demand for O_2 and production of CO_2 increase during incubation but the resistance of the shell to gaseous diffusion remains fixed, tensions of these two gases decrease and increase, respectively, inside the egg (Rahn et al., 1974; Tazawa, 1980). These changes are reflected both in the gas tensions in the air cell (Section 3.1.2) and in the blood leaving the chorioallantois (Tazawa, 1980). Air cell gas tension can be measured directly or calculated by the equation (Rahn et al., 1974):

$$P_{I_{O_2}} - P_{A_{O_2}} = \dot{M}_{O_2} / G_{O_2} \tag{7}$$

where $P_{I_{O_2}}$ (torr) is the ambient P_{O_2} corrected for the saturated water vapor pressure at the egg temperature (Paganelli et al., 1981) and $P_{A_{O_2}}$ (torr) is the P_{O_2} in the air cell. By initiation of internal pipping, air cell P_{O_2} and P_{CO_2} of about 30 species average 101 and 40 torr, respectively (Hoyt and Rahn, 1980). The range of values among these species is very small and indicates that avian embryos generally develop in similar gaseous environments, despite large differences in egg mass and incubation period (Rahn et al., 1974; Hoyt and Rahn, 1980).

As might be anticipated, the similarity in air cell gas tensions among avian eggs indicates that important relationships exist among $\dot{M}_{O_2}$, W and I. The relation formulated by Rahn and Ar (1980):

$$\dot{M}_{O_2} = 237\ W/I \qquad (8)$$

indicates that $\dot{M}_{O_2}$ is inversely proportional to I for eggs of a given mass. As in the relation among $\dot{M}_{H_2O}$, W, and I (Section 3.1.1), some exceptions exist and others are certain to be found in the future. However, in general, if $\dot{M}_{O_2}$ is plotted against W/I, the slope of the regression line approximates 1.0 (Fig. 3). Using the same example as before (Section 3.1.1), $\dot{M}_{O_2}$ of a shearwater egg would be predicted to be 50% of that of a similar sized chicken egg at the same stage of incubation; experimental evidence confirms this prediction (Ackerman et al., 1980).

Since $\dot{M}_{O_2}$ is a function of G_{O_2} [Eq. (3)], it can be shown that the incubation period is inversely proportional to G_{O_2} for a given egg mass:

$$G_{O_2} = 5.6\ W/I. \qquad (9)$$

The net effect of these relations is not only that the amount of water lost during incubation is similar in most eggs (Section 3.1.1), but also that the amount of O_2 consumed/g egg is similar (Ar and Rahn, 1978), the ΔP_{O_2} and ΔP_{CO_2} across the eggshells at comparable stages of incubation are similar (Rahn et al., 1974), and the final levels of O_2 and CO_2 in the air cell fall within narrow limits (Rahn et al., 1974). The importance of this latter observation may relate to the similarity of the tensions of O_2 and CO_2 in the air cell just prior to pipping and the levels found in the lungs of hatchlings and adults (Tazawa et al., 1971a; Wangensteen, 1972). Therefore, the conductance of the eggshell allows diffusion of appropriate levels of gases for growth and development and also prepares the embryo for the onset of aerial respiration (Wangensteen and Rahn, 1970/71; Rahn et al., 1974).

3.2.2. Tolerance of Variation in O_2 and CO_2 Exchange

If most avian embryos develop in similar gaseous environments inside the egg, it would be quite useful to know how sensitive they are

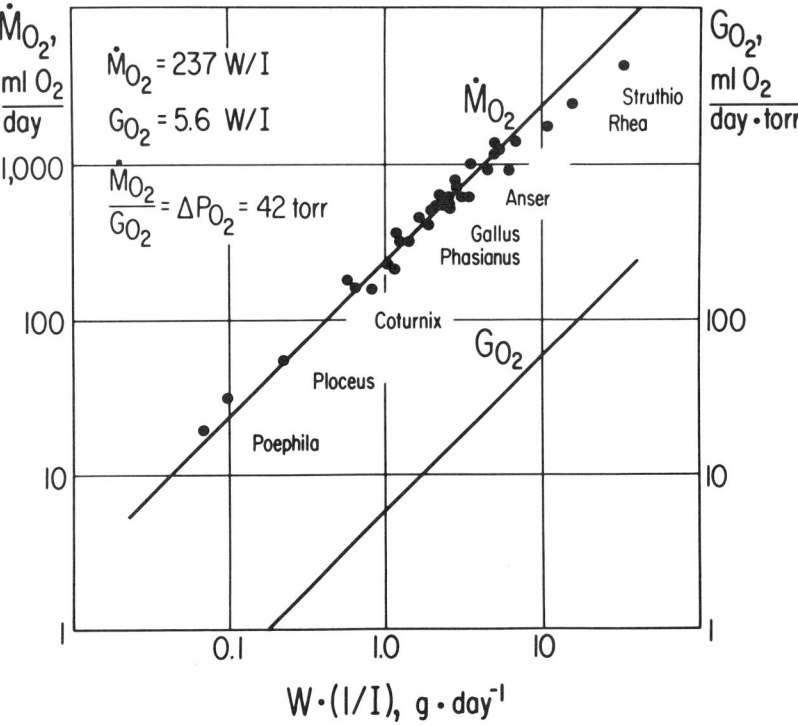

FIGURE 3. Oxygen consumption at the prepipping stage ($\dot{M}_{O_2}$) and conductance of the eggshell to oxygen (G_{O_2}) as a function of W/I. The equations for the regression lines are given in the upper left corner. The ratio of $\dot{M}_{O_2}/G_{O_2}$ gives the ΔP_{O_2} across the eggshell. Reprinted by permission of the *American Zoologist* from Rahn and Ar (1980).

to variation in external gas concentrations and what regulatory mechanisms they might possess for dealing with such variation. Unfortunately, very few experiments have tested tolerance limits of avian embryos. Most of the existing studies have involved chicken embryos that have proved highly tolerant of variation in ambient CO_2 levels (Dawes and Simkiss, 1971) but very sensitive to changes in O_2 (Tazawa, et al., 1971b). Incubation of chicken embryos in hyperoxic environments results in substantial increases in embryonic mass and oxygen consumption (Temple and Metcalfe, 1970; Visschedijk et al., 1980) but exposure to very moderate hypoxia (19.4–18.6%) results in depression of metabolic rate (Visschedijk et al., 1980).

Wild birds may prove substantially more tolerant of variation in ambient O_2 and CO_2 than embryonic chickens. Embryos of three species of wild birds maintain normal rates of growth and hatchling masses over 3600 m altitudinal gradients and one species exhibits control rates of oxygen consumption over a 2900 m gradient (Carey et al., 1982). The mechanisms by which embryos of wild birds maintain normal metabolism under hypoxic conditions at altitudes above 2000 m are unknown.

3.3. Adaptation to Diverse Gaseous Environments

The general principles concerning gaseous exchange of avian eggs that have been defined within the last decade are based on average values for a few species that generally breed in open nests with "normal" gaseous environments. It is now of interest to determine (1) if eggshell characteristics of birds breeding in habitats that pose problems for diffusive respiratory systems differ from allometric predictions, (2) if such differences could contribute to embryonic survival, and (3) if no differences exist, how do embryos develop successfully in variable gaseous environments inside the shell?

The average G_{H_2O} of eggs of American Coots (*Fulica atra*), Common Loons (*Gavia immer*), and grebes (*Podiceps cristatus* and *Podilymbus podiceps*) are 2–3 times greater than predicted on the basis of egg mass (Lomholt, 1976b; Ar and Rahn, 1978; Tullett, 1978; Ackerman and Platter-Rieger, 1979), These eggs are generally laid in nests saturated with liquid water. The very high value of G_{H_2O} may offset the high nest humidity and ensure that hydration levels of the embryo fall within ranges typical of other birds, but no measurements of water content are available to confirm this assumption.

Some eggs are incubated beneath soil, sand, or under rotting vegetation. Eggs laid by Egyptian Plovers, *Pluvianus aegyptius*, are buried during the day beneath sand which the adults moisten by carrying water in their ventral feathers. The G_{H_2O} of eggs is significantly greater than predicted on the basis of egg mass and incubation period and probably contributes to the ability of the eggs to lose normal amounts of water (17% of initial mass) in the moist, sandy environment (Howell, 1979). The eggs of Brush Turkeys (*Alectura lathami*) and Mallee Fowl (*Leipoa ocellata*) are laid in mounds of rotting vegetation. The gaseous environment of the eggs is hypoxic, hypercapnic, and fully saturated with water vapor. The eggs lose practically no water and do not form an air cell. The exceedingly high values of G_{H_2O} (35% and 120% greater than predicted for Mallee Fowl and Brush Turkey eggs, respectively) assist exchange of O_2 and CO_2 in this environment (Seymour and Ackerman, 1980).

Eggs of some species nesting in burrows can be exposed to very humid, hypoxic, and hypercapnic conditions. The average G_{H_2O} of Bank

Swallow (*Riparia riparia*) eggs is significantly greater than that of similarly sized eggs of Barn Swallows (*Hirundo rustica*) laid in open nests. This modification of bank swallow eggs is thought to maximize losses of water and CO_2 and uptake of O_2 in these burrows (Birchard and Kilgore, 1980).

Not all eggs laid in burrows experience aberrant gaseous conditions. Ambient O_2 and CO_2 tensions in 3 m deep burrows of Bonin Petrels (*Pterodroma hypoleuca hypoleuca*) are similar to those in open nests of most birds (Pettit et al., 1982) but the G_{H_2O} of these eggs is reduced to about 50% of the predicted value. The reduction in G_{H_2O} apparently ensures that petrel eggs lose about the same amount of water vapor (17% of initial mass) during the incubation period that is 1.82 times the predicted length (Grant et al., 1982).

The low ambient humidity of deserts might be expected to pose some problems for embryonic gas exchange, but Grant (1982) has found that the average P_N of nests of shorebirds breeding in hot spring and summer temperatures of California deserts does not vary from values reported for other birds (Walsberg, 1980). The average G_{H_2O} of five species fell below predicted levels, while that of one species exceeded its expected level.

These studies of eggs laid in atypical gaseous environments depend on comparisons with values predicted by allometric relationships to determine if differences exist in eggshell structure. Because of the variation inherent in allometric equations, deviations from predictions might be interpreted as adaptations to a particular environment when, in fact, they simply fall within the range of variation typical of all avian eggs.

Analysis of populational differences within the same species breeding over geographical gradients posing problems for diffusive respiratory systems avoids the problems inherent in allometric comparisons. The only existing studies of this type concern eggs of birds breeding over altitudinal gradients. Since the diffusion coefficient [D in Eq. (2)] is inversely proportional to barometric pressure (P_B, in torr; Paganelli et al., 1975), gases diffuse more rapidly at high altitudes than at sea level, all other factors in Eq. (2) held equal. While this effect might ameliorate the potential problems caused by decreasing ambient P_{O_2} (Visschedijk et al., 1980), losses of water vapor and CO_2 might increase to lethal levels above certain altitudes (Rahn and Ar, 1974). However, rates of water loss and final water contents of pipped embryos of Redwinged Blackbirds are independent of P_B over a 3050 m altitudinal gradient, despite a 31% increase in D (Carey et al., 1983). Therefore, the effect of D on diffusion of water vapor has been offset in some manner. Analysis of eggs of several wild and domesticated species breeding over at least a 3000 m altitudinal gradient indicates that G_{H_2O} of the eggshells decreases in montane habitats in approximate proportion to the decrease in P_B and increase in D (Carey, 1980b; C. Monge, unpublished data). These results suggest that conservation of water vapor and/or CO_2 has been achieved by reduction of the conductance of the eggshell, despite the fact that ambient P_{O_2} levels are reduced at high altitudes. Although conservation of water and CO_2 may be most important for embryonic survival up to about 3000 m, recent evidence indicates that the demand for O_2 may become the important priority for successful reproduction at higher altitudes (C. Monge, unpublished data). Therefore, the structure of the eggshell may reflect selection for different features depending on the altitude at which the egg is laid.

We clearly need more studies on eggshells of birds breeding in diverse gaseous environments to determine what kinds of modifications have been made in eggshell properties, what the effects of these modifications are on the gaseous environment inside the shell, and what the priorities are for embryonic survival when an optimal compromise between requirements for O_2 uptake and conservation of water vapor and CO_2 cannot be achieved. The few data we have now suggest that selection has favored adjustments in shell conductance in species breeding in diverse gaseous environments with the probable result that most embryos develop in internal gas tensions similar to those of embryos of other birds.

4. GAS EXCHANGE AND THE EGGSHELL

Since G is a function of both the pore area (Ap) and shell thickness (L) [Eq. (2)], it is now of interest to consider what is known about: (1) the structures of the shell and pores, (2) how these features are formed, (3) how pore area and shell thickness are determined for each species, and (4) whether environmental factors can select for modification of eggshell structure.

4.1. Eggshell

4.1.1. Structure of the Eggshell

The structure of the avian eggshell has aroused curiosity since Aristotle. While the chicken egg has received the most attention because of the economic problems associated with eggshell breakage (Simons,

1971), eggshells of about 150 species of wild birds have recently been studied (Tyler and Simkiss, 1959; Tyler, 1964, 1965, 1966; Hoyt et al., 1979; Becking, 1975; Board and Tullett, 1975; Tyler and Fowler, 1978, 1979; and others). Although eggshells of wild birds are interesting because of the taxonomic diversity in size, color, and pore geometry (see below), studies of these eggs are certainly important due to the need to understand how pesticides affect the physiology of eggshell formation (Cooke, 1975) and how characteristic structures of shell and pores are formed in each species. The study of eggshell synthesis may soon generate even broader interest, since cellular physiologists and biochemists are just beginning to realize the importance of transport, storage, and utilization of metal ions in cellular processes (Simkiss, 1981).

The contents of the egg are enveloped by two membranes. These are composed of protein with a polysaccharide covering (Tyler, 1969; Erben, 1970; Harris et al., 1980). They appear as a dense, fibrous mat under magnification (Simons, 1971; Becking, 1975). The size of the fibers and the spaces between fibers in membranes of wild species are directly correlated with egg mass (Becking, 1975). The membranes are loosely fused together at infrequent intervals except in the blunt end of the egg where the membranes become separated during incubation and enclose the gas volume in the "air cell" (Simons, 1971). The membranes appear to pose negligible resistance to gaseous diffusion during the major portion of incubation except in the initial stages, when the membranes are wet (Section 3.2.1). The membranes serve to contain egg contents during the process of shell formation, to anchor the seeding sites of crystal formation (Krampitz and Witt, 1979), and possibly to present a temporary barrier to larger diameter microorganisms invading the interior of the egg (Board and Fuller, 1974).

Eggshells contain about 98% crystalline calcite, a structural form of $CaCO_3$ (Simons, 1971; Krampitz and Witt, 1979). Minute amounts of aragonite and vaterite, alternative forms of $CaCO_3$, have been detected in shells or shell coverings of a few species (Erben, 1970; Tullett et al., 1976). Other components of the shell include small concentrations of magnesium and phosphorus (Romanoff and Romanoff, 1949; Board and Love, 1980) and a matrix of organic molecules distributed throughout the crystalline layer (Simkiss and Tyler, 1957; Tyler, 1969; Krampitz and Witt, 1979). The organic matter is principally a protein–mucopolysaccharide complex (Simkiss and Tyler, 1957). The function of the organic matrix is not firmly established. It may serve as an internal support structure for the growth and calcite crystals and may contribute to the mechanical strength of the shell (Board and Scott, 1980). The amino acid composition of the protein fraction appears to be species-specific, and the particular placement of specific amino acids may play a vital role in the crystallization process (Krampitz et al., 1972, 1974).

The nomenclature of the parts of the shell varies among workers and the exact microstructural arrangement of crystals is subject to considerable debate (Schmidt, 1962; Tyler, 1969; Simons, 1971; Board and Scott, 1980; Board, 1982). Typically, the shell is fastened to the outer shell membrane at the sites of organic protuberances, called mammillary cores, adhering to the membrane (Fig. 4). The mammillary cores are enclosed in calcite crystals that project downward and through the membrane fibers, forming the basal caps. Other crystals project vertically to form the cone layer. These cones are fused to form the palisade layer. In some eggs, like those of the Anatidae and Sphenisciformes, the palisade layer may be overlaid with a surface crystalline layer, a system of vertically oriented crystals about 1–10 μm thick (Tyler 1964, 1965; Becking, 1975). This layer, if present, or the terminus of the palisade layer, forms the boundary of the true shell. Measurements of shell thickness, usually made with micrometer calipers, include the distance from the basal caps, after the shell membrane is removed, to the shell surface (Tyler, 1969). Therefore, this measurement represents the total distance that gases must travel to traverse the shell. Not surprisingly, shell thickness is directly related to egg mass [Eq. (D), Appendix].

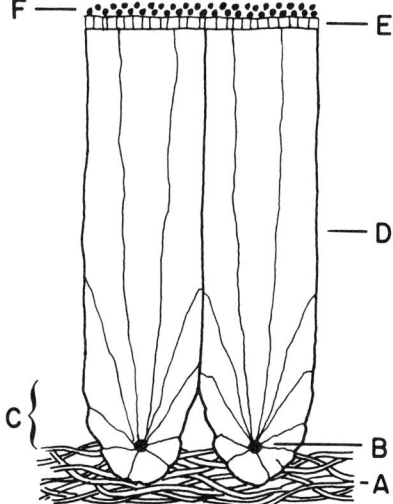

FIGURE 4. Schematic drawing of parts of the eggshell modified from Board and Scott (1980). A = outer shell membrane, B = mammillary core, C = cone layer, D = palisade layer, E = surface crystalline layer, and F = accessory coverings.

The surfaces of shells of some species are covered with a thin organic cuticle, a chalky organic cover, or an inorganic cover composed primarily of vaterite (Tyler and Simkiss, 1959; Tyler, 1964, 1969; Tullett et al., 1976). The functions of the cuticle and chalky covering are unclear and they may simply be the desiccated remnants of the lubricants and secretions of the oviduct carried out with the egg during laying. They do not appear to pose any resistance to gaseous diffusion (Tullett and Board, 1976) and they may be rubbed off during incubation by contact with the brood patch or other eggs (Board and Perrott, 1982). The cuticle might provide water repellency for eggs in open nests and the vaterite cover may serve to keep the surface of eggs clean in wet, muddy nests or open colonies where nest sanitation may be less than desirable (Board et al., 1977; Board and Perrott, 1982; Board, 1982).

4.1.2. Formation of the Eggshell

Although formation of the eggshell has been studied intensively in chickens for several decades, we are far from understanding this process completely. Moreover, very little is known about eggshell formation in wild birds due to the technical problems of sampling eggs at various stages of formation. Therefore, questions concerning how avian shell glands produce such variety of sizes, shapes, shell thicknesses, and pore geometries, and how pesticides disrupt shell formation lack answers.

The process of eggshell formation in the chicken begins when the yolk and albumen arrive in the proximal portion of the isthmus of the oviduct. The two shell membranes are secreted around these materials by glandular cells lining the lumen (Wyburn et al., 1970; Simons, 1971). Carbohydrate and about 15 g of water are added to the albumen during the process of "plumping" which requires active transport of sodium across the mucosa of the isthmus (Mongin and Carter, 1977). The plumping process appears to stretch the membranes taut by increasing the volume of egg contents. This process is hypothesized to have some crucial role in the formation of both pores and the shell thickness (Simons, 1971; Tullett, 1978; Board and Love, 1980).

As egg contents and membranes pass into the distal end of the isthmus, organic granules, the mamillary cores, are attached to the surface of the outer membrane, possibly by shrinkage and desiccation of colloidal droplets of organic matter secreted by glandular cells (Erben, 1970; Wyburn et al., 1973). The cores appear to be the site of formation of the first calcite crystals, possibly because calcium is chelated by the mucopolysaccharide or proteins in the core matrix (Simkiss and Tyler, 1958; Krampitz and Witt, 1979). The initiation of formation of cores is evident as the egg arrives in the pouch of the shell gland or uterus (Wyburn et al., 1973; Erben and Kriesten, 1974).

The shell gland produces the remainder of the calcite structure, the organic matrix, pigments, and any cuticle or cover. The calcium and carbonate are provided for crystal formation by the mucosa of the shell gland. Although appetite for calcium and the transport capacity of the small intestine for this ion increase markedly before and during laying, dietary intake must be supplemented by mobilization of calcium from medullary bone, a labile storage type of bone (Simkiss, 1975). Calcium is removed from the blood by binding proteins in the shell gland and is moved by both active transport and diffusion into the luminal fluid (Eastin and Spaziani, 1978). While some bicarbonate needed for shell construction may originate from the plasma, over 80% of the requirement is produced by carbonic anhydrase in mucosal cells by this reaction (Sauveur and Mongin, 1971):

$$H_2O + CO_2 \rightarrow H_2CO_3 \rightarrow HCO_3^- + H^+. \qquad (10)$$

The bicarbonate ion is transferred into the lumen for reaction with calcium and the hydrogen ion is transported into the blood, resulting in plasma acidosis during eggshell production (Sauveur and Mongin, 1971; Mongin and Carter, 1977).

Some calcite crystals expand from the mammillary cores downward into the meshwork of the outer membrane and fuse the growing shell to the fibers by the formation of the basal caps. Other crystals grow from the core away from the membrane to form the cones. These continue vertically for a short distance and then expand laterally until they fuse with the margins of crystals from adjacent cores and vertically to form the palisade layer.

The exact mechanism that terminates shell growth at a characteristic thickness for each species is unknown. However, the exchange of ions between the mucosal cells and luminal fluid changes during shell production (Sauveur and Mongin, 1971), and the concentrations of magnesium and phosphorus imbedded in the shell matrix increase markedly toward the shell surface (Tyler, 1969; Board and Love, 1980). Simkiss (1964) has hypothesized that secretion of phosphate into the luminal fluid could "poison" further crystal growth. Variation in the chemical composition of the luminal fluid might also result in the different pattern of calcite deposition that occurs as the palisade layer becomes overlaid by the surface crystalline pattern.

The cover or cuticle, if present, is added as the final step prior to laying (Simons, 1971). The shift from deposition of calcite to vaterite in the formation of shell coverings is also possibly the result of addition of phosphate to the luminal fluid at this terminal stage (Tullett et al., 1976).

4.1.3. Modification of Shell Structure during Incubation

While the structures of the palisade layer or surface crystalline layers do not appear to change during incubation (Simons, 1971), minerals are mobilized from the inside of the shell during incubation. The embryo obtains both about 80% of its calcium requirement and some bicarbonate, which may be used to regulate the acid–base balance (Crooks and Simkiss, 1974; Simkiss, 1975), by dissolving about 5% of the shell. Mobilization of the calcium from the mammillary cores starts about day 12 in the chicken embryo and by day 16 results in the detachment of the outer shell membrane, core matrix, and small amounts of calcium from the rest of the shell (Simkiss, 1961). It is unclear whether this detachment is necessary for some purpose or is merely a result of a weakening of the calcite structure. The constancy of G_{H_2O} during the interval between day 12 and the end of incubation when shell reabsorption occurs (Wangensteen et al., 1970/71; Rahn et al., 1974) has led to the generally accepted view that shell reabsorption has no effect on gaseous diffusion.

The mechanism of shell reabsorption is a mystery. It does not occur in the portion of the shell over the air cell nor in infertile eggs (Simkiss, 1975). The decalcifying fluid is thought to be either carbonic or citric acid (Simons, 1971; Krampitz and Witt, 1979). We have no clear idea where it is produced, how it gets to the mammillary cores, and how the calcium and bicarbonate are transported through the shell membranes to the chorioallantois. It is possible that the water that condenses on the inner shell membrane during cooling of the egg while the incubating parent leaves the eggs (Simkiss, 1974) provides a hydric environment for movement of these materials along the fibers of the membranes (A. Ar, personal communication).

4.2. Pores

4.2.1. Structure of Pores

Board and Scott (1980) and Board (1982) have recently reviewed the remarkable diversity of pore shapes and accessory coverings that have been described in avian eggs. The simplest pore is funnel-shaped (Fig. 5). The channel of the pore is often substantially narrower than the opening at the shell surface. For instance, the pore channel and mouth of the funnel of shells of the Plaintive Cuckoo (*Cacomantis merulinus*) measure 5 and 50 μm, respectively (Becking, 1975). The pore channel of eggs of many birds, such as ducks, rheas, extinct moas, and *Aepyornis* have more than one branch (Tyler and Simkiss, 1959; Tyler, 1964). The pores in shells of falconiform birds and the Emu (*Dromaius novaehollandiae*) are distinguishable only part of the distance through the shell. Small cavities in the palisade layer presumably allow gaseous diffusion through the remainder of the shell (Tyler, 1966; Board and Tullett, 1975).

Simple or branched pores may be occluded with accessory materials of unknown composition, plugged with organic or crystalline material of diverse chemical composition, or capped with organic or inorganic spheres (Board and Scott, 1980; Board, 1982; Fig. 5). Chemical analysis of these materials has revealed startling complexities. For instance, pores of four species of tinamous and the Lesser Jacana (*Microparra capensis*) are blocked with plugs containing sulfur and iron, respectively. Since these elements are not found in the adjacent shell surrounding the pore, the origin and manner of deposition of these materials are perplexing (Board and Perrott, 1979). The materials covering some eggs leave little more than minute fissures through which gases must diffuse (Board and Scott, 1980). The plugs and other accessory materials are thought to protect the pathways for gaseous diffusion from clogging by mud, nest debris, or preening oils (Board and Perrott, 1982; Board, 1982).

Much of the rapid progress over the last decade in understanding the principles governing gaseous diffusion is attributable to the simple technique suggested by Ar et al. (1974) for estimating G_{H_2O}. Although G_{H_2O} could be calculated by measuring L and Ap with direct microscopic examination, the variety of pore shapes and coverings make

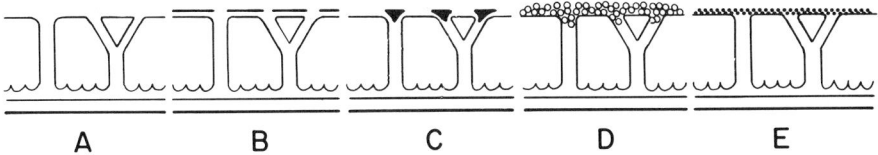

FIGURE 5. Schematic drawing of types of pores and accessory materials modified from Board and Scott (1980). A = open with branched or unbranched pores, B = occluded pores, C = plugged pores, D = capped pores, and E = reticulate pores.

accuracy of the latter method problematical. Tullett and Board (1977) and Hoyt et al. (1979) compared values of G_{H_2O} measured on the same eggs, both by the method of Ar et al. (1974) and by direct pore count. They found the values agreed within 10–15%. However, some large discrepancies were identified and were assumed to result from the inaccuracies of the direct pore counts because of the subjectivity involved in such measurements and the variation in sizes of pores.

Despite the complexity of pore shapes, sizes, and accessory materials, no differences among the G_{H_2O} of various types of structures are distinguishable when values are analyzed allometrically (Board and Scott, 1980; Board, 1982). Therefore, the performance of each type of configuration in control of gaseous diffusion appears to be similar for a given egg mass.

4.2.2. Formation of Pores

The number of pores/egg and the diameter of each pore increase and the number of pores/cm² decreases as egg mass increases (Tullett and Board, 1977). But we still have no firm idea how the shell gland creates a pore or how it makes the appropriate number of pores/egg.

Pores occur where three or more cones meet (Tyler and Fowler, 1978). Only 1:100–1:800 such junctions contain a pore; the rest are fused shut (Tyler and Fowler, 1978). Pores occupy only about 0.02% of the surface area of a typical anatid egg (Hoyt et al., 1979). Therefore, creation of a pore is a rare event in shell formation.

Tyler and Simkiss (1959) and Tullett (1978) have proposed that a pore is formed by fluid passing through the forming shell during the plumping phase. Furthermore, the pore number could be varied by variation of the initial amount of water in the albumen and the corresponding requirement for water to be added during plumping. The major problem with these proposals is that the plumping procedure overlaps with shell formation only during the first 2 hr of a 12-hr process (Mongin and Carter, 1977). Therefore, another mechanism is needed to maintain open pore channels after plumping is completed. Board (1982) has suggested that high concentrations of ions that poison shell calcification might be inserted into the stagnant fluid within the forming pore.

Krampitz et al. (1972, 1974) have noted the relation between species-specific amino acid compositions of the organic matrix of avian eggs with the formation of particular crystalline patterns. The possibility should be investigated that such a relationship could also contribute to the formation of pores.

The number of pores could be determined by genetic control of the number of mammillary cores laid down in the isthmus (Tullett, 1975). Fewer seeding sites could result in fewer pores since crystals would spread farther over the surface of the egg before forming junctions with other crystals. However, Tyler and Fowler (1978) have found no significant relation between the number of cores or cone junctions and the frequency of pores. They think that the cause of most variation in pore number is still unknown.

4.2.3. Modification of Pores during Incubation

Although Drent (1970) and Rahn and Ar (1974) have asserted that G_{H_2O} is constant throughout incubation, daily water loss and G_{H_2O} do increase in the first few days of incubation in certain species (Section 3.1.1). While several authors have proposed that additional pores may be opened during incubation (Almquist and Holst, 1931; Goryainova and Tarnovskaya, 1975), conclusive proof of this possibility has yet to be found. Shell reabsorption has been suggested as the causative factor (Sotherland et al., 1980), but the change in G_{H_2O} occurs early in incubation before shell reabsorption begins (Section 4.1.3). The drying of the shell membrane cannot be responsible for the increase in G_{H_2O} since membranes of all eggs dry but G_{H_2O} changes in only certain species. Abrasion of shell accessory materials by the brood patch or contact with other eggs might be responsible for changes in G_{H_2O}, but Tullett and Board (1976) believe that the cuticle of at least certain eggs has no effect on gaseous diffusion. This possibility might also appear to be ruled out since some artificially incubated eggs show an increase in G_{H_2O} (Hanka et al., 1979), but G. C. Packard (personal communication) now thinks that the results of Hanka et al. may be unreliable.

4.3. Modification of Eggshells in Various Environments

The G_{H_2O} of birds breeding in certain environments that pose special challenges to gaseous diffusion differ from those of conspecific populations breeding in less problematical habitats or from the values predicted on the basis of egg mass (Section 3.3). In most cases, the variation appears to result from changes in Ap and not L. For instance, eggs of loons are predicted on the basis of egg mass to have 76 pores/cm^3; instead, the actual count is 307/cm^3 (Board et al., 1977). In some cases the cross-sectional areas of the pores are more than 20-fold larger than predicted, but the number of cones per unit area is normal for eggs of this weight (Tullett, 1978).

The G_{H_2O} of eggs of six populations breeding in montane areas are significantly decreased below sea level values, but L does not vary

(Carey, 1980b). Since we cannot determine how the shell gland creates a prescribed number of pores, even in a chicken egg, we are unable to predict how selection has changed the mechanism of pore formation in these cases.

Although shell characteristics can be modified on a short-term basis by diet and pesticides (Tyler, 1969; Cooke, 1975), it would be interesting to know if environmental factors can cause a short-term modification in Ap, or if Ap is genetically fixed for the lifetime of a female. For instance, if a population of birds moved into a new habitat, such as high altitude, where a different Ap might afford greater hatchability than the Ap favored in the previous habitat, could Ap be physiologically modified or would breeding success ultimately depend on variation in Ap already present in the population? Since homing pigeons can detect alteration in P_B equivalent to change in altitude of 10 m (Kreithen and Keeton, 1974), it would be fascinating to know if females can detect changes in P_B or other associated environmental factors and adjust Ap accordingly. Chickens maintained for years at 3800 m were transferred to 1200 m and experienced a corresponding change in P_B of 37%. Within 6 weeks, the mean G_{H_2O} increased an average of 34% (Rahn et al., 1982), although considerable individual variability in response (5–70%) was apparent (Carey, 1980b). The change in G_{H_2O} was primarily due to an increase in number, rather than size of pores. A second study involving transfer of Coturnix quail (*Coturnix coturnix*) and society finches (*Lonchura domestica*) from sea level to 2897 m was unable to detect significant changes in Ap after at least six weeks at high altitude (Carey, Hoyt, Bucher, and Larson, unpublished data).

5. SUMMARY

Survival to hatching by avian embryos depends importantly on adult behavior, contents of the egg, and structural properties of the eggshell that afford appropriate levels of gas exchange, strength, and protection from invading microorganisms. The generalizations that have emerged over the last decade indicate that the eggshell conductance to gases has evolved in conjunction with the incubation period, egg mass, and external gas tension, with the result that avian embryos develop in fairly similar gaseous environments. However, since the conductance of individual eggs in natural populations varies considerably, the regulatory abilities of embryos must also contribute to their survival to hatching.

APPENDIX. ALLOMETRIC EQUATIONS DESCRIBING THE RELATION OF PROPERTIES OF AVIAN EGGS TO EGG MASS (W, in g)

In cases where equations for the same egg characteristic have been published by multiple authors, the most recent equation incorporating the largest sample size is presented.

Eq.	Property of egg	Units	Equation	Reference
A	Incubation period	day	$I = 11.64\ W^{0.221}$	Ar and Rahn, 1978
B	Water loss	mg·day^{-1}	$\dot{M}_{H_2O} = 13.24\ W^{0.754}$	Ar and Rahn, 1980
C	Conductance to water vapor	mg·day^{-1}·torr^{-1}	$G_{H_2O} = 0.384\ W^{0.814}$	Ar and Rahn, 1978
D	Shell thickness	cm	$L = 5.126 \cdot 10^{-3}\ W^{0.456}$	Ar et al., 1974
E	Shell surface area	cm^2	$A = 4.197\ W^{0.661}$	Tullett and Board, 1977
F	Functional pore area	cm^2	$Ap = 9.72 \cdot 10^{-5}\ W^{1.249}$	Ar and Rahn, 1978
G	Shell mass	g	$W_{sh} = 4.82 \cdot 10^{-2}\ W^{1.132}$	Paganelli et al., 1974
H	Shell volume	cm^3	$V_{sh} = 2.48 \cdot 10^{-2}\ W^{1.118}$	Paganelli et al., 1974
I	Pre-pipping oxygen consumption	ml·day^{-1} STPD	$\dot{M}_{O_2} = 28.9\ W^{0.714}$	Hoyt and Rahn, 1980
J	Total pores per egg	—	$N = 1041\ W^{0.504}$	Hoyt et al., 1979
H	Pores/cm^2 shell	—	$P = 279.65\ W^{-0.231}$	Tullett and Board, 1977
L	Individual pore area	cm	$P_A = 5.425 \cdot 10^{-8}\ W^{0.804}$	Tullett and Board, 1977

ACKNOWLEDGMENTS. Preparation of this review was aided by NSF-79-23403.

REFERENCES

Ackerman, R. A., and Platter-Rieger, M., 1979, Water loss by pied-billed grebe (Podilymbus podiceps) eggs, Am. Zool. **19**:921.

Ackerman, R. A., Whittow, G. C., Paganelli, C. V., and Pettit, T. N., 1980, Oxygen consumption, gas exchange, and growth of embryonic wedge-tailed shearwaters (Puffinus pacificus chlororhynchus), Physiol. Zool. **53**:210–221.

Almquist, H. J., and Holst, W. F., 1931, Variability of shell porosity in hen's eggs, Hilgardia **6**:61–72.

Ar, A., and Rahn, H., 1978, Interdependence of gas conductance, incubation length, and weight of the avian egg, in: Respiratory Function in Birds, Adult and Embryonic (J. Piiper, ed.), Springer-Verlag, Berlin, pp. 227–236.

Ar, A., and Rahn, H., 1980, Water in the avian egg: Overall budget of incubation, Am. Zool. **20**:373–384.

Ar, A., and Yom-Tov, Y., 1978, The evolution of parental care in birds, Evolution **32**:655–669.

Ar, A., Paganelli, C. V., Reeves, R. B., Greene, D. G., and Rahn, H., 1974, The avian egg: Water vapor conductance, shell thickness, and functional pore area, Condor **76**:153–158.

Becking, J. H., 1975, The ultrastructure of the avian eggshell, Ibis **117**:143–151.

Birchard, G. F., and Kilgore, D. L., 1980, Conductance of water vapor in eggs of burrowing and nonburrowing birds: Implications for embryonic gas exchange, Physiol. Zool. **53**:284–292.

Board, R. G., 1982, Properties of avian egg shells and their adaptive value, Biol. Rev. **57**:1–28.

Board, R. G., and Fuller, R., 1974, Non-specific antimicrobial defenses of the avian egg, embryo, and neonate, Biol. Rev. **49**:15–49.

Board, R. G., and Love, G., 1980, Magnesium distribution in avian eggshells, Comp. Biochem. Physiol. **66A**:667–672.

Board, R. G., and Perrott, H. R., 1979, The plugged pores of tinamou (Tinamidae) and jacana (Jacanidae) eggshells, Ibis **121**:469–474.

Board, R. G., and Perrott, H. R., 1982, The fine structure of the outer surface of the incubated eggshells of the helmeted guinea fowl (Numidia meleagris), J. Zool. Lond. **196**:445–451.

Board, R. G., and Scott, V. D., 1980, Porosity of the avian eggshell, Am. Zool. **20**:339–349.

Board, R. G., and Tullett, S. G., 1975, The pore arrangement in the emu (Dromaius novaehollandieα) eggshell as shown by plastic models, J. Microbiol. **103**:281–284.

Board, R. G., Tullett, S. G., and Perrott, H. R., 1977, An arbitrary classification of the pore systems in avian eggshells, J. Zool., Lond. **182**:251–265.

Calder, W. A., 1978, The kiwi, Sci. Am. **239**:132–142.

Carey, C., 1979, Increase in conductance to water vapor during incubation in eggs of two avian species, J. Exp. Zool. **209**:181–186.

Carey, C., 1980a, The ecology of avian incubation, BioScience **30**:819–824.

Carey, C., 1980b, Adaptation of the avian egg to high altitude, Am. Zool. **20**:449–459.

Carey, C., Rahn, H., and Parisi, P., 1980, Calories, water, lipid and yolk in avian eggs, Condor **82**:335–343.

Carey, C., Thompson, E. L., Vleck, C. M., and James, F. C., 1982, Avian reproduction over an altitudinal gradient—I. Incubation period, hatchling mass, and embryonic oxygen consumption, Auk **99**:710–718.

Carey, C., Garber, S. D., Thompson, E. L., and James, F. C., 1983, Avian reproduction over an altitudinal gradient—II. Physical characteristics and water loss of eggs, Physiol. Zool. (in press).

Cooke, A. S., 1975, Pesticides and eggshell formation, Symp. Zool. Soc. Lond. **35**:339–361.

Crooks, R. J., and Simkiss, K. 1974, Respiratory acidosis and eggshell resorption by the chick embryo, J. Exp. Biol. **61**:197–202.

Dawes, C. M., and Simkiss, K., 1971, The effects of respiratory acidosis in the chick embryo, J. Exp. Biol. **55**:77–84.

Drent, R. H., 1970, Functional aspects of incubation in the herring gull, Behavior Suppl. **17**:1–132.

Drent, R. H., 1975, Incubation, in: Avian Biology, Vol. V (D. S. Farner and J. R. King, eds.), Academic Press, New York, pp. 333–420.

Eastin, W. C., and Spaziani, E., 1978, On the mechanism of calcium secretion in the avian shell gland (uterus), Biol. Reprod. **19**:505–518.

Erben, H. K., 1970, Ultrastrukturen und Mineralisation rezenter und fossiler Eischalen bei Vogeln and Reptilien, Biomineralis. **1**:1–66.

Erben, H. K., and Kriesten, K., 1974, Mikromorphologie der Fruhstadien bei der Kristallbildung in normalen und anormalen Huhner-Eischalen, Biomineralis. **7**:28–36.

Goryainova, G. P., and Tarnovskaya, T. V., 1975, Changes in the porosity of egg shell during embryogenesis of birds, Z. Zhurn. **54**:1113–1115.

Grant, G. S., 1982, Avian incubation: Egg temperature, nest humidity, and behavioral thermoregulation in a hot environment, Ornithol. Monogr. **30**:1–75.

Grant, G. S., Pettit, T. N., Rahn, H., Whittow, G. C., and Paganelli, C. V., 1982, Regulation of water loss from Bonin petrel (Pterodroma hypoleuca) eggs, Auk **99**:236–242.

Hanka, L. R., Packard, G. C., Sotherland, P. R., Taigen, T. L., Boardman, T. J., and Packard, M. J., 1979, Ontogenetic changes in water-vapor conductance of eggs of yellow-headed blackbirds (Xanthocephalus xanthocephalus), J. Exp. Zool. **210**:183–188.

Harris, E. D., Blount, J. E., and Leach, R. M., 1980, Localization of lysyl oxidase in hen oviduct: Implications in egg shell membrane formation and composition, Science **208**:55–56.

Howe, H. F., 1976, Egg size, hatching asynchrony, sex, and brood reduction in the common grackle, Ecology **57**:1195–1207.

Howe, H. F., 1978, Initial investment, clutch size, and brood reduction in the common grackle (Quiscalus quiscula L.), Ecology **59**:1109–1122.

Howell, T. R., 1979, Breeding biology of the Egyptian plover, Pluvianus aegyptius, Univ. Calif. Publ. Zool. **113**:1–76.

Howey, P. W., Board, R. G., and Kear, J., 1977, A pulse-position modulated multichannel radio telemetry system for the study of the avian microclimate, Biotelemetry **4**:169–180.

Hoyt, D. F., and Rahn, H., 1980, Respiration of avian embryos—a comparative analysis, Resp. Physiol. **39**:255–264.

Hoyt, D. F., Vleck, D., and Vleck, C. M., 1978, Metabolism of avian embryos: Ontogeny and temperature effects in the ostrich, Condor **80**:265–271.

Hoyt, D. F., Board, R. G., Rahn, H., and Paganelli, C. V., 1979, The eggs of the Anatidae: Conductance, pore structure, and metabolism, Physiol. Zool. **52**:438–450.

Krampitz, G., and Witt, W., 1979, Biochemical aspects of biomineralization, Top. Curr. Chem. **78**:57–144.
Krampitz, G., Erben, H. K., and Kriesten, K., 1972, Über Aminosäuren-Zusammensetzung und Struktur von Eischalen, Biomineralis **4**:87–99.
Krampitz, G., Kriesten, K., and Faust, R., 1974, Über die Aminosäuren-Zusammensetzung morphologischer Eischalen-Fraktionen von Ratitae, Biomineralis **7**:1–13.
Kreithen, M. L., and Keeton, W. T., 1974, Detection of changes in atmospheric pressure by the homing pigeon, Columba livia, J. Comp. Physiol. **89**:73–82.
Kutchai, H., and Steen, J. B., 1971, Permeability of the shell and shell membranes of hen's eggs during development, Resp. Physiol. **11**:265–278.
Lomholt, J. P., 1976a, Th development of the oxygen permeability of the avian egg shell and its membranes during incubation, J. Exp. Zool. **198**:177–184.
Lomholt, J. P., 1976b, Relationship of weight loss to ambient humidity of birds eggs during incubation, J. Comp. Physiol. **105**:189–196.
Lundy, H., 1969, A review of the effects of temperature, turning, and gaseous environment in the incubator on the hatchability of the hen's egg, in: The Fertility and Hatchability of the Hen's Egg (T. C. Cater and B. M. Freeman, eds.), Oliver and Boyd, Edinburgh, pp. 143–176.
Mongin, P., and Carter, N. W., 1977, Studies on the avian shell gland during egg formation: Aqueous and electrolytic composition of the mucosa, Br. Poult. Sci. **18**:339–351.
Nice, M. M., 1962, Development of behavior in precocial birds, Trans. Linn. Soc. N.Y. **8**:1–211.
Nisbet, I. C. T., 1978, Dependence of fledging success on egg-size, parental performance and egg composition in common and roseate terns, Sterna hirundo and S. dougallii, Ibis **120**:207–215.
Paganelli, C. V., Ar, A., and Rahn, R., 1981, What is the effective ΔP_{O_2} in a gas-phase diffusion system? Resp. Physiol. **45**:9–11.
Paganelli, C. V., Ar, A., Rahn, H., and Wangensteen, O. D., 1975, Diffusion in the gas phase: The effects of ambient pressure and gas composition, Resp. Physiol. **25**:247–258.
Paganelli, C. V., Ackerman, R. A., and Rahn, H., 1978, The avian egg: In vivo conductances to oxygen, carbon dioxide, and water vapor in late development, in: Respiratory Function in Birds, Adult and Embryonic (J. Piiper, ed.), Springer-Verlag, Berlin, pp. 212–218.
Parsons, J., 1970, Relationship between egg-size and post-hatching chick mortality in the herring gull (Larus argentatus), Nature (Lond.) **228**:1221–1222.
Pettit, T. N., Grant, G. S., Whittow, G. C., Rahn, H., and Paganelli, C. V., 1982, Respiratory gas exchange and growth of Bonin petrel embryos, Physiol. Zool. **55**:162–170.
Rahn, H., and Ar, A., 1974, The avian egg: Incubation time and water loss, Condor **76**:147–152.
Rahn, H., and Ar, A., 1980, Gas exchange of the avian egg: Time, structure, and function, Am. Zool. **20**:477–484.
Rahn, H., Paganelli, C. V., and Ar, A., 1974, The avian egg: Air-cell gas tension, metabolism and incubation time, Resp. Physiol. **22**:297–309.
Rahn, H., Paganelli, C. V., and Ar, A., 1975, Relation of avian egg weight to body weight, Auk **92**:750–765.
Rahn, H., Paganelli, C. V., Nisbet, I. C. T., and Whittow, G. C., 1976, Regulation of incubation water loss in eggs of seven species of terns, Physiol. Zool. **49**:245–259.
Rahn, H., Ackerman, R. A., and Paganelli, C. V., 1977, Humidity in the avian nest and egg water loss during incubation, Physiol. Zool. **50**:269–283.

Rahn, H., Ledoux, T., Paganelli, C. V., and Smith, A. H., 1982, Changes in eggshell conductance after transfer of hens from at altitude of 3800 m to 1200 m, J. Appl. Physiol. **53**:1429–1431.

Ricklefs, R. E., 1977, Composition of eggs in several bird species, Auk **94**:350–356.

Ricklefs, R. E., 1983, Variation in the size and composition of eggs of the European starling. Condor (in press).

Ricklefs, R. E., and Montevecchi, W. A., 1979, Size, organic composition, and energy content of North Atlantic gannet Morus bassanus eggs, Comp. Biochem. Physiol. **64A**:161–165.

Ricklefs, R. E., Hahn, D. C., and Montevecchi, W. A., 1978, The relationship between egg size and chick size in the laughing gull and Japanese quail, Auk **95**:135–144.

Romanoff, A. L., 1967, Biochemistry of the Avian Embryo, Wiley, New York.

Romanoff, A. L., and Romanoff, A. J., 1949, The Avian Egg, Wiley, New York.

Sauveur, B., and Mongin, P., 1971, Étude comparative du fluide utérin et de l'albumen de l'oeuf in utero chez la poule, Ann. Biol. Anim. Biochem. Biophy. **11**:213–224.

Schmekel, L., 1960, Daten über das Gewicht des Vogeldottersackes vom Schlupftag bis zum Schwinden, Rev. Suisse Zool. **68**:103–110.

Schmidt, W. J., 1962, Liegt der Eischalenkalk der Vogel ab Submikroscopische Kristallite vor? Z. Zellforsch **57**:848–880.

Schönwetter, M., 1960–1980, Handbuch der Oologie (W. Meise, ed.), Akademie, Berlin.

Seymour, R. S., and Ackerman, R. A., 1980, Adaptations to underground nesting in birds and reptiles, Am. Zool. **20**:437–447.

Simkiss, K., 1961, Calcium metabolism and avian reproduction, Biol. Rev. **36**:321–367.

Simkiss, K., 1964, Phosphates as crystal poisons of calcification, Biol. Rev. **39**:487–505.

Simkiss, K., 1974, The air space of an egg: An embryonic "cold nose"? J. Zool., Lond. **173**:225–232.

Simkiss, K., 1975, Calcium and avian reproduction, Symp. Zool. Soc. Lond. **35**:307–337.

Simkiss, K., 1980, Eggshell porosity and the water metabolism of the chick embryo, J. Zool., Lond. **192**:1–8.

Simkiss, K., 1981, Calcium, pyrophosphate and cellular pollution, Trends in Biochem. Sci. **6**:3–5.

Simkiss, K., and Tyler, C., 1957, A histochemical study of the organic matrix of hen eggshells, Quart. J. Microbiol. Sci. **98**:19–28.

Simkiss, K., and Tyler, C., 1958, Reactions between egg-shell matrix and metallic cations, Quart. J. Microbiol. Sci. **99**:5–13.

Simons, P. C. M., 1971, Ultrastructure of the hen eggshell and its physiological interpretation, Centre for Agricultural Publishing and Documentation, Wageningen, The Netherlands.

Snyder, G. K., and Birchard, G. F., 1982, Water loss and survival in embryos of the domestic chicken, J. Exp. Zool. **219**:115–117.

Sotherland, P. R., Packard, G. C., Taigen, T. L., and Boardman, T. J., 1980, An altitudinal cline in conductance of cliff swallow (Petrochelidon pyrrhonota) eggs to water vapor, Auk **97**:177–185.

Taigen, T. L., Packard, G. C., Sotherland, P. R., Boardman, T. J., and Packard, M. J., 1980, Water-vapor conductance of black-billed magpie (Pica pica) eggs collected along an altitudinal gradient, Physiol. Zool. **53**:163–169.

Tazawa, H., 1978, Gas transfer in the chorioallantois, in: Respiratory Function in Birds, Adult and Embryonic (J. Piiper, ed.), Springer-Verlag, Berlin, pp. 274–291.

Tazawa, H., 1980, Oxygen and CO_2 exchange and acid-base balance in the avian embryo, Am. Zool. **20**:395–404.

Tazawa, H., Mikami, T., and Yoshimoto, C., 1971a, Respiratory properties of chicken embryonic blood during development, *Resp. Physiol.* **13**:160–170.

Tazawa, H., Mikami, T., and Yoshimoto, C., 1971b, Effect of reducing the shell area on the respiratory properties of chicken embryonic blood, *Resp. Physiol.* **13**:352–360.

Temple, G. F., and Metcalfe, J., 1970, The effects of increased incubator oxygen tensions on capillary development in the chick chorioallantois, *Resp. Physiol.* **9**:216–233.

Tullett, S. G., 1975, Regulation of avian eggshell porosity, *J. Zool., Lond.* **177**:339–348.

Tullett, S. G., 1978, Pore size versus pore number of avian eggshells, in: *Respiratory Function in Birds, Adult and Embryonic* (J. Piiper, ed.) Springer-Verlag, Berlin, pp. 219–226.

Tullett, S. G., and Board, R. G., 1976, Oxygen flux across the integument of the avian egg during incubation, *Br. Poult. Sci.* **17**:441–450.

Tullett, S. G., and Board, R. G., 1977, Determinants of avian eggshell porosity, *J. Zool., Lond.* **183**:203–211.

Tullett, S. G., Board, R. G., Love, G., Perrott, H. R., and Scott, V. D., 1976, Vaterite deposition during eggshell formation in the cormorant, gannet, and shag, and in "shell-less" eggs of the domestic fowl, *Acta Zool. (Stock.)* **57**:79–87.

Tyler, C., 1964, A study of the egg shells of the Anatidae, *Proc. Zool. Soc. Lond.* **142**:547–583.

Tyler, C., 1965, A study of the egg shells of the Sphenisciformes, *J. Zool., Lond.* **147**:1–19.

Tyler, C., 1966, A study of the egg shells of the Falconiformes, *J. Zool., Lond.* **150**:413–425.

Tyler, C., 1969, Avian egg shells: Their structure and characteristics, *Int. Rev. Gen. Exp. Zool.* **4**:81–130.

Tyler, C., and Fowler, S., 1978, The distribution of organic cores, cones, cone junctions and pores in the egg shells of wild birds, *J. Zool., Lond.* **186**:1–14.

Tyler, C., and Fowler, S., 1979, The size, shape and orientation of pore grooves in the egg shells of *Rhea* sp., *J. Zool., Lond.* **187**:283–290.

Tyler, C., and Geake, F. G., 1960, Studies on eggshells XIII. Influence of individuality, breed, season, and age on certain characteristics of eggshells, *J. Sci. Food Agric.* **9**:535–547.

Tyler, C., and Simkiss, K., 1959, A study of the egg shells of ratite birds, *Proc. Zool. Soc. Lond.* **133**:201–243.

Visschedijk, A. H. J., Ar, A., Rahn, H., and Piiper, J., 1980, The independent effects of atmospheric pressure and oxygen partial pressure on gas exchange of the chicken embryo, *Resp. Physiol.* **39**:33–44.

Vleck, C. E. M., Hoyt, D. F., and Vleck, D., 1979, Metabolism of avian embryos: Patterns in altricial and precocial birds, *Physiol. Zool.* **52**:363–377.

Vleck, C. M., Vleck, D., and Hoyt, D. F., 1980, Patterns of metabolism and growth in avian embryos, *Am. Zool.* **20**:405–416.

Walsberg, G. E., 1980, The gaseous microclimate of the avian nest during incubation, *Am. Zool.* **20**:363–372.

Walsberg, G. E., 1983, A test for regulation of nest humidity in two bird species, *Physiol. Zool.* (in press).

Wangensteen, O. D., 1972, Gas exchange by a bird's embryo, *Resp. Physiol.* **14**:64–74.

Wangensteen, O. D., and Rahn, H., 1970/71, Respiratory gas exchange by the avian embryo, *Resp. Physiol.* **11**:31–45.

Wangensteen, O. D., Wilson, D., and Rahn, H., 1970/71, Diffusion of gases across the shell of the hen's egg, *Resp. Physiol.* **11**:16–30.

Wyburn, G. M., Johnston, H. S., Draper, M. H., and Davidson, M. F., 1970, The fine structure of infundibulum and magnum of the oviduct of *Gallus domesticus*, *Quart. J. Exp. Physiol.* **55**:213–232.

Wyburn, G. M., Johnston, H. S., Draper, M. H., and Davidson, M. F., 1973, The ultrastructure of the shell forming region of the oviduct and the development of the shell of *Gallus domesticus*, *Quart. J. Exp. Physiol.* **58**:143–151.

CHAPTER 4

THE ORIGIN OF BIRDS AND OF AVIAN FLIGHT

LARRY D. MARTIN

1. INTRODUCTION

Birds, with their feathers, toothless bills, bipedal locomotion, and flight, form such a well-defined, uniform class, that it is hard to imagine how they were derived from some other stock. On the other hand, they provide the best known interclass transition in the vertebrata. This transitional form is *Archaeopteryx,* whose skeleton was first recognized in the Jurassic of Bavaria in 1861, only two years after Darwin's *Origin of Species* (Darwin, 1859). The timing of this discovery may have contributed to the immense public interest that has developed around the *Archaeopteryx* specimens. Various authors during this early period tried to relate birds to reptiles and a few championed specific reptilian groups (Huxley, 1867, 1868, 1870). Dinosaurs were suggested very early as potential bird ancestors (Marsh, 1877). This was probably because dinosaurs are largely bipedal reptiles and birds are also bipedal. Relationship has also been suggested with pterosaurs (Owen, 1875, Seeley, 1881); lizards (Petronievics, 1921, 1927, 1950), and mammals (Gardiner, 1982). I do not discuss pterosaurs, lizards, and mammals as possible ancestors because they have never enjoyed serious support as bird rel-

---

LARRY D. MARTIN • Department of Systematics and Ecology and Museum of Natural History, University of Kansas, Lawrence, Kansas 66045.

atives and I have not discovered anything to make me think that they deserve any. The dinosaur–bird connection was dominant until the early part of the twentieth century, when abundant remains of small diapsid reptiles with bipedal tendencies were discovered in South Africa. These reptiles were lumped together as an early stage of the archosaur radiation, the Pseudosuchia. The Pseudosuchia was thought to contain the ancestral stock of dinosaurs, crocodiles and finally birds (Broom, 1913). This theory was supported by Heilmann (1926) in his influential book, *The Origin of Birds,* and with a few exceptions (Lowe, 1933, 1944a,b; Holmgren, 1955) replaced the dinosaurian origin of birds until essentially the last decade.

2. THE NATURE OF *ARCHAEOPTERYX*

Birds are archosaurs because they have an antorbital fenestra (Fig. 1), and this is the reason that we seek their origins among the diapsid reptiles, although no known bird is diapsid. *Archaeopteryx* is considered a bird in this paper. It has feathers, a furcula and a pretibial bone. However, *Archaeopteryx* is not *ancestral* to any group of modern birds. It has specializations in its tarsometatarsus and skull that show conclusively that it is on a side-branch of avian evolution. These characters include reduction or loss of the squamosal in the skull, and the fusion of the metatarsals without the characteristic tarsal cap of modern birds. This means that *Archaeopteryx* cannot be a primitive model for all birds, as some have suggested (Ostrom, 1975a), and that we must compare it with other well-known Mesozoic birds like *Hesperornis* and *Ichthyornis,* as well as modern birds, in order to get a full picture of what the hypothetical "Proavis" was like. We should not fall into the trap of rejecting either the fossil evidence (Patterson, 1981) or the lessons we can learn from the comparative anatomy of modern animals (Ostrom, 1976b); we should try to use all such evidence.

The fact that *Archaeopteryx* is not an ancestor for modern birds does not lessen its value as a transitional form. It is the oldest and most primitive known bird and it shows a remarkable assemblage of primitive and advanced characters. This "mosaic evolution" (Simpson, 1980) seems to characterize transitions. *Archaeopteryx* has primary feathers essentially identical in their arrangement and morphology with those

FIGURE 1. Restorations of skeletons. (A) *Archaeopteryx.* (B) *Scleromochlus.* (C) *Deinonychus,* modified from Ostrom (1969, 1976c).

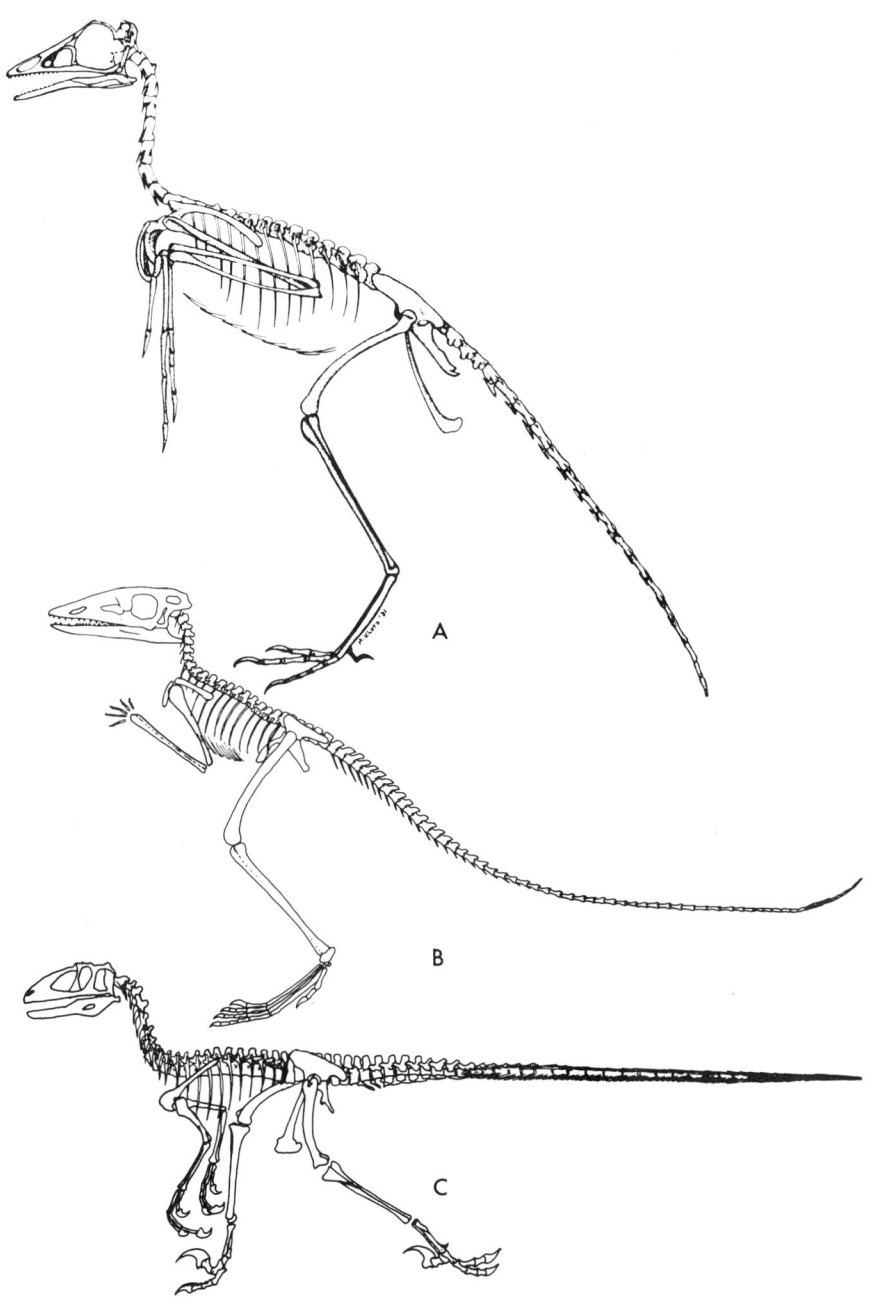

of modern birds and at the same time a reptilian jaw armed with numerous teeth.

Because *Archaeopteryx* is not a direct ancestor to modern birds we must be careful when we assume that its features are primitive. Characters shared by it and other Mesozoic birds, such as *Ichthyornis* and *Hesperornis*, were almost certainly features of the "Proavis," but characters found only in one or the other might be autapomorphies. This does not invalidate the use of unique features in either group, since we can compare them with reptiles to ascertain polarity. When we do this we find that *Archaeopteryx* provides unique information concerning the origin of the two most distinctive avian locomotor systems, bipedalism and flight.

2.1. Bipedalism

All modern birds are fully bipedal. *Archaeopteryx*, with its long hind limbs and wings, seems to correspond to modern birds in this respect. It has also been considered highly cursorial (Ostrom, 1974a, 1979). The latter interpretation should be viewed with caution, for long hind limbs and, especially, the lengthening of the tarsus can also be adaptations for jumping. *Archaeopteryx*, like other Mesozoic birds, has a considerable amount of closure of the inner wall of the pelvic acetabulum (Fig. 2). This is a primitive feature found in pseudosuchians and crocodilians, but not in saurischian dinosaurs. The theropods are highly sophisticated, obligate bipeds that have the head of the femur at a right angle to the sagittal plane and deeply inserted into a completely open pelvic acetabulum so that the weight of the body is supported with little waste of muscular energy (Charig, 1972). This posture is not possible in *Archaeopteryx*, in which the head of the femur is at an angle of about 45° to the sagittal plane and rests in an inclined attitude against the inner wall of the acetabulum (Fig. 2A and B). Like the modern lizard, *Basiliscus*, *Archaeopteryx* may have been able to sprint for short distances on the ground, but it is unlikely that either it or its predecessors could have been primarily cursorial ground dwellers.

2.2. Flight

Ostrom (1974b, 1979) thought that *Archaeopteryx* represents a pre-flight to flight transitional stage of avian evolution. Well-developed powered flight is considered unlikely because of lack of rigidity in the

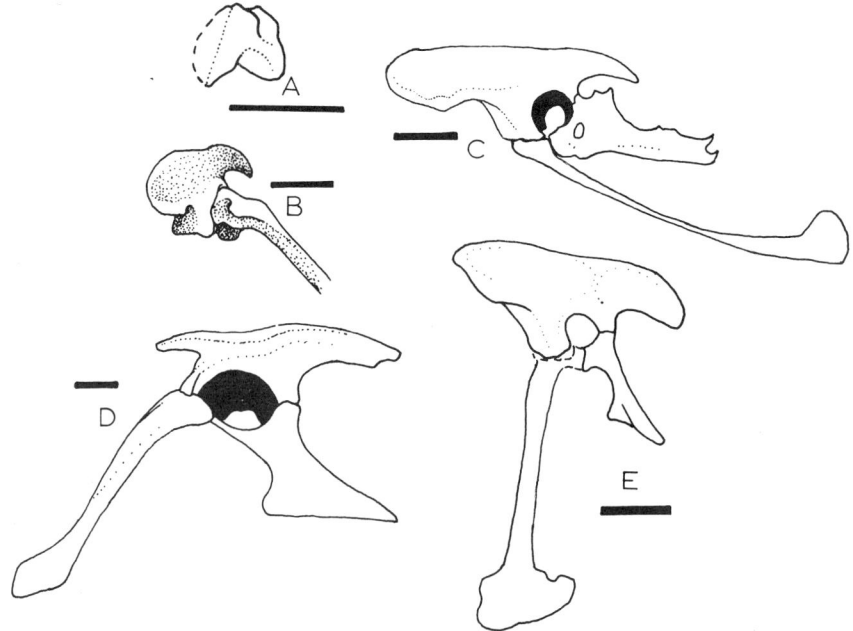

FIGURE 2. *Archaeopteryx*. (A) Proximal femur (dashed area hidden by matrix). (B) Anterior view of femur in articulation with the ilium. (C) *Archaeopteryx*. (D) *Protosuchus*, after Colbert and Mook (1951). Both C and D are lateral views of pelvi with the bone inside the acetabulum darkened. (E) *Deinonychus*, after Ostrom (1976c). Scales of A–D = 1 cm; E = 10 cm.

skeleton, lack of an ossified sternum, shortness of the coracoids, absence of quill nodes on the ulna, and absence of a pygostyle. It may well be that flight could not be as good in *Archaeopteryx* as it is in modern birds, but Ostrom's objections do not preclude powered flight. The wing is in fact organized as it is in a modern bird. The absence of an ossified sternum is mitigated by the large furcula (Olson and Feduccia, 1979) and by the rectangular coracoids, which would also be areas of origin for the flight muscles. The skeleton also seems able to sustain the forces of takeoff and landing (Yalden, 1971). The wing area seems adequate for flight when we take into consideration the lift provided by the tail (Bramwell, 1970). The most striking evidence for flight in *Archaeopteryx* has been proposed by Feduccia and Tordoff (1979), who argue that the asymmetric feathers of *Archaeopteryx* can be explained in no other way.

3. THE ORIGIN OF BIRDS

Beginning in 1970, with Galton's paper (1970) providing new arguments in favor of an avian–ornithischian dinosaur relationships, the controversy over avian origins quickly began to re-emerge. Walker (1972) suggested a "sister group" relationship between crocodilians and birds, and Ostrom (1973) reinvigorated the theropod dinosaur model. Ostrom has proven to be the most persuasive supporter of the dinosaur model since its inception and it now enjoys wide support (Cracraft, 1977). Walker's crocodilian hypothesis has recently acquired new support (Whetstone and Martin, 1979, 1981; Martin et al., 1980).

3.1. The Pseudosuchian Origin

Both the ornithischian and crocodilian hypotheses of avian origins are rooted in a common ancestor that Heilmann would probably have called a pseudosuchian. The Pseudosuchia, as it is generally utilized in the arguments concerning avian origins, is an evolutionary grade from which all of the advanced archosaurs are somehow derived. Such an argument really states that the split leading to birds is very ancient. It is thus difficult to find direct support for this model, although Tarsitano and Hecht (1980) presented a recent version of it which excludes a special relationship with crocodilians.

3.2. Ornithischian Relationship

The ornithischian model of avian origins as proposed by Galton (1970) is a sister group model suggesting that ornithischian dinosaurs share a unique common ancestor with birds but do not themselves give rise to Aves. This model has not had much acceptance. It was based primarily on the opisthopubic pelvis in both groups. The presence of a fenestra pseudorotundum in *Hypsilophodon* (Whetstone and Martin, 1981) might also support this, and I have recently found a predentary bone in *Hesperornis*, and (judging from its characteristic articular facet with the dentary) it was present in *Ichthyornis* as well. *Archaeopteryx* does not seem to have had a predentary bone, and we cannot be certain that it was present in the protobird. If it was, the ornithischian origin of birds would become more reasonable, for ornithischians form the only reptilian group considered a possible avian progenitor that has a predentary bone.

3.3. Crocodilian Relationships

No one has actually proposed that birds are directly derived from crocodilians, and all arguments relating the two groups have been based on a hypothesis of some unknown common ancestor (Walker, 1972, 1974, 1977; Whetstone and Martin, 1981). Derived characters of crocodilians that make them unlikely avian ancestors include (1) akinetic skull, (2) elongated radiale and ulnare, and (3) phalangeal formula of 2344 on the pes. Any common ancestor between birds and crocodilians would have to predate the late Triassic, because true crocodilians are well established by that time, and have an almost worldwide distribution. I agree with numerous other workers (Crompton and Smith, 1980) that *Sphenosuchus* is a sister group of the crocodilia, and is not a real crocodilian. Some of its supposed similarity to birds is based on characters that are either primitive or related to a higher degree of bipedality than is normal for the Crocodilia, and I have not found *Sphenosuchus* as similar to birds as is the Crocodilia. Considering the size constraints on flight, it is likely that birds were derived from animals of small body size, and I join most workers (de Beer, 1954; Bock, 1965, 1969; Brodkorb, 1971; Feduccia, 1980) in suggesting that the "Proavis" was arboreal. Small archosaurs that generally fit this description are not uncommon in the Triassic and include *Cosesaurus*, which was described as a protobird, and *Longisquama*, which was thought to have a furcula and feather-like scales (Sharov, 1970).

Another form that has been suggested to be related to *Cosesaurus* (Ostrom, 1976b; Ellenberger, 1977) is *Scleromochlus* from the Triassic of Scotland. *Scleromochlus* and *Cosesaurus* have elongated hind legs (not known in *Longisquama*), especially in the tibia and metatarsals, pointed snouts, and a remarkably high percentage of the features suggested to relate birds to coeleurosaurs. *Longisquama* and *Cosesaurus* also share some avian features not found in dinosaurs, including feather-like structures (Ellenberger, 1977; Sharov, 1970), a mesethmoid (Ellenberger, 1977), a furculum, and placement of the scapula parallel to the vertebral column in the avian manner. The latter character seems to be clearly present in all three genera and is of great importance, for it is not found in any known dinosaur. I suspect that it is related to climbing, in analogy to the similar placement of the scapula in primates. *Scleromochlus* has a distinct calcaneal heel, as is found in the Crocodilia. The absence of a mandibular fenestra in all Mesozoic birds for which the jaw is known is interesting, because it is a prominent feature of both crocodilian and dinosaurian lower jaws. Its lack is primitive for diapsids, as *Petrolacosaurus* does not have one. *Cosesaurus* also lacks the mandibular fenestra and it has a long retroarticular process, as do crocodilians, Archaeopteryx, and Hesperornis. Theropods have short projections in this area, but do not compare as well with Mesozoic birds such as Hesperornis as do the crocodilians. Unfortunately, the cranial features used to relate crocodilians with birds cannot be ascertained in the known material of Cosesaurus, Longisquama, and Scleromochlus, but we should take seriously the possibility that they have avian affinities. I have made comparisons with them whenever possible.

Following is a list of characters that unite crocodilians and birds. Characters 1 through 15 are Tarsitano and Hecht's (1980) summary of characters provided by Walker. The rest are cranial features that have been more recently developed (Whetstone and Martin, 1981).

1. Possession of laterosphenoids, an external mandibular foramen, and an antorbital fenestra.
2. Forward position of the quadrate head, articulating with the squamosal and prootic.
3. A kinetic skull with a streptostylic quadrate.
4. Crescentic shape of the occipital surface.
5. A short paraoccipital process projecting behind the quadrate and forming the posterior wall of the tympanic cavity.
6. Inferred possession of paired salt glands in Sphenosuchus.
7. Similar carotid circulation based on paired grooves and a pneumatic basisphenoid in Sphenosuchus.
8. An elongate cochlear duct.
9. Similarity of the manus, carpus, and elbow joint.
10. Similar pattern of digit reduction.
11. Posteriorly directed pubis of the "Jurassic crocodilian Hallopus."
12. The scapulocoracoid in Archaeopteryx can be derived from that of Sphenosuchus.
13. Similar morphology of the palatines.
14. Crocodilians were originally arboreal as evidenced by the climbing ability of juvenile crocodilians, the morphology of the tarsus, long humerus, pneumatization of the skull and limbs of fossil crocodilomorphs, marked inward and forward curvatures of the lower half of the tibia and the reduction of the first metatarsal.
15. Eustachian tube system in birds and crocodilians.
16. Fenestra pseudorotundum present in birds and crocodilians.
17. Unserrated teeth in birds and crocodilians (theropod dinosaurs normally have serrated teeth).

18. Distinct constriction between the crown and root of the teeth in birds and crocodilians.
19. Expanded bony root covered with cementum and connected to the jaw by peridontal ligaments in birds and crocodilians.
20. Circular or oval resorption pits in the teeth of birds and crocodilians formed by the tilting of a replacement tooth labially and its main development within the pulp cavity of its predecessor.
21. Implantation of the teeth of birds and crocodilians in a groove in at least the young individuals.
22. Formation of the lingual walls of the major tooth-bearing bones in birds and crocodilians by extensions of dense bone, rather than by attachment bone, as in most dinosaurs.
23. A quadrate cotylus at the anterior base of the parocciput, in birds and crocodilians.
24. A bipartite quadrate articulation in birds and crocodilians with dermal and endochondral bones—anteriorly with the prootic, squamosal, and laterosphenoid; posteriorly with the prootic and otoccipital.
25. Squamosal shelf over the ear region in birds and crocodilians.
26. Anterior-medial origin of the temporal musculature on the skulls of birds and crocodilians.
27. Periotic pneumatic cavities in the dorsal, central, and rostral positions in birds and crocodilians.
28. Two pneumatic cavities in birds and crocodilians surrounding the cerebral carotid arteries.
29. Pneumatic quadrate in both birds and crocodilians.
30. Foramen aerosum in the lower jaw in both birds and crocodilians.

Tarsitano and Hecht (1980) criticized characters 1–15 and some of this criticism seems valid. Characters 1, 3, 5, 7, 8, and 12 are either primitive diapsid features or are widely distributed in other archosaurs. Characters 3, 6, and 13 are based on interpretations of *Sphenosuchus* and early birds which are at best unclear. For instance, I could not find bird-like kinesis in the skull of *Sphenosuchus* when I examined the specimen. Characters 9 and 10 are based on only a few similarities whose importance is unclear, and character 14 is an inference which may be correct but is hard to utilize as a systematic argument. Tarsitano and Hecht (1980) argue that character 8 is a plesiomorphy because it is present in mammals, dinosaurs, and sea turtles. I gather from this that they think that the short duct in other tetrapods is derived. However, the presence of an elongated duct in birds, crocodiles, and dinosaurs renders its use as a simple character untenable. Character 11 cannot presently be confirmed with certainty (Walker, 1970, 1977). Character 15, the eustachian tube system, needs more study in birds, but, as it is presently known, it does not seem to lend much support to the crocodile–bird hypothesis.

The homology of character 16, the fenestra pseudorotunda, in crocodilians and birds is firmly supported by studies of both soft anatomy and embryology. It is an apomorphic character that is absent from many tetrapods, and in those modern forms where it does occur (lizards and mammals) it has a different embryology. It is, however, an ancient structure in birds and crocodilians, being present in Triassic crocodilians like *Eopneumatosuchus* and their relatives, such as *Sphenosuchus* (Crompton and Smith, 1980). The fenestra pseudorotunda is also present in the oldest birds (*Enaliornis* and *Hesperornis*) where the proper area is exposed. As far as I can tell it occurs throughout both the Aves and the Crocodilia.

In dinosaurs a fenestra pseudorotunda is generally absent, and when it is found it is probably best interpreted as an independent acquisition. This is the case with *Hypsilophodon*, which has a "round window," while hadrosaurs and ankylosaurs do not (Whetstone and Martin, 1979, 1981). The large carnosaurs such as *Tyrannosaurus* lack the fenestra pseudorotunda, and the material of *Dromaeosaurus* and struthiomimids are indeterminate for this structure. It is clear from the known distribution of this feature that the presence of a fenestra pseudorotunda does not unite theropod dinosaurs as a group to birds, but is either an autapomorphy or indicates a special relation of a small group, perhaps a single genus, with birds.

The tooth argument for a bird–crocodilian relationship has recently acquired new importance. Not only do the teeth of birds and crocodiles resemble each other in external morphology and differ from all known dinosaurs, but the mode of replacement and implantation is also the same. In crocodilians, the replacement tooth tilts labially early in its development and enters the tooth root through a circular resorption window. These resorption windows are also found in bird teeth, but not in dinosaurs. In dinosaurs, the replacement teeth migrate labially in a vertical position, and create a ventrally open absorption pit (Fig. 3). The septa and inner wall of the tooth-bearing bones in crocodilians and birds also seem to be formed differently from the septa and inner walls in dinosaurs. It is unlikely that the common ancestor of birds and dinosaurs was fully thecodont, and this, like many other features, suggests that any common stem must be early.

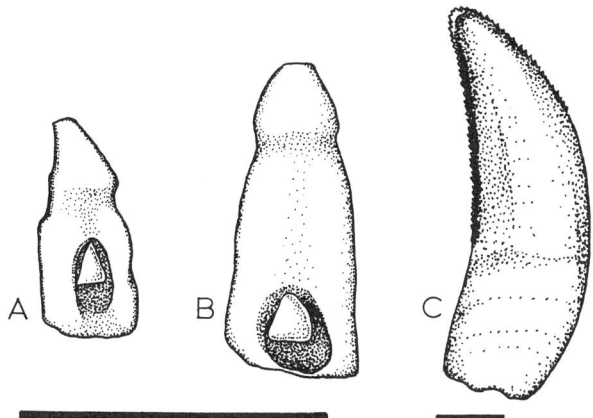

FIGURE 3. Teeth. (A) Hesperornithiform bird showing replacement tooth in its absorption pit. (B) Crocodilian. (C) Theropod dinosaur, *Megalosaurus*. Scales = 1 cm.

The quadrate articulation of birds and its relationship to the periotic pneumatic cavities presents a highly complicated complex that is uniform in its development in birds and occurs in the earliest known crocodilians. It has not been found in dinosaurs, although pneumatic cranial spaces are known in a few highly-specialized species (Osmolska, 1976). It is hard to decide what should be done if these characters should be found in some coelurosaur. There is no comparable suite of characters to unite the coelurosauria and one wonders if such a "dinosaur" might be more reasonably attached to birds and crocodiles rather than being retained in the theropoda.

3.4. The Coelurosaurian Origin

Lists of characters proposed to support a coelurosaurian origin of birds can be found in the works of Ostrom and more recently his student Padien (1982). Here are the most important of these characters:

1. A sharply tapered snout.
2. Long, elliptical external nares, bounded almost exclusively above and below by the premaxilla.
3. A large, triangular antorbital fossa, containing two small anterior openings and a large triangular posterior fenestra.
4. A slender, nearly vertical preorbital bar separating the antorbital foss and the orbit.

5. A large, circular orbit containing a large sclerotic ring.
6. A thin, straight jugal bar.
7. A stout quadrate of moderate length, which is inclined forward, i.e., descends anteriorly.
8. A lower jaw that is unusually shallow and has a conspicuous downward bend behind the tooth row.
9. A long retroarticular process.
10. Theropod-like intermandibular articulation.
11. Thoracic vertebra pleurocoelous (and probably amphicoelous).
12. 12–15? thoracics; ten cervical vertebrae.
13. Semilunate carpal articulating with metacarpals I and II.
14. Manus reduced to digits I, II, and III.
15. Metacarpal I shorter than metacarpals II and III.
16. Phalanx I longer than metacarpal I.
17. Elongated penultimate phalanges on digits I–III.
18. Well-defined deltopectoral crest on the humerus.
19. Radius and ulna shorter than the humerus.
20. Strap-like scapula.
21. Subrectangular coracoid fused to scapula.
22. Ossified gastralia.
23. Posterior process of ilium separates into ventromedial and ventrolateral blades.
24. Ischium with an anteriorly-directed pointed blade.
25. Pubis considerably longer than the ischium.
26. Acetabulum of the pelvis open.
27. Lesser trochanter well developed on the femur.
28. Femur considerably shorter than the tibia.
29. Well-developed ascending process of the astragalus.
30. Astragalus and calcaneum fused together.
31. Mesotarsal joint.
32. Two distal tarsals.
33. Pes with four digits, V being lost.
34. Metatarsal I articulated only distally.
35. Reversed hallux.
36. Phalangeal proportions of the pes.
37. Metatarsal proportions.
38. Elongations of caudal zygapophyses.
39. Caudal chevrons elongated and flattened.

This is a long list, and at first glance convincing. However, the characters are not so impressive when they are examined individually. We may begin with the skull.

Our knowledge of the skull of *Archaeopteryx* has been radically improved by the discovery of the Eichstatt specimen (Wellnhofer, 1974) and by the preparation of the cranium of the London *Archaeopteryx* (Whybrow, 1982). Unfortunately, this new information serves more to isolate *Archaeopteryx* than to provide information on the origin of birds. It is clear that *Archaeopteryx* has lost the diapsid arch and that no postfrontal is present. The squamosal is also absent or extremely reduced, and this is a derived feature that isolates *Archaeopteryx* from modern birds. A short dorsal process is present on the jugal, and the quadratojugal runs up the lateral border of the quadrate, giving a clue to the shape of the lower temporal opening in the avian progenitor. Apparently it was constricted dorsally, as in *Aetosaurus*, and would not have looked very coelurosaur-like. The orbit in *Archaeopteryx* is very large and resembles some coelurosaurs in this respect (Ostrom, 1976b), but large orbits occur in other small reptiles, including *Cosesaurus* and *Scleromochlus*, which also have the triangular antorbital fenestra and pointed snout characteristic of birds and found in some dinosaurs (Ostrom, 1976b). Well-developed sclerotic rings are present in some early crocodilians, *Cosesaurus*, *Longisquama*, and *Scleromochlus*, but are not known from most theropods. Their absence is a derived condition and would detract from the argument that birds are derived from any particular coelurosaur that lacks them. Ostrom (1976b) has suggested that the antorbital fenestra of *Archaeopteryx* contains secondary fenestrae that are a derived feature of certain theropod dinosaurs. This feature is only known from the Eichstatt *Archaeopteryx*, and my examination of this character suggests that it is open to alternate explanations. It may or may not support the coelurosauran origin of birds.

Gingerich (1979) has suggested that the intermandibular articulation of *Ichthyornis* and *Hesperornis* supports relationship to theropod dinosaurs, but *Archaeopteryx* and *Gobipteryx* do not appear to have an intermandibular joint, and the joint in *Hesperornis* and *Ichthyornis* is more like the intermandibular joint of mosasaurs than it is like the one found in theropods. Intermandibular joints may have characterized the base of the ornithurae radiation, but it seems that their common ancestor with *Archaeopteryx* lacked such a structure.

Character 13, the semilunate carpal bone, is considered by Ostrom (1969) to be the radiale in *Deinonychus*, and he describes its articulation with the radius. The ulnare in *Deinonychus* lies on the internal side of the radiale, and according to Ostrom (1969) they are tightly articulated in this position. This means that the III metacarpal lies internal to the II metacarpal and cannot be seen from an external view unless it is extended outward. This is quite different from the avian situation, where the III metacarpal lies directly in line with metacarpals I and II. The semilunate bone in birds is composed of distal carpals and articulates with the scapholunar (radiale) and cuneiform (ulnare), which then articulate with the radius and ulna. If we accept Ostrom's (1969) interpretation of the wrist of *Deinonychus*, we cannot accept the homology of the semilunate bone in the wrist of birds with that in coelurosaurs, for they articulate with different bones and form a markedly different joint. Character 15 is primitive and can be found in the oldest known diapsid, *Petrolacosaurus*. Characters 16–17 might be valid, but character 18 can be found in both *Scleromochlus* and *Sphenosuchus*. Character 19 is primitive and widely distributed.

One of the most striking features of the avian pectoral girdle is the long, narrow scapulae that lie on the back, parallel to the vertebral column, not extending diagonally across the lateral face of the rib cage, as in most reptiles. In order to achieve this relationship the bird coracoid and scapula are rotated nearly 90°, so that the former anterior margin is medial and the glenoid faces directly laterally, not posteroventrally (Fig. 4C). In this position the coracoids form a nearly continuous surface with the ventral border of the sternum. The scapula lies directly on the ribs along the lateral margins of the transverse processes. This means that they must form about a 90° angle with the coracoid. The scapulae are disarticulated from the coracoids in all but the London specimen, and in that specimen the left scapula is not in articulation. It is likely that the interpretation of the scapula and coracoid as fused is in error. The scapula of the coelurosaurs including *Deinonychus* lie against the lateral side of the rib cage in the primitive condition (Fig. 4A and B). They share a subrectangular coracoid with *Archaeopteryx*, but this shape is primitive and does not contribute to our understanding of relationship.

Character 22 is primitive and widely distributed among reptiles. Characters 23–24 are from Padien (1982) and were presented without discussion. That these features are actually present in dinosaurs as in *Archaeopteryx* is not clear. Character 25 can be found in early crocodilians, and character 26 is incorrect. All Mesozoic birds have some closure of the pelvic acetabulum, and in *Archaeopteryx* this is quite significant. In fact, the femur of *Archaeopteryx* sprawls to a fairly great degree when fitted to the acetablum (Fig. 2), and might be considered to have the semi-improved posture of Charig (1972).

One of the earliest lines of evidence developed to associate birds with coeleurosaurs was the supposed presence of a high, pointed dorsal

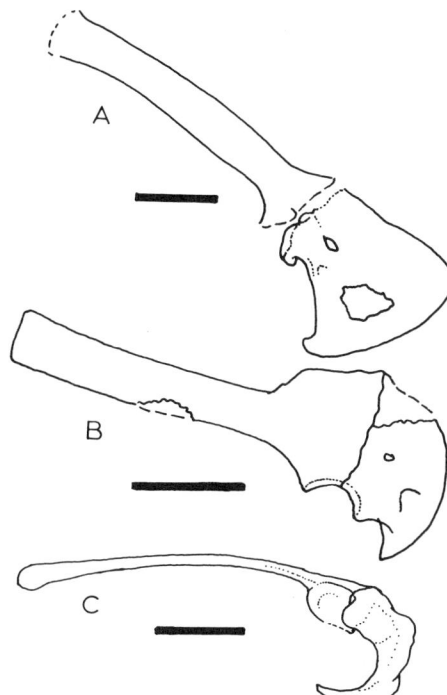

FIGURE 4. Scapulae and coracoids in lateral view as in the restored skeletons. (A) *Deinonychus*, after Ostrom (1974b, 1976c). (B) *Gallimimus*, after Osmolska, et al. (1972). (C) *Archaeopteryx*. Scales of A and C = 5 cm; B = 10 cm.

process of the astragalus in both groups (Baur 1884, 1885, 1886; Ostrom, 1975a,b). The homology of the structure in birds with that in dinosaurs was challenged by Martin et al. (1980), who pointed out that the ascending process of birds was more closely associated with the calcaneum than with the astragalus, and the Hoatzin (Martin et al., 1980) shows it attaching to the calcaneum with no connection with the astragalus.

In early stages of development *Struthio* also seems to have a special relationship between the "ascending process" and the calcaneum, as Holmgren (1955, p. 278) reports that "It is connected with the tibial corner of the fibula by means of a thin, rather diffuse strand of mesenchyme." The ascending process widens in later stages so that its base overlaps essentially all the calcaneum and about one-half the astragalus. This seems to be a general pattern, with a triangular ossification forming slightly lateral to the junction between the calcaneum and the astragalus. When this triangular ossification has a very narrow base it fuses with the calcaneum, but when the base is broad it can also develop considerable overlap on the astragalus. In other words, it may or may not overlap both the astragalus and the calcaneum, but if it only overlaps one, it overlaps just the calcaneum. This would seem to make it impossible for the ascending process of birds to be an "ascending process of the astragalus" as it is in dinosaurs, and I consider it a neomorph structure in birds, the pretibial bone (Martin et al., 1980). It appears that the base of the pretibial bone is broadest in large ground birds such as the ostrich, and those birds show the greatest overlap with the astragalus. It is therefore no accident that Baur (1884, 1885) and others have primarily used the ostrich for comparison with dinosaurs. The Mesozoic birds, *Archaeopteryx*, *Hesperornis*, and *Baptornis*, have narrow pretibial bones which are mostly associated with the calcaneum. In all birds, the calcaneum forms the outer condyle of the tibiotarsus, and the tarsal joint is stabilized by fusion of the proximal tarsal bones to each other and to the tibia early in ontogeny. In theropod dinosaurs this stability is achieved at the expense of the calcaneum by the lateral extension of the astragalus along with its ascending process until, in some genera like *Ornithomimus* (Welles and Long, 1974), the calcaneum becomes a small bone inset into the face of the astragalus, and the astragalus forms the entire functional articular surface. This is the culmination of an evolutionary trend in the tarsus, with the ascending process and astragalus both moving *laterally*. In birds, the trend is for the pretibial to widen its base *medially* over part of the astragalus. Because of these relationships, the "pretibial bone" of birds cannot be homologous to the "ascending process" of dinosaurs. That birds and dinosaurs evolved functionally similar structures is not surprising, for some sort of stop is required to prevent a mesotarsal joint from acting as a roller-bearing in bipedal animals (Charig, 1972).

The fact that dinosaurs and birds have developed ascending processes independently indicates that any common ancestor they might share would have to precede the development of that character and would probably not be an advanced biped. This would set the latest date for this common ancestor somewhere in the Triassic. The tarsometatarsi of birds and *Deinonychus* differ more in proportions than seems to be generally realized (Fig. 5), but even *Archaeopteryx* cannot be taken as the primitive condition for birds. It does not have a tarsal cap of the sort found on the tarsometatarsus of modern birds. *Archaeopteryx* probably separated from the main line of avian evolution before a fused tarsometatarsus was formed.

Finally, some dinosaurs form their foot in a manner similar to

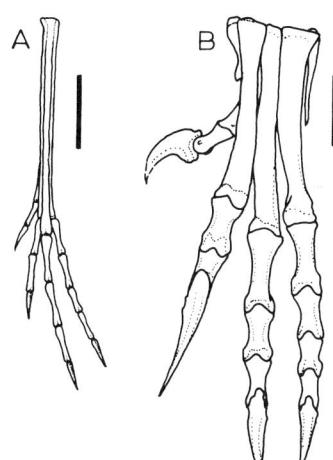

FIGURE 5. Anterior view of the left pes. (A) *Archaeopteryx* (Eichstatt specimen). (B) *Deinonychus*, after Ostrom (1969). Scales of A = 1 cm; B = 5 cm.

Archaeopteryx, and others are more similar to modern birds (Osmolska, 1981). Since birds are surely not polyphyletic, one or both of the dinosaur patterns are *convergent* to that in birds. The main feature of the foot that could be used to indicate a coelurosaurian ancestry of birds is the loss of metatarsal V and the distal articulation of metatarsal I (Ostrom, 1976b). Reduction of metatarsal V is a common evolutionary trend in reptiles and may be found in as advanced a state in the Triassic crocodilian *Protosuchus* (Colbert and Mook, 1951) as it is in coelurosaurs. The distal articulation of metatarsal I is a better character.

4. THE ORIGIN OF AVIAN FLIGHT

The features of birds that set them most apart from other living tetrapods are their feathers and bipedalism. Our understanding of how bipedalism evolved influences our hypotheses concerning the origin of birds and how they acquired flight. The early recognition that many dinosaurs were bipedal quickly made dinosaurs a prime candidate for avian origins, and it was not until the discovery of bipedal pseudosuchians that the more primitive archosaurs came into consideration. Acceptance that the ancestors of birds were bipedal before they were birds places interesting strictures on the origin of avian flight. It must be difficult for an obligate biped that cannot fly to climb vertical trees. If the ancestors of birds were cursorial bipeds highly specialized for running, as were the coelurosaurian dinosaurs, then it is hard to understand how they could also be arboreal. In fact, most who have strongly supported a dinosaurian origin have also supported an origin of flight from the ground up (Williston, 1879; Nopsca, 1907; Ostrom, 1974a, 1976a, 1979; Padien, 1982).

This is an unusual model, and it cannot be supported for any other tetrapods, although an analog can be found in flying fish. Flying fish escape predators by swimming at a high rate of speed until they break the surface of the water, at which point they spread their pectoral fins and glide. The critical difference between flying fish and some hypothetical proavis is that while both lose ground speed while gliding, the flying fish also loses visual contact with its pursuer, which might either give up the chase or not be able to relocate the particular fish it was chasing. These advantages do not pertain to a terrestrial glider, and an escape-from-predation model is unlikely. However, I like it better than Ostrom's (1979) suggestion that the proavis was a small pursuit predator that glided (or flapped) after its prey. Gliding results in a loss both of ground speed and maneuverability. It is unlikely that a predator that cannot overtake its prey on foot would be more successful gliding.

The suggestion that the feathers were an "insect net" is also hard to support. The reason that an insect net is a *net* is to permit air to pass through it. Without this special quality the insects would be blown out of the structure by the rush of air. A similar problem would have arisen for the hapless proavis as it sought to trap its dinner, unless the structure of the feathers provided a free passage of air through their fabric. However, such an open fabric would make either gliding or flight impossible. Also, low-speed flight close to the ground is difficult, and would have required more sophistication in its early stages than flight developing from transitional, arboreal-gliding stages.

Supporters of an arboreal origin of avian flight (Marsh, 1877; Heilmann, 1926; Simpson, 1946; Swinton, 1958, 1960) have in general ignored the problem of an initial bipedal stage and have emphasized the strengths of their model. These include the use of gravity to achieve airspeed, the numerous examples of transitional stages of parachuting and gliding in diverse vertebrates (Savile, 1962), the selective value in reducing predation, and the fact that the other two tetrapod groups (bats and pterosaurs) with true powered flight must surely have begun as arboreal animals. The advantage of gliding over parachuting is obvious; the gliding animal has a better chance of arriving at another tree than does the parachuter. The chief problem with gliding is the continual loss of altitude so that the incline of the glide must intersect another tree or branch to prevent a ground landing. Even a little capability to maneuver or gain altitude would have greatly enhanced the chances of remaining in the trees and these slight gains in altitude and direction would have provided the animal with a wider range of choice in its landing sites. I think this was the original selective differential for powered flight over gliding. The longer that powered flight could be maintained, the more landing choices available. Such a model has the advantage of being selectively positive even for unsophisticated flyers.

But what about the problem of an arboreal, cursorial, biped? Fortunately, part of this problem is solved by the anatomy of *Archaeopteryx* itself. It was not a highly sophisticated cursorial biped, like the coelurosaurian dinosaurs, but must have run bipedally only for short distances. It was more adapted for moving about in the trees than for a life in the open plain. Nonetheless, it has elongated and fused metatarsals, a mesotarsal joint and an elongated tibiotarsus. If these are not for running, how did they originate? The answer to this question seems to lie in a theory of bipedalism already well accepted by primatologists. The bipedal tarsiers have elongated hind feet, tibiae, and a bipedal stance; but they do not spend their time running about on the ground. They live in trees and leap from branch to branch. They launch themselves in the desired direction with their hind feet and cling to their objective with their forefeet. The anatomy of *Archaeopteryx*, with its long fingers and sharp claws on the manus, might be derived from just such vertical clinging and leaping. Such a model might also explain the absence of a stage with a broad patagium, like that found in flying squirrels and dermopterans. We should, however, be careful not to reject the idea of any patagium in the proavis. Even modern birds have a distinct patagium (Barbour, 1902; Heilmann, 1926), and in the proavis the patagium might have been more important. It would gradually be reduced as the flight feathers enlarged.

This brings us to the origin of feathers. Most have contended that downy feathers are primitive and that their original function was to conserve heat in a small endotherm (Cowen and Lipps, 1982). A few have argued that flight was their original function (Parkes, 1966). I can support the latter viewpoint. It is not completely certain that the known Mesozoic birds were endothermic, although it is reasonable to suppose that they were. They lack a system of pneumatic bones connected with airsacs. I can think of no reason for this extensive pneumatic system, which is distributed throughout modern birds, other than its use in cooling. Its absence in *Archaeopteryx*, along with the absence of an ossified sternum and probably the type of airsac breathing found in modern birds, suggests that *Archaeopteryx* did not have a serious cooling problem. It was, however, a powered flyer (Feduccia and Tordoff, 1979). The real benefit of endothermy seems to be an increase in the ability to prolong muscular activity. This would have obvious advantages to an active flying animal, and Daly (1980) has suggested that flight preceded endothermy in birds. I find this idea attractive, and it permits the development of flight feathers before that of downy and contour feathers, an arrangement that agrees with Parkes (1966).

We can develop the following scenario for the origin of flight in modern birds. We begin with a small arboreal reptile with a tendency toward bipedality. This bipedality was improved by vertical clinging and leaping, which also began the specialization of the hind legs. Bipedality was not perfected until after the acquisition of flight and was probably partially the result of the need to keep the feathers off the ground while running and walking. This also resulted in a 90° change in the plane of flexion of the radius and ulna and the manus, a rotation that was probably achieved concurrently with long primaries and was already fully developed in Archaeopteryx.

The earliest flying birds probably had reptilian breathing, and the chief area of origin for the pectoralis muscles was the large furcula (Olson and Feduccia, 1979) and the broad flat coracoids. The efficiency of the pectoralis muscles was increased by their posterior elongation onto the sternum. This would be at a stage beyond Archaeopteryx and would result in the ossification of the sternum. Further elongation of the muscles could be achieved by elongating the coracoids as is characteristic of modern birds. With the ossification of the sternum, a bellows system of breathing would be possible, and an enlarged airsac system would begin to develop at this time. The earliest known birds with ossified sterna are early Cretaceous hesperornithiform birds, although flying neornithiform birds also ought to have been present (Martin, 1980). The oldest known bird with a keeled sternum and advanced powers of flight is Ichthyornis from the early part (Turonian) of the late Cretaceous (Martin and Stewart, 1977, 1982). This is also the time that we first observe ossified uncinate processes. These processes aid with the airsac breathing of modern birds, and their ossification may be another sign that this type of breathing is beginning. Pneumatic bones have not been reported from the appendicular skeleton of Mesozoic birds, and this may indicate that they operated at lower body temperatures than do modern birds. Pneumatic bones must be effective heat exchangers, for they lie at the centers of the large muscle masses and have air regularly circulated through them. A pneumatic appendicular skeleton seems to be an early Tertiary phenomenon and seems to be one of the last major innovations related to powered flight.

Feathers must have been fairly early inventions, but they may have first appeared as tracts of enlarged scales along the margin of well-developed patagia. They may have developed because the patagium could not be enlarged by extending it down the back to the hind legs and still retain the flexibility of the forelimbs needed for vertical clinging and leaping. As powered flight was added to gliding, the need for sustained activity promoted the development of a constant high body temperature and with it an insulative covering of downy and contour feathers.

5. CONCLUSIONS

The coelurosaurian model, as it is presently constructed, implies a direct ancestor–descendant relationship between some coelurosaur and the first bird. The characters that support it probably do not have great antiquity in the coelurosauria, so the transition to birds must have been relatively late. Some workers have gone so far as to argue that *Archaeopteryx* might better be classified as a coelurosaur (Padien, 1982). Lowe extended this idea by arguing that several groups of coelurosaurs independently gave rise to different groups of birds. Osmolska (1981) argued that the structure of the tarsometatarsus of some late Cretaceous dinosaurs placed them closer to modern birds than is *Archaeopteryx*, another hypothesis resulting in a polyphyletic Aves; it also implies that feathers had already evolved in the coelurosaurs.

It is interesting to note that the characters supporting a bird–crocodilian relationship are almost all cranial and are distributed throughout the two groups, except in clear cases of secondary loss (for instance, the loss of teeth in modern birds). The coelurosaurian argument is based primarily on features of the postcranial skeleton. The pseudosuchian and ornithischian models, like the crocodilian, seem to demand an early origin of the Aves. The characters of the teeth and their implantation, pelvis, femur, and ankle joint agree with each other, that any bird–dinosaur ancestor must have existed before the late Triassic. It is difficult to reconcile this with a model proposing a small, cursorial coelurosaur as the hypothetial proavis. The crocodilian model is at least a viable alternative, and we should not reject ornithischian dinosaurs without further consideration of the new evidence.

The use of feathers for thermoregulation may postdate the development of powered flight. It is possible that feathers originally evolved for their aerodynamic qualities. *Archaeopteryx* must have had limited powers of flight, but it did not have a fully-improved bipedal posture.

Fully-improved bipedal posture seems to have appeared relatively late in avian evolution, perhaps not before the Cretaceous. It is unlikely that avian flight originated in a cursorial-terrestrial biped. The bipedality of birds may have originally been for arboreal leaping rather than for fast running. The development of the typical bellows-type of breathing using the airsac system must postdate *Archaeopteryx*, and the development of postcranial, skeletal airsac systems must have occurred in the late Cretaceous or early Tertiary.

ACKNOWLEDGMENTS. For allowing me to examine specimens, I thank J. P. Lehman, D. Goujet, F. Poplin, and D. E. Russell (Museum National d'Histoire Naturelle, Paris); A. J. Charig, A. Milner, and C. A. Walker (British Museum of Natural History, London); G. S. Cowles and C. J. O. Harrison (British Museum of Natural History, Ornithological Department, Tring); A. D. Walker (University of Newcastle Upon Tyne); G. Viohl (Jura Museum, Eichstatt); H. Jaeger and H. Fischer (Humboldt Museum für Naturhunde, Berlin); P. Wellnhofer (Bayerische Staatssammlung, Munich); Z. Kielan-Jaworowska, A. Elzanowski, and H. Osmolska (Polaska Akademia Nauk, Warsaw); P. Ellenberger (Laboratoire de Paleontologie des Vertebres, Montpellier); and J. H. Ostrom and M. Turner (Yale Peabody Museum, New Haven).

I have benefited from many stimulating conversations with S. Olson, J. Cracraft, P. Rich, A. Feducia, R. Mengel, P. Brodkorb, C. Harrison, C. A. Walker, A. Milner, J. D. Stewart, and K. N. Whetstone. J. D. Stewart has in particular made important suggestions concerning avian teeth. M. A. Klotz prepared the figures. M. A. Jenkinson, R. Mengel and J. D. Stewart critically read the manuscript. Funding was provided by the University of Kansas (sabbatical leave) and University General Research Grant 3251-5038, NSF DEB 7821432, and National Geographic Grant 2228-80.

REFERENCES

Barbour, E. H., 1902, President's Address—The progenitors of birds, *Proc. Nebr. Ornithol. Union, Third Annu. Meeting*, pp. 8–39.

Baur, G., 1884, Dinosaurier und Vogel, *Morphol. Jahrb.* **10**:446–454.

Baur, G., 1885, Bemerkungen uber das Becken der Vogel and Dinosaurier, *Morphol. Jahrb.* **10**:613–616.

Baur, G., 1886, Zur Vogel-Dinosaurier-Frage, *Zool. Anz.* **8**:441–443.

de Beer, G., 1954, *Archaeopteryx Lithographica*, A study based upon the British Museum specimen, British Museum (Natural History), London.

Bock, W. J., 1965, The role of adaptive mechanisms in the origin of higher levels of organization, *Syst. Zoo.* **14**:272–287.

Bock, W. J., 1969, The origin and radiation of birds, Ann. N.Y. Acad. Sci. **167**:147–155.
Bramwell, C. D., 1970, Quantitative assessment of the flight of Archaeopteryx, Nature **228**:185–186.
Brodkorb, P., 1971, Origin and evolution of birds, in: Avian Biology (D. S. Farner and J. R. King, eds.), Academic Press, New York and London, pp. 19–55.
Broom, R., 1908, On the early development of the appendicular skeleton of the ostrich with remarks on the origin of birds, Trans. S. Afr. Phil. Soc. **16**:355–368.
Broom, R., 1913, On the South African pseudosuchian Euparkeria and allied genera, Proc. Zool. Soc. Lond. **1913**:619–633.
Charig, A. J., 1972, The evolution of the archosaur pelvis and hindlimb: An explanation in functional terms, in: Studies in Vertebrate Evolution (K. A. Joysey and T. S. Kemp, eds.), Oliver and Boyd, Edinburgh, pp. 121–155.
Colbert, E. H., and Mook, C. C., 1951, The ancestral crocodilian Protosuchus, Bull. Am. Mus. Nat. Hist. **97**(3):143–182.
Cowen, R., and Lipps, J. H., 1982, An adaptive scenario for the origin of birds and flight in birds, Third N. Am. Paleo. Conv. Proc. **1**:109–112.
Cracraft, J., 1977, Special review. John Ostrom's studies on Archaeopteryx, the origin of birds and the evolution of avian flight, Wilson Bull. **39**:488–492.
Crompton, A. W., and Smith, K. K., 1980, A new genus and species of crocodilian from the Kayenta Formation (Late Triassic?) of northern Arizona, in: Aspects of Vertebrate History (L. L. Jacobs, ed.), Museum of Northern Arizona Press, Flagstaff, pp. 193–217.
Daly, E., 1980, The origins of homiothermy, Trans. Kansas Acad. Sci. **83**(4):247.
Darwin, C., 1859, On the Origin of Species by Means of Natural Selection, or the Preservation of Favoured Races in the Struggle for Life, 2nd ed., John Murray, London, 1869.
Ellenberger, P. P., 1977, Quelques precisions sur l'anatomie et la place systematique tres speciale de Cosesaurus aviceps (Ladinien superieur de Montral, Catalogne), Caudernos Geologia Iberica **4**:169–188.
Feduccia, A., 1980, The Age of Birds, Harvard University Press, Cambridge, Massachusetts and London.
Feduccia, A., and Tordoff, H. B., 1979, Feathers of Archaeopteryx: Asymmetric vanes indicate aerodynamic function, Science **203**:1021.
Galton, P. M., 1970, Ornithischian dinosaurs and the origin of birds, Evolution **24**:448–462.
Gardiner, B. G., 1982, Tetrapod classification, Zool. J. Linn. Soc. **74**:207–232.
Gingerich, P. D., 1979, The stratophenetic approach to phylogeny reconstruction in vertebrate paleontology, in: Phylogenetic Analysis and Paleontology (J. Cracraft and N. Eldredge, eds.), Columbia University Press, New York, pp. 41–77.
Haeckel, E. H. P. A., 1866, Generelle Morphologie der Organismen, G. Reimer, Berlin.
Heilmann, G., 1926, The Origin of Birds, Witherby, London.
Holmgren, N., 1955, Studies on the phylogeny of birds, Acta Zool. **36**:243–328.
Huxley, T. H., 1867, On the classification of birds and on the taxonomic value of the modifications of certain of the cranial bones observable in that class, Proc. Zool. Soc. Lond. **1867**:415–472.
Huxley, T. H., 1868, On the animals which are most nearly intermediate between the birds and reptiles, Ann. Mag. Nat. Hist. Lond. **2**(4):66–75.
Huxley, T. H., 1870, Further evidence of the affinity between the dinosaurian reptiles and birds, Quart. J. Geol. Soc. Lond. **26**:12–31.
Lambrecht, K., 1933, Handbuch der Palaeornithologie, Borntraeger, Berlin.
Lowe, P. R., 1933, On the relationships of the Struthiones to the dinosaurs and to the

rest of the avian class, with special reference to the position of *Archaeopteryx*, *Ibis* **13**:398–432.

Lowe, P. R., 1944a, Some additional remarks on the phylogeny of the Struthiones, *Ibis* **86**:37–43.

Lowe, P. R., 1944b, An analysis of the characters of *Archaeopteryx* and *Archaeornis*. Were they reptiles or birds? *Ibis* **86**:517–543.

Marsh, O. C., 1877, Introduction and succession of vertebrate life in America, *Proc. Am. Assoc. Adv. Sci.* **1877**:211–258.

Martin, L. D., 1980, The foot-propelled diving birds of the Mesozoic, *Acta XVII Congressus Int. Ornithol.* **2**:1237–1242.

Martin, L. D., and Stewart, J. D., 1977, Teeth in *Ichthyornis* (Class: Aves), *Science* **195**:1331–1332.

Martin, L. D., and Stewart, J. D., 1982, An ichthyornithiform bird from the Campanian of Canada, *Can. J. Earth Sci.* **19**:324–327.

Martin, L. D., Stewart, J. D., and Whetstone, K. N., 1980, The origin of birds: Structure of the tarsus and teeth, *Auk* **97**:86–93.

Nopsca, F., 1907, Ideas on the origin of flight, *Proc. Zool. Soc. Lond.* **1907**:463–477.

Olson, S. L., and Feduccia, A., 1979, Flight capability and the pectoral girdle of *Archaeopteryx*, *Nature* **278**:247–248.

Osborn, H. F., 1900, Reconsideration of the evidence for a common dinosaur-avian stem in the Permian, *Am. Natural.* **34**:777–799.

Osmolska, H., 1976, New light on the skull anatomy and systematic position of oviraptor, *Nature* **262**:683–684.

Osmolska, H., 1981, Coossified tarsometatarsi in theropod dinosaurs and their bearing on the problem of bird origins, *Palaeontol. Pol.* **42**:79–95.

Osmolska, H., Roniewicz, E., and Barsbold, R., 1972, A new dinosaur *Gallimimus bullatus* n. gen., n. sp. (Ornithomimidae) from the Uppermost Cretaceous of Mongolia, *Palaeontol. Pol.* **27**:95–143.

Ostrom, J. H., 1969, Osteology of *Deinonychus antirrhopus*, an unusual theropod from the Lower Cretaceous of Montana, *Bull. Yale Peabody Mus. Nat. Hist.* **30**:1–165.

Ostrom, J. H., 1973, The ancestry of birds, *Nature* **242**:136.

Ostrom, J. H., 1974a, *Archaeopteryx* and the origin of flight, *Quart. Rev. Biol.* **49**:27–47.

Ostrom, J. H., 1974b, The pectoral girdle and forelimb function of *Deinonychus* (Reptilia: Saurischia): A correction, *Postilla* **165**:1–11.

Ostrom, J. H., 1975a, The origin of birds, *Annu. Rev. Earth Plan. Sci.* **3**:55–57.

Ostrom, J. H., 1975b, On the origin of *Archaeopteryx* and the ancestry of birds, *Colloques Int. Centre Nat. Rech. Sci.* **218**:519–532.

Ostrom, J. H., 1976a, Some hypothetical anatomical stages in the evolution of avian flight, *Smithson. Contrib. Paleobiol.* **27**:1–21.

Ostrom, J. H., 1976b, *Archaeopteryx* and the origin of birds, *Biol. J. Linn. Soc.* **8**:91–182.

Ostrom, J. H., 1976c, On a new specimen of the lower Cretaceous theropod dinosaur *Deinonychus antirrhopus*, *Brevoria* **439**:1–21.

Ostrom, J. H., 1979, Bird flight: How did it begin? *Am. Sci.* **67**:46–56.

Owen, R., 1875, Monograph of the fossil reptiles of the Liassic formations II. Pterosauria, *Palaeontol. Soc. Monogr.* **1875**:41–81.

Padien, K., 1982, Macroevolution and the origin of major adaptations: Vertebrate flight as a paradigm for the analysis of patterns, *Third N. Am. Paleo. Conv. Proc.* **2**:387–392.

Parkes, K. C., 1966, Speculations on the origin of feathers, *Living Bird* **5**:77–86.

Patterson, C., 1981, Significance of fossils in determining evolutionary relationships, *Annu. Rev. Ecol. Syst.* **12**:195–223.

Petronievics, B., 1921, Uber das Becken den Schultergurtel und einige andere Teile der Londoner Archaeopteryx, Georg, Genf. Buchhandl.

Petronievics, B., 1927, Nouvelles recherches sur l'osteologie des Archaeornithes, Ann. Paleo. **16**:39–55.

Petronievics, B., 1950, Les deux oiseaux fossiles les plus anciens (Archaeopteryx et Archaeornis), Ann. Geol. Pen. Balkan **18**:89–127.

Savile, D. B., 1962, Gliding and flight in the vertebrates, Am. Zool. **2**:161–166.

Seeley, H. G., 1881, Prof. Carl Vogt on the Archaeopteryx, Geol. Mag. **8**(2):300–309.

Sharov, A. G., 1970, An unusual reptile from the Lower Triassic of Fergana, Paleo. Z. **1970**:127–130.

Simpson, G. G., 1946, Fossil penguins, Bull. Am. Mus. Nat. Hist. **87**:1–95.

Simpson, G. G., 1980, Fossil birds and evolution, Contrib. Sci. Nat. Hist. Mus. Los Angeles County **330**:3–8.

Swinton, W. E., 1958, Fossil Birds, British Museum (Natural History), London.

Swinton, W. E., 1960, The origin of birds, in: Biology and Comparative Physiology of Birds, Volume 1 (A. J. Marshall, ed.), Academic, New York, pp. 1–14.

Tarsitano, S., and Hecht, M. K., 1980, A reconsideration of the reptilian relationships of Archaeopteryx, Zool. J. Linn. Soc. Lond. **69**(2):149–182.

Walker, A. D., 1970, A revision of the Jurassic reptile Hallopus victor (Marsh), with remarks on the classification of crocodiles, Phil. Trans. Roy. Soc. Lond. (B) **257**:323–372.

Walker, A. D., 1972, New light on the origin of birds and crocodiles, Nature **237**:257–263.

Walker, A. D., 1974, Evolution, organic, McGraw-Hill Yearbook Sci. Tech. **1974**:177–179.

Walker, A. D., 1977, Evolution of the pelvis in birds and dinosaurs, in: Problems in Vertebrate Evolution (S. M. Andrews, R. S. Miles, and A. D. Walker, eds.), Academic Press, London, pp. 319–357.

Welles, S. P., and Long, R. A., 1974, The tarsus of theropod dinosaurs, Ann. S. Afr. Mus. **64**:191–218.

Wellnhofer, P., 1974, Das funfte Skelettexemplar von Archaeopteryx, Palaeontographica (A) **147**:169–216.

Whetstone, K. N., and Martin, L. D., 1979, New look at the origin of birds and crocodiles, Nature **279**:234–236.

Whetstone, K. N., and Martin, L. D., 1981, Common ancestry for birds and crocodiles? A reply, Nature **289**:98.

Whybrow, P. J., 1982, Preparation of the cranium of the holotype of Archaeopteryx lithographica from the collections of the British Museum (Natural History), N. Jb. Geol. Palaontol. Mh. **3**:184–192.

Williston, S. W., 1879, "Are birds derived from dinosaurs?" Kansas City Rev. Sci. **3**:457–460.

Yalden, D. W., 1971, Flying ability of Archaeopteryx, Nature **231**:127.

CHAPTER 5

THE GREAT PLAINS HYBRID ZONES

J. D. RISING

1. INTRODUCTION

About 130 taxa of birds reach distributional limits in the central Great Plains, and 28 of these are replaced there by closely related taxa that are apparently their ecological counterparts. In many cases, it is reasonable to assume that the presence of a similar, competing form is a factor limiting distribution. Hybridization between the counterparts is known to occur at least occasionally in 22 (11 pairs of taxa) of these 28 (14 pairs) taxa, and many of the commonly hybridizing taxa are common in the Plains. In terms of numbers of individuals they comprise a sizable proportion of the bird life of that region. Although species specialized to live in grasslands are numerous in the Plains, only two, the Eastern and Western meadowlarks (*Sturnella magna* and *S. neglecta*), are known to hybridize in the Plains. The other hybridizing taxa are adapted to deciduous thicket or woodland edge habitats. For the most part, an eastern taxon is replaced by and hybridizes with a western taxon, but in some examples replacement is northeast–southwest, and in one case north–south. For east–west distributional replacements, the counterpart taxa seemingly differentiated in allopatry during the Pleistocene and are presently in secondary contact (Short, 1965; Remington, 1968; Mengel, 1970). For the northeast–southwest

---

J. D. RISING • Department of Zoology, University of Toronto, Toronto, Ontario, Canada M5S 1A1.

replacements, it has been suggested that woodland species became isolated in the Edwards Plateau region of central Texas. Many of the forms isolated there seem to have come from the southwest at a time when there was a continuous corridor of woodland habitat from Mexico to central Texas (Remington, 1968).

In the late 1950s, C. G. Sibley and his students examined in detail the distributions and interactions of several of the east–west complementary taxa, namely the "Yellow-shafted" and "Red-shafted" flickers (*Colaptes auratus, s. l.*; Short, 1965), Baltimore and Bullock's orioles (*Icterus galbula* and *I. bullockii*; Sibley and Short, 1964), Indigo and Lazuli buntings (*Passerina cyanea* and *P. amoena*; Sibley and Short, 1959), the Rose-breasted and Black-headed grosbeaks (*Pheucticus ludovicianus* and *P. melanocephalus*; West, 1962), and "plain" and "spotted" towhees (*Pipilo erythrophthalmus, s. l.*; Sibley and West, 1959). Although these workers examined specimens from throughout the Plains region, the best samples available to them were from Nebraska and Colorado, and usually had been obtained from along the Platte River, which runs from the Rocky Mountains eastwards across northeastern Colorado and central Nebraska. In all of these taxa they found evidence that hybridization was occurring in the central Plains region. There were individuals intermediate in plumage pattern and, in some cases, size between the nominal forms.

Also in the 1950s, Dixon (1955) examined the hybridization between eastern "gray-crested" titmice and the southwestern "black-crested" titmice (*Parus bicolor, s. l.*) in Texas. These titmice show northeast–southwest replacement in the Great Plains.

These studies raised a number questions. Is assortative mating occurring, or does selection against intermediate individuals reduce gene flow between populations? Are these "steps" stable through time in both geographic position and in width? Have these step clines developed as a consequence of hybridization, following secondary contact, or are they examples of geographic variation?

One approach to differentiating between allopatric divergence followed by secondary contact and geographic variation is to look for interspecific similarities in distributions among taxa. Remington (1968) designated so-called "suture-zones," or regions in which several different taxa show patterns of step-clinal variation that are more or less congruent. The "Rocky Mountain Eastern Suture-Zone" and "Central Texas Suture-Zone" are among those he designated. They are defined not only by the hybridizing birds mentioned above, but also on the basis of putative hybrid zones between several complementary pairs of non-avian taxa. It is unlikely that several different taxa would have highly similar distributions unless they had been similarly influenced by the same historical events. Remington, therefore, argues that suture-zones describe geographical regions in which there have been major rejunctions of biotas that had previously diverged while in allopatry. By inference, the individual species showing this pattern are hybridizing in secondary contact.

In recent years, these "hybridizing" birds have been the subject of new research. In addition to summarizing this research, I shall examine the distributions of the taxa involved to clarify whether or not the data on bird distributions support the interpretation of the Great Plains as a region where there are suture-zones.

1.1. Delimiting the Plains

The Great Plains physiographic province extends from the Rocky Mountains east to the eastern "Plains States" (Texas to North Dakota), and from Great Slave Lake to the Balcones Fault and Rio Grande River in Texas (Hunt, 1974). For practical purposes, and also for reasons that make good ecological sense, I take the "Great Plains" as equivalent to the prairies, the grasslands and associated woodland edge, from southern Canada south through the Plains States. This includes the Osage Plains and the Saskatchewan Plains from the Central Lowland province (Hunt, 1974). I have, therefore, omitted discussion of such species as the Yellow-bellied Sapsuckers (*Sphyrapicus varius, s. l.*), the Myrtle and Audubon's warblers (*Dendroica coronata, s. l.*) and the MacGillivray's and Mourning warblers (*Oporornis tolmiei* and *O. philadelphia*), which breed in habitats that are not characteristic of the prairies.

1.2. What is Hybridization?

Biologists sometimes differentiate between "hybridization," interbreeding between individuals of two different species, and "intergradation," interbreeding between individuals from two phenotypically different populations of the same species. Inasmuch as it is presently popular to define species on the basis of reproductive isolation, it is tautologically impossible for hybridization in the strict sense (as defined above) to occur commonly. In this paper, hybridization is used in a broad sense, i.e., interbreeding between individuals from two different well-marked populations that are in secondary contact (Mayr, 1970). In most cases, secondary contact cannot be proven, but it probably has occurred between distinct, but closely related taxa that have essentially parapatric distributions. In some of the cases discussed, the two taxa "blend" into each other in a step-cline; in others, intermediate individuals ("hybrids") are rare or unknown. In the general discussion I shall review the evidence for secondary contact. Strictly speaking, without such evidence we should refer to these taxa as "perhaps hybridizing," but for simplicity I use words like "hybridizing" and "hybrids" even when there is no evidence for secondary contact.

2. ACCOUNTS OF HYBRIDIZING TAXA

2.1. *Otus*

Among North American screech owls, eastern and western screech owls (*O. asio, s.s.,* and *O. "kennicottii"*) are clearly differentiated, differing in song, plumage coloration, and bill color (Marshall, 1967). In the Great Plains, western screech owls are found east along the Cimarron River to Kenton and Boise City, in the Oklahoma Panhandle. Along the Cimarron, at Elkhart, in extreme southwestern Kansas, apparently only eastern screech owls occur. Along the Arkansas River, western screech owls breed from Fountain Creek, south of Colorado Springs, east to Rocky Ford, Colorado. The vegetation along the Arkansas River thins out in eastern Colorado, and the next specimen available along the Arkansas River to the east is from Coolidge, in extreme western Kansas; it is an eastern screech owl, and eastern screech owl songs have been recorded there (Marshall, 1967). Along the Canadian River, Marshall (1967) recorded eastern screech owl songs northwest of Tascosa, in the Texas Panhandle, and found no habitat suitable for *Otus* west of there. Similarly, lack of suitable habitat doubtless prevents contacts along the Pecos River. There are possible contacts between these owls along the Cimarron River, between Elkhart, Kansas, and Boise City, Oklahoma, and along the face of the Rocky Mountains, between Denver and Colorado Springs, Colorado (Marshall, 1967). To the north of the region around the headwaters of the Arkansas River, the ranges of these two screech owls are widely separated. In the extreme southwestern Plains, however, they are sympatric along the Rio Grande River for a distance of ca. 185 km, between Langtry and Boquillas. Marshall (1967) found two mixed pairs in that region in the early 1960s, and concluded that the "two forms interbreed as freely as their sparse populations permit. . . ."

2.2. *Colaptes*

Of all the hybrid contacts in the Great Plains, that between the "red-shafted" and "yellow-shafted" flickers (*C. "auratus"* and *C. "cafer"*) was the earliest to be described by ornithologists. Audubon, in his Missouri River Journal (Audubon, 1897), describes a mixed pair of these flickers and their brood, from Fort Union, near the present boundary of North Dakota and Montana. "Hybrid" flickers predominate in that region today (Short, 1965; personal observations). Although hybridization between the flickers has been documented extensively in the Great Plains, from the Oklahoma Panhandle north to Alberta, it has been studied carefully only along the Platte River in Nebraska and Colorado (Short, 1965) and in South Dakota (Anderson, 1971). In both South Dakota and Nebraska, yellow-shafted flickers are replaced by red-shafted flickers in the far west. In the middle parts of both states, many individuals are phenotypically intermediate, and populations on average, grade from yellow-shafted to red-shafted, east to west (Fig. 1). The red-

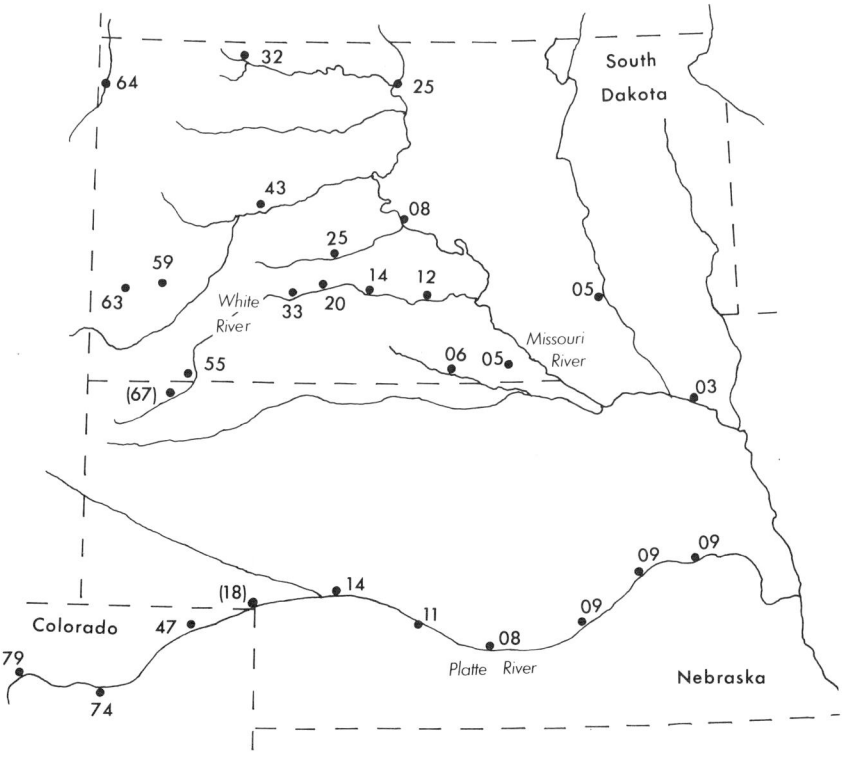

FIGURE 1. Average character index scores of flickers from the Central Plains region. To pool sexes, scores are scaled to range from 0 ("typical yellow-shafted") to 100 ("typical red-shafted"). Numbers are averages for samples reported in Short (1965) and Anderson (1971).

shafted influence is especially noticeable in the Black Hills Region, and at the base of the Rocky Mountains, e.g., north-central Colorado. The majority of the flickers from the southern Canadian prairies, from southwestern Saskatchewan (Cypress Hills Region) westward, are also intermediate in plumage characteristics (Short, 1965; personal observations).

In Kansas and Oklahoma, hybrid flickers are not as evident as in Nebraska and South Dakota. Although there are hybrid specimens from western Kansas (Rising, 1974) and the Oklahoma Panhandle (Sutton, 1967), over 80% of the breeding specimens from western Kansas appear to be yellow-shafted (personal observations).

In order to ascertain whether or not the flicker hybrid zone is stable through time, we can compare samples taken from the same sites at different times by Short (1965) and Anderson (1971). To quantify variation in plumage colors and patterns, both Short and Anderson used a "Character Index" (or "Hybrid Index") that was devised by Short (1965). For this index, a given character state like that of a "typical" yellow-shafted flicker is given a value or score of "0," and the state of that feature typical of a red-shafted flicker a score of "2" or "4," depending on the characteristic. A total score is calculated for each individual by summing the scores of each of the several plumage variables that differ between the two taxa. A bird with a total score of "0" is a pure yellow-shafted flicker, and one with a total score of "24" (for males) or "20" (for females) is a red-shafted flicker. The totals differ because the males possess one additional feature (the malar stripe) that varies between the taxa. An average score for a sample can be calculated. If that average is close to "0," the sample is composed of birds that by and large resemble yellow-shafted flickers. Table I compares the average character index scores of flickers from three sites in South Dakota. To pool data for both sexes, and thus to increase the sample sizes, the character index scores are expressed on a scale of 0–100, i.e., a male

TABLE I
Comparisons of Flicker Character Index Scores from South Dakota[a]

Locality	1955[b]	N	1965–1968[c]	N
Howes	42	7	43	19
Kadoka	42	6	33	14
Custer	61	18	63	18

[a]Scores on a scale of 0–100, with 0 typical "yellow-shafted" and 100 typical "red-shafted."
[b]From Short (1965).
[c]From Anderson (1971).

with a score of "15" is "63" (15/24) on this scale, and a female with the same score is "75" (15/20). Although samples sizes are small, these data indicate no change in flicker hybridization in South Dakota between 1955 and 1965. Recall also that the flickers found and described by Audubon in western North Dakota were like ones that can be found there today.

Anderson (1971) points out, however, that the flooding of the Missouri River Valley in South Dakota, and consequently the destruction of much of the habitat suitable for flicker nesting in the central part of that state, may restrict the flow of "yellow-shafted" genes from the east into western South Dakota. If such gene flow is important in determining the phenotypic characteristics of flickers in western South Dakota, this could have the effect of increasing the relative frequency of red-shafted flickers in that area in the future.

2.3. Centurus

Golden-fronted and Red-bellied woodpeckers (*C. aurifrons* and *C. carolinus*) meet in Texas in a zone that extends from the Rockport area, on the Gulf Coast, through Austin to the extreme southwestern corner of Oklahoma (Selander and Giller, 1959; Sutton, 1967; Oberholser, 1974). At Austin, the two co-occur, apparently without interbreeding (Selander and Giller, 1959), and there is no indication of any recent changes in their ranges in that region. Their overlap in the Austin area is only ca. 32 km in width, but in some places there the two species are regularly seen together (Selander and Giller, 1959). On the other hand, records indicate that *C. aurifrons* has only recently moved into southwestern Oklahoma, where they occur regularly today (Sutton, 1967). There are several specimens in the Stovall Museum (University of Oklahoma) from southwestern Oklahoma that are intermediate in plumage features and it is probable that hybridization is occurring there (Scott Wood, personal communication). Perhaps hybridization between these two taxa occurs only where contact between them is recent.

2.4. Myiarchus

The Great Crested and Ash-throated flycatchers (*M. crinitus* and *M. cinerascens*) are locally sympatric along the Cimarron River in southwestern Kansas (Meade County; Rising, 1974), in southwestern Oklahoma (Comanche and Tillman counties; Sutton, 1967), and southeastward across central Texas (Oberholser, 1974). However, their interactions have not been studied, and hybrids are unknown.

2.5. Contopus

Two factors complicate elucidation of the distributions and interaction of the Western and Eastern pewees (or wood pewees; *C. sordidulus* and *C. virens*), their scarcity in the western Great Plains, where they are sympatric, and their great phenetic similarity. It is probable that they are locally sympatric in North Dakota (Stewart, 1975), as they are in southwestern Manitoba (Knapton, 1979; Rising and Schueler, 1980). Rising and Schueler (1980) collected *Contopus* of both species in north-central Montana in early June. Using multivariate analyses (of both plumage characters and body measurements) they found no evidence of hybridization. However, since only a few of the specimens available to them were birds known to be breeding in areas of sympatry, the interactions between the two pewees remain obscure.

2.6. Cyanocitta

The Blue Jay (*C. cristata*) is found virtually throughout the prairies where there are trees, although they are scarce in extreme southwestern Kansas (Rising, 1974) and south of there. They are probably extending their range westward in southwestern Kansas and the Oklahoma Panhandle, and have only comparatively recently appeared as a breeding bird in the foothills of Colorado (Williams and Wheat, 1971). Williams and Wheat describe at least four individuals that are apparent hybrids between the Blue Jay and the Steller's Jay (*C. stelleri*) seen at feeders at Boulder, Colorado in the fall of 1969; a brood of probable backcross individuals was seen during mid-August, 1970. The Steller's Jay is primarily a bird of the montane coniferous forests, and it seems likely that hybridization between these taxa will be confined to ecotonal areas between deciduous woods and pine forests.

2.7. Parus (Chickadees)

The ranges of the Black-capped and Carolina chickadees (*P. atricapillus* and *P. carolinensis*) are parapatric from New Jersey westward across southern Illinois and into southern Kansas. In Kansas, Black-capped Chickadees are found throughout the state in the north, but are rare in the southwest along the Cimarron River where they might co-occur and interbreed with Carolina Chickadees, which are also found in low density in that region (Rising, 1968, 1974). In southeastern Kansas, there is some indication of limited hybridization (Rising, 1968), although the great similarity of the taxa makes it difficult to definitely ascertain the extent to which this occurs. Outside the Great Plains region, Brewer (1963) found that the ranges of these two chickadees were separated by a region where none occurred. "Brewer's Gap" may be caused by selection against hybridization.

2.8. *Parus* (Crested Titmice)

The eastern, "gray-crested," Tufted Titmouse (*P. bicolor, s.s.*) is found westward into eastern Oklahoma (Sutton, 1967) and Texas (Dixon, 1955; Oberholser, 1974), where it is replaced by the smaller, "black-crested" titmouse (*P. "atricristatus"*). In addition to differing in crest color and size (wing length), black-crested birds have longer crests and shorter bills than gray-crested ones. Where the two are sympatric, however, mating is apparently random and hybrids are produced from such mixed pairs (Dixon, 1955). At present, there is a zone of contact that extends from Port Lavaca and Refugio, near the Gulf coast, north through Waco and Fort Worth, to Gainesville, on the Red River. Hybrids occur frequently along this zone (Dixon, 1955), as well as in southwestern Oklahoma (Jackson County; Sutton, 1967). Between 1886–1940, the weather in Texas was warmer and drier than earlier in the 1800s, and many southern and southwestern birds, including the black-crested titmouse, moved northeastward across Texas. This movement has apparently been facilitated not only by this change in climatic conditions, but also because of the extensive clearing of eastern forests that occurred during that time, and continues today (Oberholser, 1974). The northward extension of black-crested titmice has been accompanied by a corresponding decrease in the range of gray-crested birds, apparently as a consequence of competition with the black-crested ones (Oberholser, 1974), which have similar habitat preferences in Texas (Dixon, 1955). Black-crested titmice were unknown in Oklahoma until recently, but since the 1960s have occurred regularly along the Red River in the extreme southwestern corner of the state. They now hybridize with, and are probably displacing, gray-crested birds there. There are disjunct populations of black-crested titmice in the Texas Panhandle (Palo Duro Canyon) and in southwestern Texas. The western Plain Titmouse (*P. inornatus*) breeds in the Black Mesa country of the western Oklahoma Panhandle (Sutton, 1967) and in the Guadalupe Mountains of southwestern Texas (Oberholser, 1974), but apparently is not sympatric with other titmice in either place.

2.9. *Sialia*

The Eastern and Mountain bluebirds (*S. sialis* and *S. currucoides*) co-occur locally in the western Dakotas (Stewart, 1975), central Saskatchewan (Houston and Street, 1959), and southern Manitoba (Knapton, 1979). However, the only hybrid individual known was collected in western Manitoba (Lane, 1968). This individual was mated to females of both nominal species, and presumably fathered a brood of seven young via his *S. currucoides* mate, whereas his *S. sialis* mate produced five infertile eggs. In short, although overlap between these taxa occurs at several sites, hybridization is apparently rare, and the above observation indicates some degree of inviability between them.

2.10. Sturnella

The phenotypic similarity between the Western and Eastern meadowlarks (*S. neglecta* and *S. magna*) makes it difficult to determine the extent to which hybridization between them occurs. Nonetheless, in a multivariate study, Rohwer (1972a) demonstrated that only a small percentage (ca. 3%) of meadowlarks from areas of sympatry in the central prairies were phenetically intermediate between reference samples of meadowlarks from both east and west. Throughout the prairies of Kansas, Nebraska, and Oklahoma, *S. neglecta* tend to occur in the relatively xeric grasslands, and increase in abundance in the west. *S. magna*, on the other hand, is found in the more mesic grasslands, and to the west becomes increasingly restricted in its distribution to riparian areas. Along most trans-Plains rivers, Eastern Meadowlarks are more or less continuously distributed. However, along the Platte River, in Nebraska, the distribution of Eastern Meadowlarks is patchy (Rohwer, 1972b). The highest percentage of intermediate birds comes from along the Platte River (6.7%) where "colonies" of Eastern Meadowlarks are surrounded by an "ocean" of Western Meadowlarks (Rohwer, 1972a). The scarcity of intermediate individuals indicates that selection is against hybridization.

2.11. Icterus

The present distributions of Baltimore and Bullock's orioles are shown in Fig. 2. Hybridization between these has been studied in detail in Oklahoma (Sutton, 1968), Kansas (Rising, 1970, 1983), Nebraska (Sibley and Short, 1964; Corbin and Sibley, 1977), South Dakota (Anderson, 1971), and the Canadian Prairies (Rising, 1973).

Throughout the Great Plains, from the southern Canadian prairies south to western Oklahoma, intermediate individuals ("hybrids") occur wherever populations of these birds are sympatric. In the Canadian prairies, at the turn of the Century, Bullock's orioles bred along the

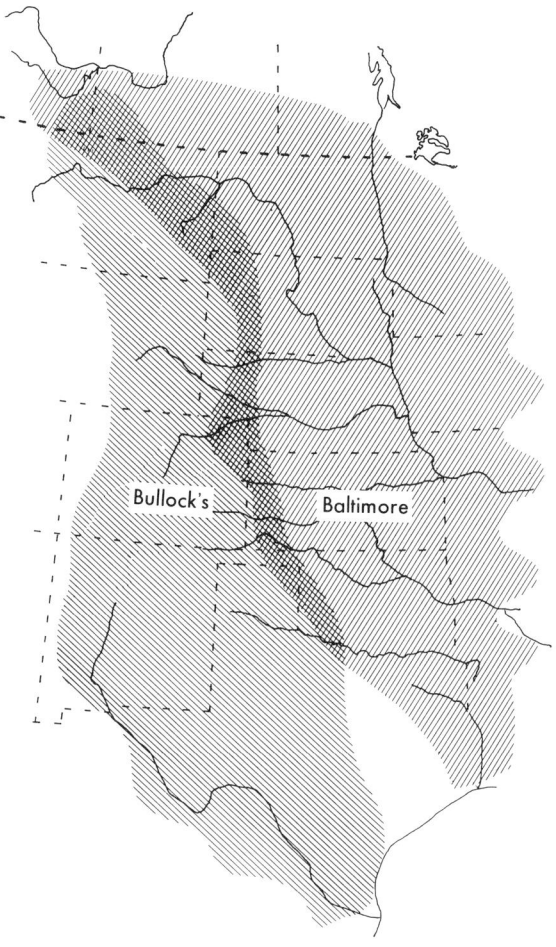

FIGURE 2. The ranges of the Baltimore and Bullock's orioles in the Great Plains. Where the distributions overlap, hybridization commonly occurs. (Both taxa are found outside of the Great Plains as well.)

South Saskatchewan River in southeastern Alberta, near Medicine Hat, and hybrids occurred in the Cypress Hills region (southeastern Alberta and southwestern Saskatchewan). In the early 1970s, hybrid or Baltimore-like individuals predominated in this region, suggesting that Baltimore orioles are displacing Bullock's orioles in this region (Rising, 1973). In the late 1960s, when Anderson collected in South Dakota, hybrid and Bullock's orioles occurred along the Little Missouri River, in northwestern South Dakota, and along this river in North Dakota (Stewart, 1975), and there were hybrid and mixed populations along the White, Bad, Cheyenne, and Grand rivers. On the Missouri River in southeastern South Dakota, the orioles were predominately of Baltimore plumage. In the past, the Missouri River was probably an important corridor for the dispersal of woodland birds, such as these orioles. Dam building in the 1950s has resulted in the destruction of most of the riparian vegetation along the Missouri in South Dakota, and Anderson (1971) predicts that this will lead to an increase in Bullock's orioles in the western Dakotas because that region is now more accessible to western woodland birds than to eastern ones.

In Nebraska, hybridization was common in the central to western part of the state in the 1950s and 1960s (Sibley and Short, 1964; Rising, 1970). In the 1950s, the "center" of the hybrid zone in Nebraska apparently ran through Valentine along the Niobrara River (samples from the northern part of the state are small) and through Big Spring along the Platte River. At Big Springs, 15 of the 18 specimens collected by Sibley and Short were intermediate between Baltimore and Bullock's orioles in plumage pattern characteristics (Fig. 3A; three looked like Bullock's orioles). In 1967, Rising (1970) collected seven specimens there, and all were hybrids. In 1974, however, many of the 25 specimens collected by Corbin at Big Springs appeared to be Baltimore orioles, indicating that Baltimore orioles were displacing Bullock's and hybrids at this site (Corbin and Sibley, 1977). Samples from the western Platte, in northeastern and north-central Colorado, were predominately Bullock's orioles in the mid-1950s as well, but were of mixed composition in the late 1960s (Corbin and Sibley, 1977; Fig. 3B). At Crook, Colorado, in the 1960s, Corbin and Sibley found a bimodal distribution of birds, suggesting that either assortative mating or reduced hybrid viability were occurring. At the very least, it appears that Baltimore orioles are extending their range westward at the expense of Bullock's orioles along the Platte River in Nebraska and Colorado, as they are apparently doing in the Canadian prairies.

In Kansas, Bullock's orioles are restricted to the western extremes of the state where they breed commonly in riparian trees along the Smoky Hill, Arkansas, and Cimarron rivers (Rising, 1970, 1983). Mixed populations, comprised largely of intermediate individuals, occur in a zone approximately 150 km wide that runs from the northwest corner southward through Scott City, Garden City, and Liberal (Rising, 1970, 1983). Rising (1983) found no evidence of any change in the hybrid zone along the Smoky Hill River (northern Kansas) between the mid-1960s and the late 1970s. However, along the Cimarron River (southern

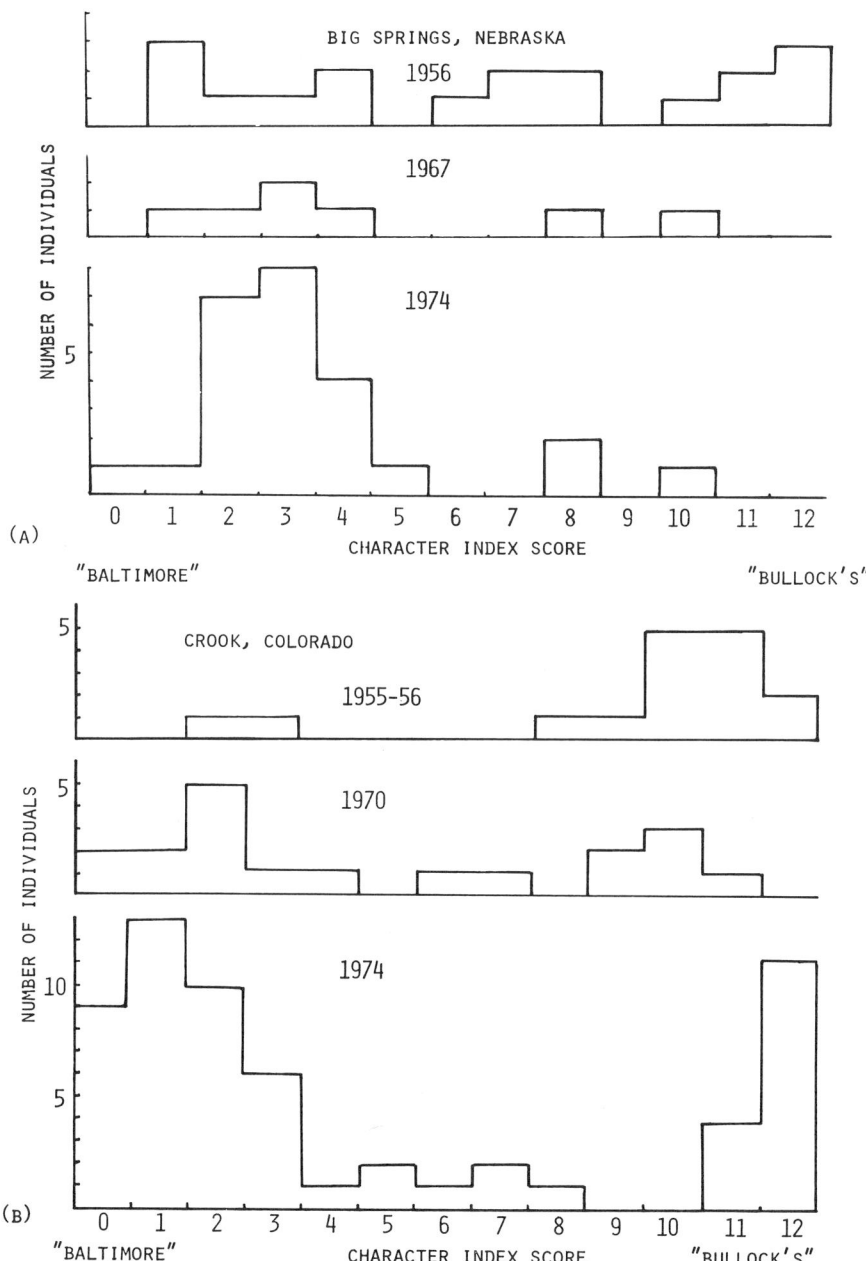

FIGURE 3. Histograms showing the change in oriole phenotype at two localities along the Platte River. (A) The change at Big Springs, Nebraska, (B) the change at Crook, Colorado.

Kansas), female Bullock's-like orioles are now prevalent farther east than they were in the 1960s (Fig. 4). Thus, in contrast to what is happening along the Platte River in Nebraska and Colorado, and probably in the Canadian prairies, Bullock's orioles may be extending eastward across southern Kansas at the expense of Baltimore orioles.

Rising (1983) found no tendency for males and females of the same taxon to associate preferentially with each other, so it is unlikely that assortative mating occurs.

In Oklahoma, Bullock's orioles occur commonly in the western Panhandle along the Cimarron River (Sutton, 1968; Rising, 1983), and populations from the western edge of Oklahoma are composed mostly of intermediates (Sutton, 1968). Overlap in Texas is apparently minimal (Oberholser, 1974), although hybridization probably occurs in the eastern Panhandle.

Throughout the Plains, Bullock's orioles average larger than Baltimore orioles in weight and wing length (Rising, 1970), as well as in many postcranial skeletal features (Rising, 1983). The pattern of size variation of orioles collected in western Kansas in 1976–1978 parallels the pattern of plumage pattern variation. Populations from the center

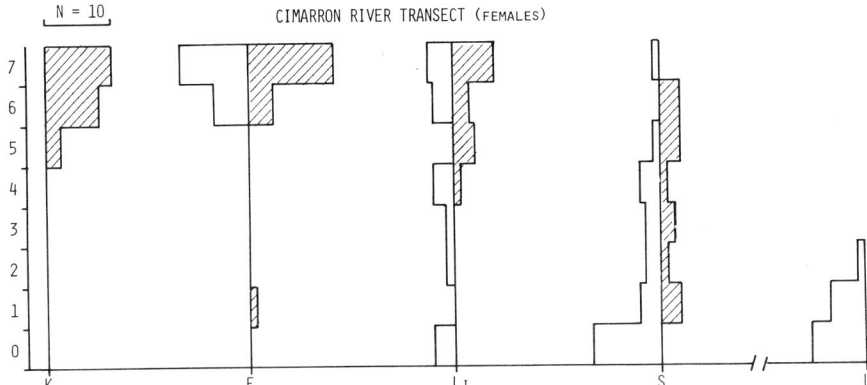

FIGURE 4. Histograms showing the change in the phenotype of female orioles from along the Cimarron River in southwestern Kansas (from Rising, 1983). The vertical scale ranges from 0 ("typical Baltimore oriole") to 7 ("typical Bullock's oriole"). The hatched bars are scores of orioles collected in 1976 and 1978, the open bars are scores of birds taken in the mid-1960s. The localities are arranged from west to east, and the letters refer to the following localities: K, Kenton, Oklahoma; E, Elkhart, Kansas; Li, Liberal, Kansas; S, Sitka, Kansas; and L, Lawrence, Kansas. The distances along the horizontal axis are proportional to the geographical distances between the localities. For scale, Liberal is about 105 km east of Elkhart. Lawrence is in eastern Kansas, about 400 km east of Sitka.

of the hybrid zone are more variable in size as in plumage, but there was no correlation between a bird's size and its plumage pattern, i.e., a bird that was the size of a Bullock's oriole might not look like one. The center of the hybrid zone is defined by populations with great variation in plumage, but the existence of a parallel pattern of variation in an independent character set supports the hypothesis that these birds are interbreeding in a zone of secondary contact, not merely showing geographic variation in plumage pattern (Schueler and Rising, 1976).

Corbin et al. (1979) examined allozymic variability in populations of Bullock's, Baltimore, and hybrid orioles from along the Platte River. Polymorphisms occur at two esterase loci, but the clines of allelic frequencies at these loci do not parallel those defined by plumage or size in any obvious manner. There is linkage disequilibrium at one locus, which could mean that assortative mating is occurring.

In summary, hybridization between these two types of orioles apparently occurs at any site where the two nominal forms come into contact. There is some indication that there is either a certain amount of assortative mating occurring in phenotypically mixed populations along the Platte River in northeastern Colorado, or that selection is against hybrid individuals there. There is no indication of either assortative mating or selection against hybrids in Kansas. In the northern prairies, Baltimore orioles appear to be displacing Bullock's orioles. This is seemingly occurring quite rapidly along the Platte River (the "center" of the hybrid zone has moved westward some 200 km in the past 20 years). In southern Kansas, on the other hand, Bullock's orioles seem to be displacing Baltimore orioles to the east in a similarly rapid fashion (ca. 100 km in 10 years; Rising, 1983). There is no evidence that the "hybrid zone" is increasing in width. Patterns of size variation parallel those of plumage variation, supporting the hypothesis that these birds are hybridizing in a zone of secondary contact.

2.12. Pheucticus

The interactions between Black-headed and Rose-breasted grosbeaks (P. melanocephalus and P. ludovicianus) have been carefully documented in Nebraska (West, 1962), South Dakota (Anderson and Daugherty, 1974), and North Dakota (Kroodsma, 1974a,b). In the Dakotas, Black-headed Grosbeaks nest along the Missouri River and its tributaries. In South Dakota, Rose-breasted Grosbeaks are common in the southeastern corner of the state, and in North Dakota they are fairly common in the Turtle Mountains in the north-central part of the state and in woodlands throughout the Agassiz Lake Plain (approximately the eastern third of the state). In the Dakotas, hybridization appears to be relatively infrequent, with hybrids coming from central North Dakota (where Rose-breasted Grosbeaks that nest along the Souris River breed near Black-headed Grosbeaks from along the Missouri River), and southeastern South Dakota (in the Missouri River Valley). In Canada, Rose-breasted Grosbeaks breed south in low density to southwestern Manitoba and southeastern Saskatchewan (Knapton, 1979), and to central Saskatchewan (Houston and Street, 1959; Godfrey, 1950). They may breed sympatrically with Black-headed Grosbeaks in the Cypress Hills Region (Godfrey, 1950) and southern Alberta (Rand, 1948), but this has not been substantiated. No hybrids are known from Canada. On the other hand, hybridization seems to occur fairly frequently in central Nebraska, especially along the Platte River (Fig. 5). In Kansas and Oklahoma, Rose-breasted Grosbeaks breed in the east. Black-headed Grosbeaks breed locally and uncommonly in western Kansas, especially northwestern Kansas, and at least one hybrid specimen is known from that region (Rising, 1974). Black-headed Grosbeaks might breed in low density in extreme western Oklahoma, but breeding there is not documented (Sutton, 1967).

In studies of mated pairs, Anderson and Daugherty (1974) found no evidence of assortative mating, but mixed pairs produced smaller clutches than like pairs.

2.13. Passerina

Although the Indigo and Lazuli buntings (*P. cyanea* and *P. amoena*) are parapatric in the central prairies, overlap is minimal in the northern and southern prairies where one or both of the taxa are rare. Indigo Buntings breed south to about the Nueces River in southern Texas, and west to the central part of the state (eastern Edwards Plateau; Oberholser, 1974), north to western Oklahoma (Sutton, 1967), and possibly to southwestern Kansas (Rising, 1974) and western Nebraska (Sibley and Short, 1959; Emlen et al., 1975). In the Dakotas, *P. cyanea* breed in low density along the Missouri River and its tributaries, decreasing in frequency relative to *P. amoena* westward, in the James River drainage in the eastern parts of those states, and (uncommonly) along the Souris River in north-central North Dakota. They are uncommon in southern Manitoba. Lazuli Buntings breed rarely in west Texas (no records since 1903; Oberholser, 1974), western Oklahoma and the Panhandle, and extreme southwestern Kansas. They breed more commonly to the north, in western Nebraska and the western Dakotas, as far west as the Rocky Mountains, and north to southern Alberta and Saskatch-

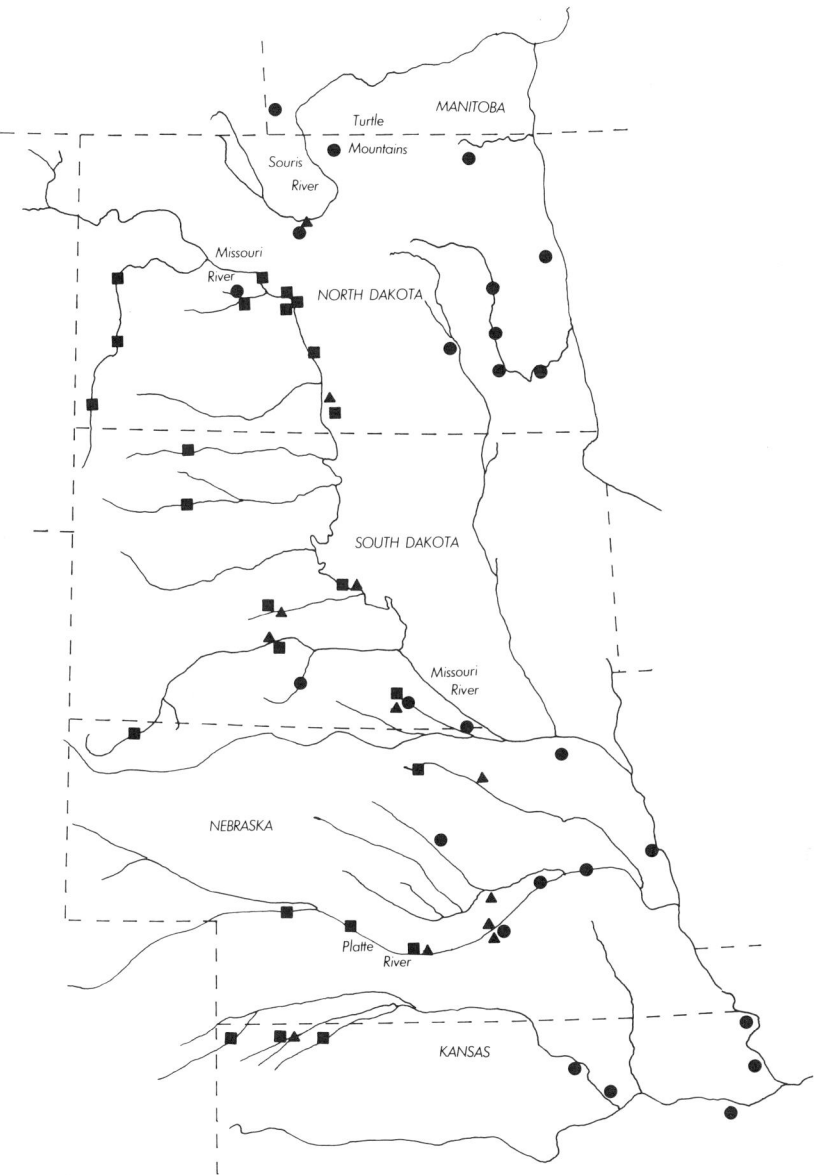

FIGURE 5. The distribution of grosbeaks (*Pheucticus*) in the Central Plains region. The symbols are on sites from which breeding grosbeaks have been collected or reported. Circles indicate sites for Rose-breasted, squares sites for Black-headed, and triangles for hybrid grosbeaks. (Data from West, 1962; Anderson and Daugherty, 1974; Kroodsma, 1974b; Rising, 1974; Knapton, 1979; personal observations.)

ewan (Rand, 1948; Godfrey, 1950). The two probably breed sympatrically, but uncommonly, in western Oklahoma (Roger Mills County; Sutton, 1967) and southwestern Kansas (Morton County; Rising, 1974). They breed sympatrically, and hybridize, along the Platte and Niobrara rivers in Nebraska (Sibley and Short, 1959; Emlen et al., 1975), the Missouri River in the central Dakotas (Sibley and Short, 1959; Kroodsma, 1975), and the Souris River in north-central North Dakota (Kroodsma, 1975). In their description of the hybrid zone in Nebraska and South Dakota, Sibley and Short (1959) argued that there was extensive hybridization in those states. Sixty-six of 95 male specimens, from 21 different localities, were designated by them as either "hybrids" or "backcrosses." Emlen et al. (1975), however, found hybridization to be somewhat less extensive, at least along the Niobrara River in northern Nebraska and eastern Wyoming. They found that really "definite" hybrids comprised only a small part of their samples (only six individuals of 101 from the Plains were clearly intermediate). Similarly, Kroodsma (1975) found hybridization to be relatively infrequent in North Dakota. The apparent differences in the implied frequency of hybridization in the studies of Kroodsma and Emlen et al., on the one hand, and that of Sibley and Short, on the other, are in part a function of the way that the specimens are scored on the character index scale [Kroodsma (1975) and Emlen et al. (1975) defined the "pure" taxa more broadly]. Despite these differences in technique, it is clear that buntings do hybridize regularly (perhaps as many as one-third of the individuals where they are sympatric and of approximately equal abundance), but backcrosses are seemingly rare (suggesting hybrid inviability). Indeed, Emlen et al. (1975) found evidence of reduced viability in mixed broods.

Along the Niobrara River, in northern Nebraska, the range of the Indigo Bunting has expanded westward by about 200 km in the 15-year interval, 1955–1969, and it appears to be rapidly replacing the Lazuli Bunting there (Emlen et al., 1975). The Indigo Bunting's range is also expanding westward in western Kansas (Rising, 1974).

The songs of the two buntings are different, but apparently substantial portions of both the utterance and response to song are learned. Thus, where the two are sympatric on a regular basis, song switching and interspecific territoriality occur (Sutton, 1967; Emlen et al., 1975).

2.14. *Pipilo*

Rufous-sided Towhees (*P. erythrophthalmus*, s. l.) in the Great Plains are restricted to deciduous thickets. In the breeding season, they are common from southern Saskatchewan and Alberta south to the Platte River in Nebraska and northeastern Colorado, but are uncommon as far south as northwestern Kansas (Sibley and West, 1959; Rising, 1974). The "hybridization" between the eastern "plain" and the western "spotted" towhees has been carefully studied only in Nebraska and South Dakota (Sibley and West, 1959). The males of these taxa differ only in "back coloration." In eastern towhees the coverts, scapulars, and interscapulars are unspotted, and there is a large, white patch at the base of the primaries. In western towhees there is extensive spotting in the scapulars, and no white at the base of the primaries. The females also show these differences, and eastern females have a rich brown head and back, whereas spotted females have a gray head and back. No differences in size have been described.

Along the Platte River, in southern Nebraska and northeastern Colorado, the east-to-west transition in towhee phenotype is fairly gradual. Populations from the central part of the state of Nebraska are comprised primarily of intermediate individuals. Along the Niobrara River, in northern Nebraska, and from along the Missouri River in South Dakota, the towhees are predominately spotted (Fig. 6). In Saskatchewan and Alberta, the towhees from along the South Saskatchewan River are, for the most part, spotted (personal observation) as are those from the Souris River in north-central North Dakota (Sibley and West, 1959). Towhees are uncommon from the southwestern Manitoba-southeastern Saskatchewan region, but one bird netted in that region was spotted (Knapton, 1979). On the other hand, towhees from along the Pembina River in northeastern North Dakota and south-central Manitoba are unspotted.

To summarize, spotted towhees are found east along the Missouri river into northeastern Nebraska, and in the Canadian prairies across Saskatchewan and into extreme southwestern Manitoba. Eastern towhees are found into eastern Kansas, Nebraska, and north through the eastern Dakotas into south-central Manitoba. Along the Platte River, the transition between typical forms is gradual, but along the Missouri and Souris rivers, where it has not been well documented, it must be a great deal more abrupt.

3. DISCUSSION

3.1. Stability of Zones

It has been argued that it is important to determine whether or not a cline is stable in geographic position, so as to differentiate between clines caused by secondary contact and hybridization, and those caused by clinal variation and selection (Schueler and Rising, 1976). If the

FIGURE 6. The average character index scores of towhees from collections in the Great Plains. The scores for individuals range from 0 ("typical eastern") to 4 ("typical western"), averages based on fewer than five individuals are in parentheses. (Data from Sibley and West, 1959; Rising, unpublished.)

step-cline has developed as a consequence of interbreeding after secondary contact ("hybridization") there are three possibilities: (1) individuals that hybridize are under negative selection, (2) hybrids are selectively equal to parental types, or superior to them in all habitats, or (3) hybrids are selectively equal or superior in the hybrid zone, but under negative selection outside the zone. In the first case, we would expect the step-cline to disappear, in the second, we would expect it to broaden, flatten, and eventually disappear, in the third, the step-cline should be stable in width and configuration, but might move geographically (shift one direction or the other, but not increase in width; Endler, 1977). In the case of geographic variation, (i.e., the stepcline developed without secondary contact and interbreeding between differentiated populations), the cline is (presumably) maintained by selection and should be stable through time (and hence inseparable from the third alternative above). However, an *apparently* stable cline may not be evidence that selection is operating because, unless the birds disperse at rates greater than those that have been calculated (Barrowclough, 1980a), it would be virtually impossible to detect any change in a hybrid zone in even a few score of generations. Such "neutral" hybrid zones (after initial contact and hybridization) increase in width so slowly that changes would be obvious only after many generations (probably thousands; Barrowclough, 1980b). For example, as mentioned in the preceding section, Audubon and his party collected "hybrid" flickers along the Missouri River in 1842, some 115 years before Short (1965) studied flicker hybridization in detail. This shows that the flickers were in contact and hybridizing in the Plains before Europeans modified the environment there, and since the climate there has not changed greatly in the past 3000 years or so (Mengel, 1970), there is no reason to doubt but that they have been in contact there for at least that long. Using a single locus model (Endler, 1977), and given the width of the hybrid zone at 250 km (the approximate width of the flicker zone), if 3000 generations had passed since the initial contact, in the absence of selection, i.e., assuming the passive diffusion of alleles, the root mean squared dispersal distance is estimated to be 2.7 km/generations, a figure that is somewhat higher than those calculated from dispersal studies by Barrowclough (1980a). Given the above figures (i.e., contact 3000 generations old, rate of dispersal 2.7 km/generation, one generation/year), a "neutral" zone would increase in width by about 0.04 km/generation. It would have been about $4\frac{1}{2}$ km wider when Short studied them in the 1950s than in Audubon's time. If the contact were older, the rate of increase in the width of the hybrid zone would be lower. We have no idea how old the contact is. Remington (1968) suggests that some western species might have crossed the Rocky Mountains and moved into the Great Plains since 1850. But, unless the contact is very recent and dispersal distances large, "neutral" clines formed by hybridizing taxa could appear to be stable.

There are two ways to assess directly the stability of hybrid zones, firstly, by determining former distributional patterns from the literature, and secondly, by repeated sampling along a transect across the zone. In the Great Plains, repeated sampling has been done only for flickers in South Dakota (Anderson, 1971), orioles in Nebraska (Corbin and

Sibley, 1977) and Kansas (Rising, 1983), and buntings in Nebraska (Emlen et al., 1975). The flicker zone appears to be stable at three sites, but the zones of orioles and buntings have shifted in position, although they seemingly have not become wider. With the exception of the orioles in southern Kansas, where changes are indicated, eastern forms are pushing western forms out of the Plains. We would expect the opposite, it would seem, if it were the western taxa that had just recently gained access to the Great Plains (see above).

The literature indicates that the "black-crested" titmice are displacing "gray-crested" ones in eastern Texas and southwestern Oklahoma. These changes in distribution feature the displacement of one taxon by the counterpart and hence can be taken as evidence that selection is favoring one morph over the other. Even if there were no direct hybrid inviability, we would not anticipate a blending of the morphs in such cases. The rapidity with which some of these displacements appear to be occurring also indicates that we might underestimate rates of gene flow, perhaps specifically between populations of migratory birds (that a priori would seem to have high dispersal rates because young birds would tend to "get lost" while attempting to return to their natal sites).

The Golden-fronted Woodpecker's range may be expanding northward, but this is not certain. For the other species discussed here, I can find no reason to suggest that range changes are taking place.

3.2. Increased Variability in Zones

Generally, we infer that hybridization is occurring between two taxa when we find (1) a step in a pattern of geographic variation of a characteristic and (2) increased variability in that characteristic in populations that have intermediate character states. Endler (1977), however, has shown that steps can develop in "ramp-clines" via selection, without secondary contact. Thus "geographic variation" can appear to be "hybridization." Schueler and Rising (1976) reasoned that in the case of hybridization (interbreeding after secondary contact) (1) all characteristics that differ between the presumed hybridizing taxa should show approximately the same patterns of clinal or step-clinal variation, and (2) there should also be an increase in the variability of all of the features that differ between the taxa as a consequence of the general mixing of their genomes. A weakness of the "general increase in variability" criterion is that in genetically simple clines, such as a cline in the frequency of two alleles at a single locus, sampling variance will necessarily be higher at intermediate frequencies. However, it seems unlikely that this statistical artifact would cause increases in variances for characteristics that have complex genetic bases (such as "wing length," or some multivariate measure such as a score on a discriminate function axis; Rising, 1983). In Rising's (1983) reexamination of Bullock's and Baltimore orioles in western Kansas, the patterns of plumage variation, which have led ornithologists to argue that these birds are hybridizing, as well as patterns of size variation, were examined. The variation in size (measured multivariately) was in general greatest where the variation in plumage was greatest. Both the parallel patterns of size and plumage pattern variability, and the increased variability in intermediate populations support the hypothesis that the orioles are hybridizing in a zone of secondary contact.

3.3. Suture-Zones in the Great Plains

Of the 14 pairs of birds discussed, 11 show east–west replacement. The distributional patterns of the Golden-fronted and Red-bellied woodpeckers and the titmice are quite similar. In both cases, there is a zone of contact that runs from the Gulf Coast through east-central Texas and into southwestern Oklahoma, a northeast–southwest pattern. In Texas and Oklahoma, the range overlap of the Great Crested and Ash-throated flycatchers is like that between the woodpeckers and titmice, but the flycatchers occur farther to the north as well, and are sympatric in southwestern Kansas. The contact between the Black-capped and Carolina chickadees, from Kansas to New Jersey, is north–south (also altitudinal in the Appalachians).

There are several different east–west patterns of replacement. The ranges of the grosbeaks, towhees, and bluebirds show one pattern (Figs. 5 and 6). In these, the western counterpart is found in the southwestern Canadian prairies, and in the Missouri River drainage southeast into southern South Dakota. The eastern form is found in the eastern Dakotas. In the Dakotas, the ranges of the buntings show a similar pattern, but they are found farther to the south in the Plains. North of Nebraska the major rivers tend to run more north–south than east–west, and there is a "treeless" region to the east of the Missouri River. For the bluebirds, grosbeaks, buntings, and towhees, this region appears to be a significant barrier. For towhees, hybridization is common along the east–west flowing Niobrara and Platte rivers in Nebraska. Grosbeak and bunting hybridization also seems to be commoner there than in the Dakotas. In the Plains, the Mountain Bluebird does not breed as far south as Nebraska. The distributions of the orioles and flickers in the Plains region are remarkably similar (Figs. 1 and 2), and illustrate a second east–west pattern. The eastern forms of these taxa are found northwest along the Missouri River nearly to the Rocky Mountains and

Black Hills. The screech owls, pewees, jays, and meadowlarks all show east–west replacement, but their distributional patterns are not alike, nor are they like either the "grosbeak–towhee" or "flicker–oriole" patterns. In screech owls, pewees, and jays the complementary forms are uncommon in the vicinity of contact and it seems unlikely that their ranges are limited by competitive exclusion. On the other hand, exclusion probably limits the ranges of the meadowlarks in the Great Plains.

The two basic patterns of distribution found, the northeast–southwest and east–west replacement, more or less correspond to two of Remington's (1968) suture-zones, The Rocky Mountain Eastern Suture-Zone and the Central Texas Suture-Zone. These suture-zones, however, are broadly delimited and it is unclear how close the correspondence among species' ranges must be in order to support the general contention that these are regions where different biotas are in contact. Coincidence may be involved in some cases, and the similarities among these 14 taxon pairs of Great Plains birds are not great.

Remington characterizes the Central Texas Suture-Zone as a "mature zone" because many of the forms apparently in secondary contact there do not hybridize. Data on the two *Centurus* woodpeckers from the Austin, Texas region support that contention, but the crested titmice, which seem to interbreed randomly when in contact, do not. Both the black-crested titmouse and the Golden-fronted Woodpecker seem to be more like their "eastern" counterparts, e.g., the Tufted Titmouse and the Red-bellied Woodpecker, than comparable taxa from the southwest, e.g., the Plain Titmouse (*Parus inornatus*) and the Gila Woodpecker (*C. uropygialis*). This suggests that these taxa, which we assume differentiated in isolation on the Edwards Plateau of Texas, were derived from the east rather than the west.

A range of interactions occurs in the birds in the Rocky Mountain Eastern Suture-Zone. For example, the two towhees seem to be only slightly differentiated, whereas the pewees may be completely separate, and any generalizations about the age of the contact, based on the extent to which the daughter taxa have differentiated, would be biased by the *a priori* selection of taxa. For most of these east–west counterparts, there are, as well, Mexican (or southwestern) representatives. In some, the Mexican taxon is more like the western taxon than is the western like the eastern, e.g., the orioles. In others, the relationships are unclear, e.g., flickers, buntings, and in still others, the east–west relationships seem closer than the west-Mexican ones, e.g., the screech owls, pewees, grosbeaks, and towhees. The southwestern meadowlarks are probably derived from the eastern ones. Thus, while the east–west–Mexican pattern is repeated over and again, there does not appear to be a single "historical" pattern of invasion and differentiation, e.g., Mexico to the west, then to the east).

3.4. Taxonomic Comments

Because we visualize speciation in sexually reproducing animals as generally taking place via extrinsic isolation and divergence, most biologists prefer to combine taxa that, where sympatric, regularly interbreed and backcross (Mayr, 1970). It is felt that in such cases differentiation is incomplete (because it is assumed that otherwise hybridization will decrease in extent). Commonly, however, it is difficult to ascertain whether selection is operating against hybrids or individuals that tend to hybridize. In the Great Plains birds, there is fairly good direct evidence that selection is against hybrid bluebirds, grosbeaks, and buntings. The scarcity of intermediate meadowlarks suggests that selection is against hybridization. The bimodal distribution of orioles at Crook, Colorado, if not anomalous, is indirect evidence of selection against hybrids as well. There is mutual exclusion between the two chickadees in some parts of their ranges, but probably there is occasional hybridization between them in southern Kansas. The screech owls hybridize where their ranges overlap, but this is in a very limited area and hybrid viability (two broods?) is not known. In the other taxa discussed, hybridization is either unknown or rare. Thus, even if one rigorously supports the "biological species concept," cases can be made to "split" all these taxa except the created titmice, flickers, and towhees.

There are no entirely satisfactory or objective criteria for delimiting species and inasmuch as these 28 taxa are all easily distinguished in the field, essentially allopatric in distribution, and possibly evolutionarily independent, there is virtue in splitting them all. Among other things, this would result in a maximal input of information about these from the millions of birdwatchers in North America who usually do not keep notes on subspecies.

ACKNOWLEDGMENTS. I thank Nancy Flood, Trudy Rising, and Richard Snell for their many helpful comments.

REFERENCES

Anderson, B., 1971, Man's influence on hybridization in two avian species in South Dakota, *Condor* **73**:342–347.

Anderson, B., and Daugherty, R. J., 1974, Characteristics and reproductive biology of grosbeaks (*Pheucticus*) in the hybrid zone in South Dakota, *Wilson Bull.* **86**:1–11.

Audubon, M. R., 1897, *Audubon and His Journals, With Zoological and Other Notes by Elliot Coues*, Chas. Scribners Sons, New York.
Barrowclough, G. F., 1980a, Gene flow, effective population sizes, and genetic variance components in birds, *Evolution* **34**:789–798.
Barrowclough, G. F., 1980b, Genetic and phenotypic differentiation in a wood warbler (Genus *Dendroica*) hybrid zone, *Auk* **97**:655–668.
Brewer, R., 1963, Ecological and reproductive relationships of Black-capped and Carolina chickadees, *Auk* **62**:49–69.
Corbin, K. W., and Sibley, C. G., 1977, Rapid evolution in orioles of the genus *Icterus*, *Condor* **79**:335–342.
Corbin, K. W., Sibley, C. G., and Ferguson, A., 1979, Genic changes associated with the establishment of sympatry in orioles of the genus *Icterus*, *Evolution* **33**:624–633.
Dixon, K. L., 1955, An ecological analysis of the interbreeding of crested titmice in Texas, *Univ. Cal. Publ. Zool.* **54**:125–206.
Emlen, S. T., Rising, J. D., and Thompson, W. L., 1975, A behavioral and morphological study of sympatry in the Indigo and Lazuli buntings of the Great Plains, *Wilson Bull.* **87**:145–179.
Endler, J. A., 1977, *Geographic Variation, Speciation, and Clines*, Princeton University Press, Princeton, New Jersey.
Godfrey, W. E., 1950, Birds of the Cypress Hills and Flotten Lake regions, Saskatchewan, Bull. 120 Nat. Mus. Canada, Ottawa, pp. 1–96.
Houston, C. S., and Street, M. G., 1959, The birds of the Saskatchewan River, Carlton to Cumberland, Special Publ. No. 2, Sask. Nat. Hist. Soc., Regina, pp. 1–205.
Hunt, C. B., 1974, *Natural Regions of the United States and Canada*, W. H. Freeman and Company, San Francisco, California.
Knapton, R. W., 1979, Birds of the Gainsborough-Lyleton region (Saskatchewan and Manitoba), Special Publ. No. 10, Sask. Nat. Hist. Soc., Regina, pp. 1–72.
Kroodsma, R. L., 1974a, Species-recognition behavior of territorial male Rose-breasted and Black-headed grosbeaks (*Pheucticus*), *Auk* **91**:54–64.
Kroodsma, R. L., 1974b, Hybridization in grosbeaks (*Pheucticus*) in North Dakota, *Wilson Bull.* **86**:230–236.
Kroodsma, R. L., 1975, Hybridization in buntings (*Passerina*) in North Dakota and eastern Montana, *Auk* **92**:66–80.
Lane, J., 1968, A hybrid Eastern Bluebird X Mountain Bluebird, *Auk* **85**:684.
Marshall, J. T., Jr., 1967, Parallel variation in North and Middle American screech-owls, Monogr. No. 1, Western Foundation of Vert. Zool., Los Angeles, pp. 1–72.
Mayr, E., 1970, *Populations, Species, and Evolution*, Harvard University Press, Cambridge, Massachusetts.
Mengel, R. M., 1970, The North American Central Plains as an isolating agent in bird speciation, in: *Pleistocene and Recent Environments of the Central Great Plains* (W. Dort, Jr., and J. K. Jones, Jr., eds.), Special Publ. No. 3, Univ. Kansas, Dept. Geology, University of Kansas Press, Lawrence, pp. 279–340.
Oberholser, H. C., 1974, *The Bird Life of Texas*, University of Texas Press, Austin.
Rand, A. L., 1948, Birds of southern Alberta, Bull. 111, Nat. Mus. Canada, Ottawa, pp. 1–105.
Remington, C. L., 1968, Suture-zones of hybrid interaction between recently joined biotas, in: *Evolutionary Biology*, Volume 2 (T. Dobzhansky, M. K. Hecht, and W. C. Steere, eds.), Appleton-Century-Crofts, New York, pp. 321–428.
Rising, J. D., 1968, A multivariate assessment of interbreeding between the chickadees, *Parus atricapillus* and *P. carolinensis*, *Syst. Zool.* **17**:160–169.
Rising, J. D., 1970, Morphological variation and evolution in some North American orioles, *Syst. Zool.* **19**:315–351.

Rising, J. D., 1973, Morphological variation and status of the orioles, Icterus galbula, I. bullockii, and I. abeillei, in the northern Great Plains and in Durango, Mexico, Can. J. Zool. **51**:1267–1273.
Rising, J. D., 1974, The status and faunal affinities of the summer birds of western Kansas, Univ. Kansas Sci. Bull. **50**:347–388.
Rising, J. D., 1983, The progress of oriole "hybridization" in Kansas, Auk **100**:(in press).
Rising, J. D., and Schueler, F. W., 1980, Identification and status of wood pewees (Contopus) from the Great Plains: What are sibling species? Condor **82**:301–308.
Rohwer, S. A., 1972a, A multivariate assessment of interbreeding between the meadowlarks, Sturnella, Syst. Zool. **21**:313–338.
Rohwer, S. A., 1972b, Distribution of meadowlarks in the central and southern Great Plains and in the desert grasslands of eastern New Mexico and west Texas, Trans. Kansas Acad. Sci. **75**:1–19.
Schueler, F. W., and Rising, J. D., 1976, Phenetic evidence of natural hybridization, Syst. Zool. **25**:283–289.
Selander, R. K., and Giller, D. R., 1959, Interspecific relations of woodpeckers in Texas, Wilson Bull. **71**:107–124.
Short, L. L., Jr., 1965, Hybridization in the flickers (Colaptes) of North America, Bull. Am. Mus. Nat. Hist. **129**:307–428.
Sibley, C. G., and Short, L. L., Jr., 1959, Hybridization in the buntings (Passerina) of the Great Plains, Auk **76**:443–463.
Sibley, C. G., and Short, L. L., Jr., 1964, Hybridization in the orioles of the Great Plains, Condor **66**:130–150.
Sibley, C. G., and West, D. A., 1959, Hybridization in the Rufous-sided Towhees of the Great Plains, Auk **76**:326–338.
Stewart, R. E., 1975, Breeding Birds of North Dakota, Tri-College Center for Environmental Studies, Fargo.
Sutton, G. M., 1967, Oklahoma Birds, University of Oklahoma Press, Norman.
Sutton, G. M., 1968, Oriole hybridization in Oklahoma, Bull. Oklahoma Ornithol. Soc. **1**:1–7.
West, D. A., 1962, Hybridization in grosbeaks (Pheucticus) of the Great Plains, Auk **79**:399–424.
Williams, O., and Wheat, P., 1971, Hybrid jays in Colorado, Wilson Bull. **83**:343–351.

CHAPTER 6

SPECIES CONCEPTS AND SPECIATION ANALYSIS

JOEL CRACRAFT

1. INTRODUCTION

Systematic biologists have directed much attention to species concepts because they realize that the origin of taxonomic diversity is the fundamental problem of evolutionary biology. Questions such as, What are the units of evolution? and, How do these units originate? thus continually capture the attention of many. It is probably no exaggeration to say that most believe the "systematic" aspects of the problem have been solved to a greater or lesser extent, whereas the task before us now is to understand the "genetic" and "ecologic" components of differentiation, i.e., those aspects often perceived to constitute the "real mechanisms" of speciation:

> A study of speciation is, to a considerable extent, a study of the genetics and evolution of reproductive isolating mechanisms (Bush, 1975, p. 339).
> ... a new mechanistic taxonomy of speciation is needed before population genetics, which deals with evolutionary mechanisms, can be properly integrated with speciation theory; that is, the various modes of speciation should be characterized according to the various forces and genetic mechanisms that underly [sic] the evolution of isolating barriers (Templeton 1980, p. 720).

JOEL CRACRAFT • Department of Anatomy, University of Illinois, Chicago, Illinois 60680; and Field Museum of Natural History, Chicago, Illinois 60605.

Thus, in terms of the systematic component of speciation, it is now widely accepted, especially within vertebrate zoology, that the units of evolution are "biological" species, characterized by their *reproductive discontinuity* from other such units, and that their origins can be analyzed in terms of their geographic pattern (see summaries in Mayr 1963; Bush, 1975). Ornithologists have contributed substantially to the development of this view. One only has to invoke the pivotal contributions of Stresemann, Rensch, Mayr, Miller, and others to appreciate the impact of ornithological systematics on questions of species concepts and speciation analysis.

The interests of classification theory and those of systematic analysis have long been viewed as being in conflict with each other (Selander, 1971, p. 99), with "species" as classificatory units functioning in ways different from "species" as units of evolution. In fact, because these two conceptions of "species" often do not correspond, and this can be shown to be true in ornithology as well, it can be argued that the functions of both are being compromised. The problem, Selander (1971, p. 99) believes, is that "our typologically based system of nomenclature cannot adequately deal with speciation phenomena. . . ." But is the "typologically based system of nomenclature" the real problem? I suggest the answer is no, the problem lies elsewhere. Indeed, the problem can be posed as the mirror image of that identified by Selander. Our classificatory system is reasonably capable of dealing with the results of speciation, i.e., the hierarchical arrangement of taxic interrelationships; rather, it is speciation analysis which is incapable of accomodating the prevailing concept of species that is being forced upon it. In particular, the idiosyncratic nature of the polytypic species concept and the *definition* of species as *discontinuous reproductive units* rather than as *phylogenetic units* prevents an adequate assessment of the patterns and processes of speciation.

Speciation can be viewed as the phylogenetic deployment of differentiated taxonomic units through space and time. If reproductive discontinuity does not precisely correlate with this deployment, then any definition of these taxonomic units solely in terms of that discontinuity will logically result in phylogenetic history (speciation) being reconstructed incompletely, at best, or incorrectly, at worst. There is abundant evidence, of course, that the pattern of reproductive disjunction among taxa does not necessarily correlate with the history of their differentiation (see especially Rosen, 1979), yet the implications of this have not been fully appreciated by many workers interested in the broad area of speciation analysis. Moreover, allegiance to the polytypic-biological species concept hinders our understanding of speciation by focusing attention on unproductive problems (Is taxon X a species or a subspecies? How much genetic change must occur before a new species is formed? are two examples).

In this paper, I first want to explore concepts of species. Unlike many previous discussions, I will be less concerned with definitions than with the analytical consequences of these definitions. I subscribe to the view that definitions within science are only meaningful to the extent that they can lead to growth in knowledge (Popper, 1959, 1972), and thus my criterion of judgment is whether a particular species concept facilitates our understanding of the history of taxonomic diversification. Following this, I want to consider the research strategy we might adopt in developing hypotheses about the phylogenetic differentiation of taxa. The emphasis, therefore, is placed on how we might reconstruct the *pattern* of taxonomic diversification. The discussion focuses almost entirely on the literature of avian systematics, and I will suggest that pattern analysis has unfortunately been neglected by many previous workers. Because of this, ornithologists understand the history of avian diversification much less than heretofore realized.

2. SPECIES CONCEPTS

2.1. The "Biological Species" Concept

The polytypic-biological species concept (hereafter, BSC), as summarized by Mayr (1942, 1948, 1957a 1957b, 1963, 1969), is without question the dominant view of species within ornithology. Indeed, it would be difficult to find any serious opposition to its acceptance, at least in principle, within the ornithological community.

According to this conception, species are defined as "groups of interbreeding natural populations that are reproductively isolated from other such groups" (Mayr, 1969, p. 26). If two or more groups are sympatric or parapatric, the presence or absence of reproductive isolation can usually be determined unambiguously. If groups are allopatric, on the other hand, then biological data must be evaluated in terms of whether it is thought individuals of these populations could interbreed were they to come into contact. Within the framework of the BSC, data on the extent of hybridization between populations are critical in determining their taxonomic status as species (Sibley, 1957; Mayr, 1969; Short, 1969).

Polytypic species include allopatric populations (often designated subspecies) assumed to be capable of interbreeding. It is important to realize that the polytypic species concept arose (Kleinschmidt, 1900; Stresemann, 1936; Mayr, 1942, pp. 110–113), not only to facilitate our understanding of taxonomic differentiation, but also as an attempt to come to grips with a classificatory problem, namely the propensity of many nineteenth century systematists to apply species names to every variation within local populations. Faced literally with many thousands of species names, the classificatory problems appeared real indeed, and a reaction against this situation developed. Yet, it is instructive to remember that classificatory "convenience" remains a major justification for the polytypic species concept (Mayr, 1969, p. 38).

There is a tendency for those in vertebrate systematics to assume that the BSC is widely accepted and applied within biology. Speaking of the "morphological" species concept, Mayr (1963, pp. 16–17), for example, remarks that "In recent years most systematists have found this typological-morphological concept inadequate and have rejected it... Where the taxonomist applies morphological criteria, he uses them as secondary indications of reproductive isolation." Many biologists consider this assessment to be false, for not only have the vast majority of plant systematists ignored or rejected the BSC, but most zoological systematists probably have done likewise, at least when dealing with the practical problems of describing the earth's biota. The list of the critics of the BSC is long (see summaries by Ehrlich, 1961; Sokal and Crovello, 1970; Raven, 1976; Rosen, 1978; Cronquist, 1978; Levin, 1979; Paterson, 1981), and it is fair to say that their criticisms have not been answered satisfactorily. Importantly, the basis for this criticism does not lie entirely with the problem of classifying the taxonomic units of nature; in fact, the rejection of the BSC stems primarily from the realization that it does not function well in helping us to understand the pattern and process of taxonomic differentiation. If species are defined strictly in terms of known or presumed reproductive disjunction, these workers point out, then it compels biologists to consider the origin of reproductive isolation as the major problem of taxonomic differentiation, and such is not the case. A solution to the "species problem" does not lie, however, in adopting many different kinds of "species" (Scudder, 1974), for to do so would be to abandon the search for general patterns of biotic diversification.

2.2. Are "Biological Species" the Units of Evolution?

Scientific theories direct attention to the relationships and interactions among things, or entities, thought to have an ontological status in nature. Theories themselves sometimes impart or predict the expectation of reality to things not known to exist. One only has to recall the predictions of particle physics or astronomy to appreciate this. Likewise, theories of evolution are theories about the descent and modification of entities. Biologists have traditionally called these entities "species," but it is essential to remember that the term "species" has been used in many different ways. Moreover, it is widely recognized that taxonomic species as such have not always been accepted as the "units of evolution" by many biologists. Indeed, these "units" have been postulated to be genes, gametes, individual organisms, local demes, populations, varieties, subspecies, "biological species," and even higher taxa (Lewontin, 1970; Hull, 1980). If this vast nomenclature signifies anything, perhaps it is the diversity of opinion that exists about the nature of the evolutionary process itself, for conceptions or theories about how nature is organized and has developed do influence opinions about the ontological status of the evolving entities. All of this is by way of introduction to the question of whether "biological species" (as in the BSC) can be defended as the "units of evolution."

Although it might be generally admitted that the BSC arose primarily as a classificatory concept, its contemporary role in evolutionary analysis developed later (Mayr, 1942, 1963, 1970; Cain, 1954). "Biological species" are viewed by many biologists (in particular Mayr, 1963, 1982) as the units of evolution, but this naturally implies that the origin of reproductive isolation is the most important component of taxonomic diversification. If, however, reproductive isolation is not the central issue of taxonomic differentiation, then logically a species concept based on isolation may not be necessary or desirable for systematic and evolutionary analysis. This would not imply, of course, that the phenomenon of reproductive isolation is uninteresting or lacking in general significance, only that its role in the description and explanation of patterns of evolutionary diversification has been exaggerated or misinterpreted.

A common theme runs through discussions of species concepts and speciation analysis, namely that the origin of taxa involves the *differentiation* of groups of individual organisms. The systematic status of these groups depends upon this differentiation first and foremost, and only secondarily upon the observation that individuals of each group interbreed with one another. Thus, it is the possession of unique phenotypes that permits us to recognize taxonomic units (Platnick, 1977; Rosen, 1978, 1979). Naturally, characters diagnostic of these differentiated units do not have to be morphological, but can include any intrinsic attribute, whether biochemical, physiological, or behavioral (in this sense, then, *a broad concept of "phenotype" is adopted through-*

out this paper). Importantly, it is these intrinsic attributes of taxa that prevent interbreeding with other taxa, no matter how much they might generally resemble one another. All basic taxa (call them species) of birds, for example, can be distinguished from their close relatives by intrinsic characters alone, and I do not know of a single example in which data on reproductive cohesion or disjunction are the sole factors establishing taxonomic limits. Indeed, even with sibling species, phenotypic differences of some kind, e.g., behavioral or biochemical, are always the primary data that lead to their recognition as distinct taxa.

Given an isolated population that has evolved one or more apomorphous (derived) characters, most biologists would call it a "unit of evolution," in that the primary characteristic of evolution, i.e., the origin of a new taxon and the evolutionary modification of primitive characters, is satisfied. Following this line of thought, can this new taxon be equated with a "biological species" as conceived by advocates of the BSC? It might be, but avian systematics is replete with examples in which such taxa are not considered to be "biological species," and as was mentioned earlier, the BSC arose as a methodological construct specifically designed under some circumstances to reject many of these differentiated taxa as "species." If so, then clearly many evolutionary taxonomic units would not be "biological species" to advocates of the BSC. Conversely, designated "biological species" (1) may be equivalent to a single evolutionary taxonomic unit, (2) may comprise two to many such taxonomic units, or (3) may contain a collection of evolutionary taxonomic units that, for one reason or the other, does not comprise a (strictly) monophyletic assemblage of related forms. Such a potpourri of possibilities is an inevitable outcome of the BSC itself and of the methodology recommended to identify "biological species" (see Mayr, 1942, 1963, 1969). "Biological species" are, by and large, identified purely on a subjective assessment of phenotypic (usually morphological) distinctness (Sokal and Crovello, 1970; Rosen, 1978, 1979). If allopatric populations are not considered "sufficiently" distinct, they are lumped in the same polytypic species. If one or more of these populations are viewed as being distinct "enough," they are assigned to their own "biological species." In this manner, then, the BSC becomes a servant to classificatory philosophy (particularly that advocated by evolutionary systematists), rather than an instrument to analyze the pattern and process of taxonomic differentiation.

One might conclude from the foregoing that "biological species" are equivalent to the "units of evolution" only in cases in which they are monotypic. In the North American avifauna, approximately 59% of the "species" are monotypic (Mayr and Short, 1970, p. 96), and according to Keast (1961, p. 393) about 44% of the Australian "species" are monotypic. But what about the polytypic species of these faunas? In actuality, both studies demonstrate that a substantial number of differentiated evolutionary units go unrecognized when "biological species" are accepted as the basis for systematic analyses of this kind. A polytypic "biological species" cannot logically constitute the lowest-level taxonomic "unit of evolution" because these "species" may be composed of a variable number of evolutionary units, each possessing their own geographic, phenotypic, and (presumably) genetic integrity.

Differentiated taxonomic units as recognized by the methods of systematics constitute our only evidence about the kinds of evolutionary entities that exist in nature. At any one time, of course, our hypotheses may overestimate or underestimate their numbers. Nevertheless, comparative evolutionary analysis will be impossible without some general agreement among biologists over which taxonomic unit is to be considered equivalent to the "unit of evolution." In fact, biologists have traditionally adopted the concept of "species" as fulfilling that role. Because the BSC fails to designate the numbers of differentiated taxonomic units correctly, evolutionary analysis based on the BSC will inevitably lead to incorrect conclusions about evolutionary history. If the BSC is to be maintained, its defense will have to be based solely on an argument about taxonomic "convenience," i.e., its original rationale and an argument already rejected by many systematists, rather than on one about its efficacy for evolutionary analysis.

2.3. A Proposed Species Concept for Ornithology

This section will pursue the problem of formulating a species concept with widespread applicability within biology, including ornithology. Such a concept should be compatible with the two primary aims of systematic biology, namely the taxonomic recognition, description, and historical analysis of all potential evolutionary units, and then the expression of this information within the context of Linnaean hierarchical classifications. Clearly, the BSC does neither entirely satisfactorily. As noted above, "biological species" often do not represent evolutionary units, and as will be discussed shortly, assumptions underlying the identification and taxonomic ranking of "biological species" have the potential to obscure the analysis of taxonomic diversification when those "species" are then used as the elements of classification or as units of evolution.

2.3.1. Taxonomy: Traditional Methodology and Its Problems

Once differentiated taxonomic units have been recognized, a systematist is faced with the question of how to treat them within a Linnaean classification scheme, and, furthermore, how to interpret the pattern of differentiation historically. Traditional solutions to the former question will be considered here, and the latter will be postponed until the next section.

Our discussion can depart from three observations: (1) in terms of the formal rules of naming taxa, the International Code of Zoological Nomenclature recognizes only two categorical ranks, species and subspecies, which might be used when discussing speciation analysis, (2) there is no general agreement among avian systematists over whether species (of whatever kind) or subspecies are "units of evolution," and (3) in modern systematic practice, the distinction between species and subspecies ranking is almost always based on some subjective measure of the degree of phenotypic (in the broad sense) differentiation.

Traditionally, speciation analysis within ornithology has focused on a detailed description of geographic variation and its correlation with available knowledge about geography and climate, both of the present and in the past (Vuilleumier, 1980). Many studies have elaborated the complexities of spatial variation in birds, and nearly all workers have confronted the difficulty of expressing these patterns of variation within the framework of conventional Linnaean classification. Because the International Code of Zoological Nomenclature recognizes only taxa of specific and subspecific rank, workers have typically found two taxonomic levels inadequate in actual systematic practice. Further complicating matters are not only different philosophies about what species and subspecies should mean but also how taxonomic rank should be assigned. Until recently a prevailing attitude has been that *somehow* all geographic variation must be expressed within classifications, and this has led to many taxa (especially subspecies) being defined arbitrarily. This may be one of the contributing factors to the decline in the perceived importance of the subspecies over the last several decades (see especially Wilson and Brown, 1953). As Mayr notes:

> The taxonomist is an orderly person whose task it is to assign every specimen to a definite category (or museum drawer!). This necessary process of pigeonholing has led to the erroneous belief among nontaxonomists that subspecies are clear-cut units ... Such situations exist occasionally ... But subspecies intergrade almost unnoticeably in nearly all cases in which there is distributional continuity (Mayr, 1942, p. 106).

Two systematic practices, therefore, have been particularly fundamental in shaping current attitudes toward the analysis of taxonomic differentiation. One is the long-standing ambivalence over the ontological status of subspecies. Are they merely subjective partitions of continuous variability, i.e., a taxonomic convenience, or do they represent real units of evolution? Logically, they cannot be both. For those viewing subspecies as units of evolution, i.e., as incipient species, there is an inclination to consider them as discrete taxa. But if subspecies are envisioned as mere tools of the descriptive taxonomist without any necessary connection to speciation, then they can become convenient, often arbitrary, designations of perceived patterns of variation. Because of the bewildering array of patterns of variation found in birds, it is not unusual to see systematists adopting both views.

The most influential book on speciation analysis was certainly Mayr's *Systematics and the Origin of Species* (1942). At one point he clearly assigns evolutionary status to subspecies:

> Geographic speciation is thinkable only, if subspecies are incipient species. This, of course, does not mean that every subspecies will eventually develop into a good species. Far from it! All this statement implies is that every species that developed through geographic speciation had to pass through the subspecies stage (p. 155).

That there is an evolutionary continuum, expressed in terms of taxonomic rank, from isolated population to subspecies to species, and then to genus, is a reflection of Mayr's transformationist view of taxonomic differentiation (and his classificatory philosophy; see below); evolution generally tends to be slow and gradual (Mayr, 1940, 1942, p. 159 et seq.; Mayr, 1963, p. 24; see also below). Yet this attitude also permits the acceptance of subspecies as being entirely subjective:

> Every subspecies that was ever carefully analyzed was found to be composed of a number of genetically, i.e., phenotypically, distinct populations. It is, in many cases, entirely dependent upon the judgment of the individual taxonomist how many of these populations are to be included in one subspecies. The limits of most subspecies are therefore subjective ... (Mayr, 1942, p. 106).

In his later writings, Mayr's attitude toward the meaning of subspecies seems to harden, and they become taxonomic conveniences rather than a means to investigate taxonomic differentiation:

> The difficulties of the subspecies concept are intensified by persistent attempts to consider the subspecies not merely as a practical device of the taxonomist, but also as a "unit of evolution" ... the subspecies, which conceals so much of the inter- and intrapopulational variability, is an altogether unsuitable category for evolutionary discussions; the subspecies as such is not one of the units of evolution (1963, pp. 347–348; see also p. 349).

> The primary use of subspecies is as a sorting device in collections, that is, as an index to populations that differ from each other "taxonomically" (1982, p. 595).

Opinions among other ornithologists also vary widely, and although some agree with Mayr that subspecies are primarily taxonomic conveniences, most recent commentary suggests a growing desire to see the subspecies concept applied to discrete taxonomic entities having status as evolutionary units (Gill, 1982; Barrowclough, 1982; Lanyon, 1982; Johnson, 1982; Zusi, 1982; Monroe, 1982; O'Neill, 1982). Sympathy for subspecies as pigeon-holing devices seems to be waning.

The second major factor influencing speciation analysis is the practice of ranking geographical isolates according to their degree of differentiation. Evolution is thus taken to be relatively gradual, with the degree of differentiation paralleling the age of the taxon (Mayr, 1942, pp. 158–167, 173–176, 218–219; Mayr and Phelps, 1967, p. 290). Ranking of these isolates is determined solely by a subjective assessment of the degree of differentiation. Have the isolates differentiated sufficiently to be classified as a distinct species?

It is customary now within much of vertebrate systematics, and certainly within ornithology, to allocate forms (allopatric or parapatric) showing low levels of differentiation to taxa of subspecific rank. Well-differentiated allopatric forms are generally given the rank of species and often are placed in a superspecies along with their hypothesized close relatives (Mayr, 1942; Amadon, 1966).

Although everyone would probably agree that the amount of differentiation must be related to age to some extent, to adopt this criterion for interspecific comparisons either within a genus or between genera necessitates the assumption that evolutionary rates among these taxa are equal. Few, if any, systematists would suggest that evolutionary rates of phenotypic characters are equal. Indeed, all would surely agree that the presence of primitive and derived characters in a taxon is *prima facie* evidence for unequal rates. Yet, the underlying assumption of equality permeates discussions of taxonomic differentiation within the ornithological literature, from those about the age of isolates on islands (Mayr, 1942, pp. 158–162; Diamond and Marshall, 1977) or mountains (Mayr and Phelps, 1967) to the question of analyzing the relative ages of double invasions (Keast, 1961, pp. 396–398; Mayr, 1963, pp. 504–506).

At a theoretical level, at least, the errors that can be introduced by the assumption of equal evolutionary rates are easy to appreciate, but basically it confounds our attempts to understand the genealogical relationships of the taxa being studied. The assumption logically demands that the degree of similarity among taxa correlate with their recency of common ancestry. One could not claim, of course, that the workers cited above believe rates of character evolution are equal, and in fact many have explicitly denied this. Yet, the widespread adoption of equality in rates as an underlying working assumption cannot be denied. One major consequence of having taxonomic rank based on the degree of differentiation is that one cannot then use these systematic data to investigate the relative ages of lineages or faunas, two extremely important problems within speciation analysis and historical biogeography. An alternative approach to species is therefore essential.

2.3.2. The Phylogenetic Species Concept

In order to study evolution, we must have a hypothesis about the identity of those entities thought to be evolving. And, if we want to develop theories about that evolution, then these entities must be named because scientific theories are class statements about processes acting on named entities (Ghiselin; 1974; Hull, 1976, 1977, 1978, 1980). The basic taxonomic units of evolution are those populations characterized by one or more evolutionary novelties, and, given the structure of Linnaean classifications, it follows that these basic taxonomic units have to be provided names with specific or subspecific rank. Most present approaches to the species question are in need of revision, because within the context of the BSC we do not now have a scientifically defensible alliance between classification theory on the one hand and evolutionary analysis on the other. Thus, if it is argued that subspecies should not be considered units of evolution, as recommended by some systematists, then we are left with only taxa of species-rank to designate those units. But this logically requires that the BSC as it is now conceived and applied be abandoned. A "biological species" simply does not refer to a single unit of evolution in many cases.

A solution to the above dilemma is forthcoming when we realize that a species concept is best formulated from the perspective of the *results* of evolution rather than from one emphasizing the processes thought to produce those results. Biologists have now come to believe that these processes are highly variable and often depend upon the group being studied. To have a number of species concepts, each possibly applying to a different group, obscures the potential discovery of common phylogenetic and evolutionary patterns from one taxon to another. The results of evolution appear to be more or less the same in all groups. Evolution produces taxonomic entities, defined in terms of their evolutionary differentiation from other such forms. These entities should be called species. By emphasizing the results of evolution, i.e., differentiated taxonomic units (species), comparisons among diverse groups of organisms become possible, even when different "processes" are thought to have produced those species. Thus, by defining species in terms of the resulting *pattern*, it allows us to investigate these processes, unbiased by a species concept that is derived from our preconceptions of those processes.

Accordingly, one possible definition of a species might be: *A species is the smallest diagnosable cluster of individual organisms within which there is a parental pattern of ancestry and descent.*

This *phylogenetic species concept* is not significantly different in content from those recommended by Eldredge and Cracraft (1980, p. 92) or Nelson and Platnick (1981, p. 12). It differs, however, from the definition of Eldredge and Cracraft (1980) in eliminating reference to reproductive disjunction from other species-level taxa (see below). Although most species will be defined by uniquely derived characters, this cannot be a component of a species definition (Rosen, 1978, 1979), otherwise it would not be possible to recognize ancestral species, which must have primitive characters relative to their descendants (see Eldredge and Cracraft, 1980, Chapter 4). Species possess, therefore, only unique combinations of primitive and derived characters, that is, they simply must be diagnosable from all other species. Finally, as recognized by all workers, diagnostic characters must be passed from generation to generation, and must be taken to define a reproductive community, not simply males, females, parts of life cycles, or morphs. Hence, a species definition must include some concept of "parental ancestry and descent." This does not mean, however, that such a definition is predicated on reproductive *disjunction* as is the BSC, but only on an acknowledgment that *all* species definitions must have some notion of reproductive cohesion *within* some definable cluster of individual organisms.

The phylogenetic species concept has important advantages over the BSC, including:

1. The known diagnosable taxonomic units are by definition equivalent to the known evolutionary units. There no longer exists the problem of evolutionary units being arbitrarily ranked as either subspecies or as species. Consequently, evolutionary problems within groups are clarified, and two of the major questions of speciation analysis, What are the units of evolution? and, What are their relationships? are not encumbered by a subjective approach to taxonomic ranking.

2. The concept clarifies the distinction between recognizing species-taxa and the analysis of geographic variation. Because species are

now defined in terms of diagnostic characters, their taxonomic status is seen to be independent of the patterns of variability that might be observed for nondiagnostic characters. Hence, incongruent patterns of clinal variation exhibited by different characters should no longer be considered a serious taxonomic problem (see Wilson and Brown, 1953, pp. 100–102). Because the phylogenetic species concept focuses attention on patterns of taxonomic differentiation, the concept thereby places a new perspective on the analysis of variation within a species. It is no longer necessary to interpret such variation from a taxonomic point of view. Because it is variation within the smallest taxonomic unit, it has no immediate taxonomic relevance. Instead, emphasis can be placed on investigating the dynamics of that variation, including elucidating the question of how character change takes place within populations.

3. The phylogenetic species concept clarifies the status and systematic role of subspecies. Subspecies cannot have ontological status as evolutionary units under a phylogenetic species concept. While it therefore can be argued that this renders subspecies superfluous for systematic and evolutionary analysis, their continued use might be thought defensible under some circumstances. If one so chose, subspecies names could be applied to populations showing clinal variation, and subspecies boundaries could then be determined by sharp gradients in character variation. In this context, however, subspecies are merely descriptors of variation seen in sometimes subjectively chosen, nondiagnostic characters and do not represent taxa having independent ontological status. Because of this, I would recommend that subspecies names not be used. Only objective taxonomic entities should be classified.

4. A major advantage of the phylogenetic species concept is that it places a new interpretation on the question of reproductive isolation, an interpretation not shared with any other species concept. The degree of reproductive isolation observed or presumed to exist between populations has always presented systematists with difficulties because (1) species are said to be defined in terms of reproductive disjunction, (2) species, however, are recognized only rarely by reproductive criteria, and (3) biologists have long realized that a definition based on reproductive criteria often obscures the pattern of differentiation in plants and many groups of animals. Unlike the BSC and the evolutionary species concept (Wiley, 1978, 1981), the phylogenetic species concept does not use data on reproductive isolation, e.g., hybridization, in recognizing species taxa. Species are recognized strictly in terms of their hypothesized status as diagnosable evolutionary taxa, which itself is revealed by shared character distributions within and between populations. Thus, even if two sister-taxa broadly hybridize, both can still

be considered to be species if each is diagnosable as a discrete taxon (of course, it may not be possible to assign hybrids to one or the other species). The critical point is that both species have had a distinct phylogenetic and biogeographic history prior to hybridization, and the phylogenetic species concept merely acknowledges that history.

The phylogenetic species concept does not deny the importance of reproductive cohesion or disjunction when discussing ideas about the evolutionary process. It simply claims that incorporating reproductive criteria in a species concept, and using those criteria to determine the taxonomic status of populations, not only obscures the analysis of historical pattern but also impedes our understanding of the reproductive relationships themselves. Patterns of hybridization among taxa are discernible only when the taxa are defined independent of that hybridization and when we have a phylogenetic hypothesis for all those diagnosable taxonomic units (see Nelson and Platnick, 1981).

Because the phylogenetic species concept emphasizes diagnosable taxa rather than those delimited by reproductive relationships, it will have broad applicability in botany and zoology alike. Rather than focusing on the processes producing taxa (Raven, 1976, p. 293), we first need to identify the *results* of these processes (taxa), determine their interrelationships, and then attempt to decipher the possible causes for their phylogenetic and biogeographic pattern.

5. Another advantage of the phylogenetic species concept is that it directs more attention to the geographical history of species than does the BSC (discussed in more detail in the next section). The phylogenetic concept demands the recognition of all differentiated taxonomic (evolutionary) units, and as a consequence also leads us to ask where they are distributed, in other words, where they are endemic. The BSC, on the other hand, does not recognize all evolutionary taxonomic units, which might mean that some areas of endemism are not recognized, or only incompletely so. And, if this is the case, it is very easy to argue that common patterns of speciation will be hidden or lost altogether (next section). The phylogenetic species concept avoids these difficulties.

6. A phylogenetic species concept places a strong emphasis on character analysis, in particular, the search for diagnostic characters. Much of present-day speciation analysis is concerned with the quantitative description of variation at the expense, sometimes, of delimiting taxa on the basis of discrete characters. Both approaches are necessary, but an interpretation of quantitative variation would seem possible only within the context of a hypothesis about species limits.

7. Finally, because the phylogenetic species concept seeks to identify all evolutionary taxonomic units, a much more accurate assessment of intra- and intercladal diversity patterns can be obtained than with

the BSC. Many questions in species diversity analysis, or in the study of macroevolutionary patterns (Eldredge and Cracraft, 1980), depend upon having a measure of the numbers of evolutionary taxa. It is clear that the BSC can greatly underestimate the numbers of these evolutionary units, whereas such will not be the case with a phylogenetic species concept.

Given these advantages we might ask what might be some potential difficulties with adopting a phylogenetic species concept. Clearly, one of the outcomes would be the recognition of many more species-level taxa. To some ornithologists, having a large number of species is apparently a disadvantage. Mayr (1969, p. 38, 1982, p. 290) believes, for example, that the BSC "led to a great clarification and simplification" as the numbers of avian species were reduced in various groups over the past 60 years or so. Precisely why avian species taxonomy should be considered "clarified" with this reduction has never been explained in detail, but I believe this perception is more illusory than real. We certainly do not have a better understanding of the numbers of evolutionary units as a consequence of the BSC. If we know more about the limits of avian taxa, surely it is merely a result of increased systematic work. Moreover, the number of taxonomic names in ornithology has not been reduced significantly, for most of those taxa no longer recognized as species are now named subspecies. And given the fact that some systematists (Mayr, 1969, 1982) do not believe subspecies should be units of evolution, then the argument could be made that present species taxonomy is actually less "clarified," at least as far as that taxonomy reflects the results of evolution.

For conventional systematic practice within ornithology, the phylogenetic species concept should not increase the numbers of taxa already recognized. In general, the major effect will be to elevate some subspecies to species. In fact, depending upon one's philosophy toward the recognition of subspecies, under the phylogenetic species concept the total number of specific and subspecific taxa might actually decrease in some instances. This will certainly be true if efforts are made to eliminate subspecies based on subjectively chosen quantitative variation.

Why ornithologists should be uncomfortable with a family having 125 species under a phylogenetic species concept compared to that family having, say, 85 species under the BSC (to take an arbitrary example) is not clear. If those 125 species represent our best estimate of the number of real evolutionary taxa in the family, then 40 additional species seems a small price to pay, especially when most, if not all, of those taxa will already have been described and have valid names available.

The question will naturally arise about potential situations in which small isolated populations or demes can be defined by discrete biochemical characters. Are these groups to be treated taxonomically as distinct species? One approach to answering this question is first to make clear the structure of the pattern of variation. Are there really taxonomic characters diagnostic of these populations, and are they congruent or incongruent with other biochemical characters? If the populations are truly distinct, do they also possess a spatial unity that is distinct from other such groups? Finally, are we interested merely in describing these populations, or are we interested in their phylogenetic and distributional history? If the latter, then it seems clear that distinct units of evolution must be delimited, and naming them would not necessarily be inappropriate.

There is no evidence at present to conclude that cases such as the preceding are at all common in birds (Barrowclough, 1983; personal communication). Indeed, local populations of birds do not seem diagnosable by present methods of genetic analysis. The conclusion therefore seems to be that presently known patterns of variation within avian populations are compatible with the phylogenetic species concept.

One, possibly unwelcome, outcome of adopting a phylogenetic species concept will be the necessity of revising much of our current species-level systematics. It is difficult to view this as a disadvantage to ornithology in that the evolutionary status of the species and subspecies of different groups would be re-evaluated.

In summary, the phylogenetic species concept would be beneficial to ornithology because it places emphasis on the description and recognition of evolutionary taxonomic units. It would help bring species-level systematics of birds more in line with the practices of systematic botany and zoology as a whole, where tendencies to identify diagnosable taxa as species have always predominated thinking (Rosen, 1978; Cronquist, 1978). Finally, a phylogenetic species concept would encourage a more rigorous approach toward the phylogenetic and biogeographic analysis of avian diversification.

3. SPECIATION ANALYSIS

3.1. Introduction

In ornithology, the methodology of speciation analysis has lain almost exclusively within the research program of evolutionary systematics (Mayr, 1942, 1969). It has adopted the BSC, thereby emphasizing polytypic species and superspecies, and has viewed speciation primarily in terms of the ecological and genetic processes presumed to result in reproductive isolation. One could perhaps find no better characterization of this research program than that given by Vuilleumier (1980, p. 1298), who listed the following protocol:

1. Analyze population structure; document isolates and secondary contact zones.
2. Review literature on Plio-Pleistocene climatic and vegetational history; from this "reconstruct a spatio-temporal sequence of eco-geographical events."
3. Correlate these events with the postulated course of speciation.
4. Assign relative or absolute ages to the events of speciation "as a function of the amounts of differentiation of populations."
5. Describe the best speciation sequence.

Vuilleumier (1980, pp. 1298–1299) also identified four assumptions found in most speciation papers: (1) relative amounts of morphologic and genetic differentiation are correlated with each other, (2) the amount of differentiation is proportional to age, (3) competition often determines the distribution of closely related species, and (4) following sympatry, competition often leads to character displacement. Although some might question details of Vuilleumier's protocol and assumptions, an examination of the literature would show that his is an accurate assessment (see Mayr, 1942, 1963, 1969; Keast, 1961; Selander, 1971; Haffer, 1969, 1974; Mayr and Short, 1970). Yet another observation could be made. With few exceptions (notably Keast, 1961; Haffer, 1974), speciation analyses within birds have been noncomparative. Thus, most analyses attempt to reconstruct the speciation history of single groups (usually genera, occasionally families) rather than compare patterns of two or more groups sharing common areas of distribution. Studies of single groups are needed, but as will be discussed shortly, there are limits to the kinds of historical inferences that can be made when studies are not comparative.

The discussion of species in the preceding section partly exposes the conceptual interaction between viewpoints about the ontological status of species and the ways in which species are thought to evolve. And, within contemporary speciation analysis, the intellectual tension between these two has been especially apparent. The BSC, for instance, has obviously influenced methods of studying taxonomic differentiation. Witness the emphasis on the description of reproductive relationships among taxa. Likewise, one of the major goals of most speciation analyses has been to decide whether taxa are "good" species or not. I believe the adoption of the BSC has been an impediment to speciation analysis, and the problems inherent in the BSC, which have been documented here and by many other workers, should lead us to re-examine the goals and methods of contemporary speciation analysis. The questions we seek to ask of these investigations need to be reformulated, and alternative methods to answer these new questions need to be explicated. It is the purpose of this section to explore some of these questions and methods.

Some major questions about taxonomic differentiation can be derived from four established observations:

1. The organisms of nature can be clustered into diagnosable taxonomic units (called species).
2. These species have a hierarchical relationship with one another, as evidenced by congruence in their shared derived characters.
3. Each species is endemic to an area.
4. These areas of endemism are often seen to be congruent from group to group.

These observations pose a series of parallel questions, and taken together they form the core of a research program for speciation analysis: What species are there?, What are the interrelationships of these species?, Which areas of endemism exhibit a significant degree of congruence?, and What are the historical interrelationships of these areas? Such questions certainly do not exhaust the subject matter of taxonomic differentiation, but without some answers to each of them we cannot expect to understand the speciation of any group or be able to pursue other questions in any important detail. Unfortunately, except perhaps for the first of these questions, none have figured very prominently (if at all) in the current primary literature on speciation analysis (Mayr, 1942, 1963, 1969; Cain, 1954; Grant, 1971; Bush, 1975; White, 1978). This is not to say, of course, that problems of phylogenetic relationships, or of endemism, have not received attention from systematics as a whole or from ornithology in particular. Nevertheless, their significance for speciation analysis, in ornithology as well as in other disciplines, has not been entirely appreciated.

Why do the above questions form the core of speciation analysis? The answer lies in the observations themselves. Taxonomic differentiation appears to be a spatial phenomenon, i.e., all taxa arise in a restricted area, and the geographical histories of different groups have elements in common, that is to say, they exhibit congruence. Accordingly, there are two ways to approach speciation studies. First, either by ignoring that congruence, in which case we can focus our attention on one group, define its taxa, formulate hypotheses about their interrelationships, and then interpret these results geographically in terms of earth history, climate, and/or ecology. Or we can investigate that congruence by directing attention to common patterns of spatial history across taxa and likewise interpret these results. Both approaches are necessary, but a comparative analysis of speciation should be our primary objective. Only through comparison can knowledge about the common elements of pattern shared from group to group be obtained, and thus only through comparison can those components of pattern that are unique to each group be isolated. It is the primary task of science to reveal and explain common patterns calling for general explanations. Unique events of history experienced by individual groups, if those events are considered of interest at all, must be explained by causes less general. Furthermore, it is likely that explanations based on non-comparative studies will require revision once they are placed in a comparative context.

3.2. How Are Areas of Endemism Determined?

As used here, areas of endemism are hypotheses about areas of origin. The area of endemism for a species is determined by mapping its known distribution, both that of the present and of the past. Because the area of origin itself is a changing entity (most species do not appear to arise spontaneously but over some period of time), areas of endemism, in that they represent those areas of origin, are abstractions. Fossils can rarely, if ever, help us locate the area of origin, because they are themselves nearly always distributed in only part of the range of a species, and thus they give us only a point in space and time. Fossils are, therefore, primarily useful in widening the known area of endemism. Whether the area of endemism is the same as the area of origin never can be determined for any species. Nevertheless, once the geographic ranges of many species sharing roughly the same area are known, it is possible to identify that which is unique in the distribution of each species (see below).

Many birds exhibit long-distance, seasonal migrations. What is their area of endemism, the breeding area, the non-breeding area, or both? Until we understand the history of migratory movements, we may not be able to answer this question. Nevertheless, upon examining these distributions, the breeding areas of species exhibit a marked degree of congruence, and the little data that are available suggest these areas have had common histories as well (Mengel, 1964). The breeding ranges, therefore, seem more important as estimators of area of origin than do nonbreeding ranges (especially in those species in which the two ranges are very widely separated), but this is a problem needing much more investigation.

Biologists have known for a very long time that areas of endemism exhibit congruence (see historical review in Nelson, 1978; Nelson and Platnick, 1981). To say that two or more species exhibit congruence does not mean their ranges are precisely the same. Each species has its own autecological characteristics and so ranges would not be expected to coincide exactly (except perhaps in cases in which obligatory coevolution might produce such coincidence). Congruence, then, simply means that species share some, usually significant part of their distributions. General areas of endemism are also abstractions of sorts, because determining the general (commonly shared) area of endemism will depend upon which species are examined and their distributions. Nevertheless, constructing hypotheses about general areas of endemism has not proved overly difficult for biologists.

What is the significance of these general areas of endemism? It is often suggested they represent coherent ecological entities, but this is true only when the ecological characteristics of species are described in superficial terms. Virtually all general areas of endemism are highly diverse ecologically, and species endemic in those areas are typically unlike one another in their detailed ecologies. General areas of endemism signify much more than common ecology, they signify common history. Areas of endemism represent biotas that are separated from other such biotas by geological barriers and/or by climatic–environmental extremes. These areas have significance for speciation analysis because the species inhabiting them comprise a biota and might be expected to share a common history of differentiation.

3.3. How Is the History of Areas of Endemism Determined?

The methods used to reconstruct the history of areas of endemism have been thoroughly discussed in the literature (Platnick and Nelson, 1978; Rosen, 1978; Nelson and Platnick, 1981; Wiley, 1980, 1981; Patterson, 1981; Cracraft, 1982, 1983), and it is therefore unnecessary to repeat them in detail here. Instead, a brief case study will be described to illustrate the relevance of a historical analysis of areas of endemism for deciphering avian speciation patterns.

The example concerns avian speciation patterns in northern and eastern Australia (Cracraft, 1982, 1983). The method of analysis will be comparative, emphasizing the history of areas of endemism, and the results will then be used to identify and interpret some unique aspects of the speciation events in the taxa being studied.

In northern and eastern Australia, and in New Guinea, a number of well-defined areas of endemism can be recognized (Fig. 1). The question is whether the biotas of these areas have any general pattern of interrelationships with each other. Such a pattern would be an expression of the congruence observed in the phylogenetic relationships of different clades distributed in these areas. Accordingly, the phylogenetic relationships of some of these clades must be investigated (using methods outlined in Eldredge and Cracraft, 1980; Nelson and Platnick, 1981; Wiley, 1981).

Figure 2 presents phylogenetic hypotheses for four lineages each with species endemic in three or more of these areas (detailed systematic data bearing on these hypotheses will be presented elsewhere, all differentiated forms are here treated as phylogenetic species): (A) some grassfinches (*Poephila*) with six species (see also Goodwin, 1982, pp. 221–230; Cracraft, 1983), (B) three species of wrens in the genus *Malurus*, (C) four species of robins, *Eopsaltria* (*Tregellasia*), and (D) four species of rifle-birds, genus *Ptiloris*.

Assuming the correctness of the phylogenetic hypotheses of Fig. 2, what information do they convey about the interrelationships of the areas of endemism? Based on these four clades, a single general area-cladogram can be hypothesized (Fig. 3). The simplest explanation of the area-cladogram itself is that a biota, once widespread in northern and eastern Australia and in New Guinea, became progressively subdivided (discussed below). Because these avian lineages represent components of this biota, and have relationships exhibiting congruent clad-

FIGURE 1. Areas of endemism for the birds of northern and eastern Australia (Keast, 1961; Ford, 1978; Cracraft, 1982). These areas include: (1) a moist forest track in coastal southeastern Queensland–eastern New South Wales, (2) the Cairns-Atherton rainforest area, (3) the Cape York Peninsula (the historical relationships between the northeast corner, here designated 3a, and the remainder of Cape York is uncertain), (4) the Arnhem Land plateau, and (5) the Kimberley plateau. For simplicity, New Guinea is considered as a single area of endemism (area 6), which is clearly not the case.

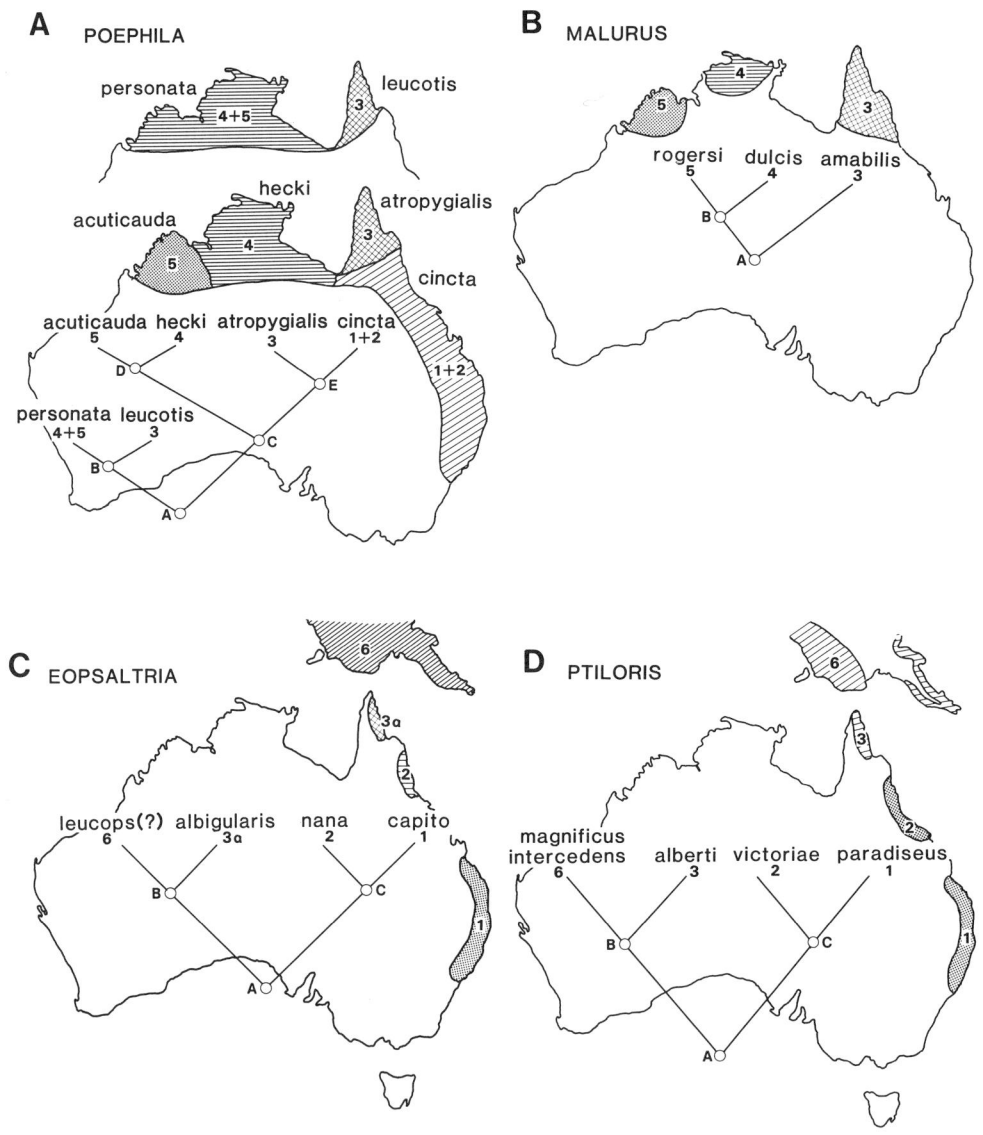

istic pattern, they can be hypothesized to have shared a common history. Knowledge of this history is necessary if we are to explain the speciation patterns of individual taxa in any detail. The genus *Poephila* is a case in point. If a study had been restricted to the grassfinches, the pattern obtained in Fig. 2A would not be informative about the history of areas 1, 2, 3a, or 6. Based on comparisons with other clades, however, we might conclude that *P. leucotis*, *P. atropygialis*, and *P. cincta* failed to differentiate in these areas, whereas other taxa did respond to geologic and/or climatic events. In the absence of a comparison, we would not have an indication that the evolutionary history of *Poephila* is different in this regard. Consequently, we would not be led to ask why species of *Poephila* might be less responsive to isolating barriers than are some other species.

3.4. How Might General Area-Cladograms Be Explained?

To the extent that the history of differentiation of a group is congruent with the history of other groups, then an explanation for the

---

FIGURE 2. Phylogenetic hypotheses for the (phylogenetic) species of four genera having distributions in the areas of endemism shown in Fig. 1. (A) *Poephila*, postulated derived characters include: A, brown back, breast and belly buff, flanks with black band; B, yellow bill, black face pattern; C, extensive black throat, gray head; D, tail elongated, pointed; and E, bill black. Diagnostic characters of the species include: *personata* (face without white patch), *leucotis* (face with white patch), *acuticauda* (bill pink to red), *hecki* (bill yellow), *atropygialis* (upper tail coverts black), *cincta* (upper tail coverts white). (B) *Malurus*, postulated derived characters include: A, female with bluish back; B, males with lavender patch on flank. Diagnostic characters of the species include: *rogersi* (in females, lores and feathers near eye, white), *dulcis* (in females, lores and feathers near eye, chestnut), *amabilis* (males without lavender patch). (C) *Eopsaltria*, postulated derived characters include: A, small size, short broad bill; B, face white, head black; C, breast and belly pale yellow. Diagnostic characters of the species include: *leucops* (no white eye-ring), *albigularis* (white eye-ring), *nana* (lores and eye-ring buff), *capito* (lores and eye-ring white). These species are often placed in the genus *Tregellasia*, which may prove more closely related to *Poecilodryas* than to *Eopsaltria*, species limits within "*albigularis*" and "*leucops*" are uncertain (to be discussed elsewhere), and "*albigularis*" may also occur in New Guinea where "*leucops*" is restricted. (D) *Ptiloris*, postulated derived characters include: A, males with green, blue, or blue-green crown, throat, and upper breast, and tail black with blue-green central feathers; B, in males, flank plumes filamentous, females vermiculated below; C, in males, feathers of abdomen and flanks black but edged with green. Diagnostic characters of the species include: *magnificus-intercedens* (in males, underparts with pink-purple gloss), *alberti* (in males, underparts with little or no pink-purple gloss), *victoriae* (much smaller than *paradiseus*), *paradiseus* (much larger than *victoriae*). Note: species taxa are not required to be characterized by derived characters (see text).

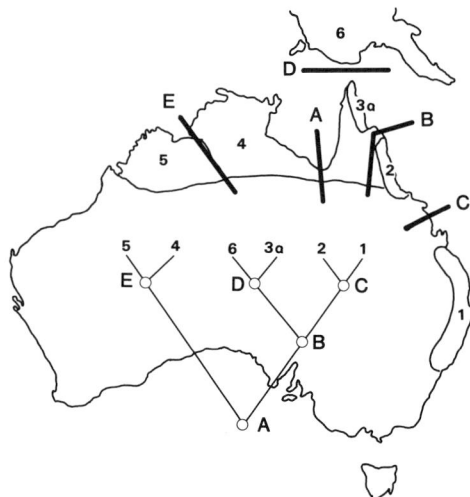

FIGURE 3. The general area-cladogram for the six areas of endemism shown in Fig. 1, based on the results shown in Fig. 2. Thick bars refer to geographic isolating events (see text).

general area-cladogram is also an explanation for those individual groups. In the preceding example it is evident that a remarkable amount of congruence exists, therefore developing a general explanation becomes especially important.

How might an area of endemism arise? Two answers can be readily suggested. First, a widespread biota can be subdivided into two or more units by development of geographic barriers, with subsequent differentiation defining them as areas of endemism (a vicariance explanation). A second possibility is that repeated dispersal across pre-existing barriers, already separating two areas, could lead to differentiation and an area of endemism (a long-distance dispersal explanation). There seem to be few other potential explanations. As Patterson (1981, p. 270) has expressed it, either the biota existed before the area of endemism did (vicariance) or the area existed before the endemic taxa (dispersal across a barrier). A choice between these two explanations in any given instance will depend upon the nature of the pattern itself. If patterns are highly congruent as in the Australian example, then vicariance of a widespread biota is the simplest explanation. If patterns are not highly congruent and if only a few taxa are endemic in the area in question, then perhaps those endemics arose following dispersal across a barrier.

Dispersalist biogeography has traditionally played a major role in avian speciation analysis. Part of this stems from the belief that birds, being winged creatures, are capable of reaching any isolated area given sufficient time. Dispersalist thinking also is characteristic of those methodologies which have not stressed the need for precise phylogenetic hypotheses and the search for congruence in the historical patterns of different groups. Moreover, the complex nature of dispersal has not always been understood, especially by those reacting negatively to vicariance explanations. Hence, the two major kinds of dispersal, range-expansion and long-distance, are not usually distinguished when vicariance biogeographers are incorrectly accused of dispensing with dispersal altogether. That species disperse and thereby increase or maintain their ranges is well known. Range-expansion dispersal is a phenomenon of populations, governed by the internal population dynamics of the species, by habitat availability, and by physical parameters of the environment. Long-distance dispersal, in contrast, is almost always a fortuitous event, typically involving very few individuals. One would not expect, therefore, that long-distance dispersal would play a significant role in establishing an area of endemism, particularly if (1) that area housed endemics from a broad taxonomic spectrum, and (2) the endemics of that area exhibited a high degree of congruence in their relationships to taxa endemic in other areas. This is not to say that long-distance dispersal might not establish an area of endemism, but demonstrating this in any instance is difficult and requires a phylogenetic analysis of each clade with endemic species in the area in question (Platnick and Nelson, 1978; Nelson and Platnick, 1981).

The areas of endemism in northern and eastern Australia are all defined by geologic or ecologic (climatic) barriers (Fig. 3). We can therefore postulate a historical sequence for the origin of these areas: (1) the northern area (4 + 5) was isolated from the eastern area (1 + 2 + 3 + 6), probably as a result of aridification in the region of the Gulf of Carpentaria (barrier A), (2) subsequent to this, areas 4 and 5 were isolated from each other by the development of the Victoria and Daly river valleys and by a more arid climatic regime (barrier E; see Ford, 1978), (3) also subsequent to barrier A, the Cape York Peninsula and apparently New Guinea also were separated from eastern Australia (barrier B), possibly as a result of changes in topographic relief and aridification of the climate, (4) following barrier B, New Guinea and the Cape York Peninsula were isolated by a rise in sea-level (barrier D), and (5) following barrier B, areas 1 and 2 were separated by the origin of more arid, savannah-like vegetation (Keast, 1961; barrier C). Data on the ages of these isolating barriers are not available, but when they are they can be used to date the origin of the areas and establish the maximum ages of their endemics.

3.5. Conclusions

This example suggests that one cannot expect to reconstruct accurately the history of speciation for a single group unless comparisons are made directly with historical patterns of other taxa. A comparative analysis is essential in order to reveal (1) whether areas of endemism are congruent and to what extent they differ from other species, (2) by the analysis of phylogenetic relationships, whether the areas of endemism are related in a nonrandom pattern, and (3) whether the origin of congruent areas of endemism can be explained by common geologic or ecologic factors. Moreover, one cannot recognize whether dispersal has led to endemism until it can be shown that a particular endemic is not part of a more general vicariance pattern.

Comparative studies are also necessary if we want to compare rates of differentiation from one group to the next (Cracraft, 1981, 1982). The general area-cladogram is a hypothesis about the relative ages of areas of endemism. As a consequence, taxa endemic in the same area can be hypothesized to be the same age. Only in this way can rates of differentiation be studied. The traditional assumption that age is correlated with the degree of differentiation (and usually taxonomic rank) should be abandoned. It can lead to spurious conclusions about the ages of taxa and areas of endemism and cannot serve as a methodological yardstick to compare evolutionary rates. Inasmuch as avian systematists have barely begun to define areas of endemism and understand their history, we can safely predict that speciation analysis will be an important subject for continuing research in ornithology.

ACKNOWLEDGMENTS. I wish to thank David Hull, V. A. Funk, N. K. Johnson, Richard F. Johnston, and Norman Platnick for their helpful comments on the manuscript. I am also grateful to the National Science Foundation (through grant DEB79-21492) for supporting this research.

REFERENCES

Amadon, D., 1966, The superspecies concept, *Syst. Zool.* **15**:245–249.
Barrowclough, G. F., 1982, Geogeographic variation, predictiveness, and subspecies, *Auk* **99**:601–603.
Barrowclough, G. F., 1983, Biochemical studies of microevolutionary processes, in: *Per-*

spectives in Ornithology (A. H. Brush and G. A. Clark, Jr., eds), Cambridge University Press, New York.

Bush, G. L., 1975, Modes of animal speciation, *Annu. Rev. Ecol. Syst.* **6**:339–364.

Cain, A. J., 1954, *Animal Species and Their Evolution*, Hutchinson University Library, London.

Cracraft, J., 1981, Pattern and process in paleobiology: The role of cladistic analysis in systematic paleontology, *Paleobiol.* **7**:456–468.

Cracraft, J., 1982, Geographic differentiation, cladistics, and vicariance biogeography: Reconstructing the tempo and mode of evolution, *Am. Zool.* **22**:411–424.

Cracraft, J., 1983, Cladistic analysis and vicariance biogeography, *Am. Sci.* **71**:273–281.

Cronquist, A., 1978, Once again, what is a species? *Beltsville Symp. Agric. Res.* **2**:3–20.

Diamond, J., and Marshall, A. G., 1977, Niche shifts in New Hebridean birds, *Emu* **77**:61–72.

Ehrlich, P. R., 1961, Has the biological species concept outlived its usefulness? *Syst. Zool.* **10**:167–176.

Eldredge, N., and Cracraft, J., 1980, *Phylogenetic Patterns and the Evolutionary Process*, Columbia University Press, New York.

Ford, J., 1978, Geographical isolation and morphological and habitat differentiation between birds of the Kimberley and the Northern Territory, *Emu* **78**:25–35.

Ghiselin, M. T., 1974, A radical solution to the species problem, *Syst. Zool.* **23**:536–544.

Gill, F. B., 1982, Might there be a resurrection of the subspecies? *Auk* **99**:598–599.

Goodwin, D., 1982, *Estrildid Finches of the World*, Cornell University Press, Ithaca, New York.

Grant V., 1971, *Plant Speciation*, Columbia University Press, New York.

Haffer, J., 1969, Speciation in Amazonian forest birds, *Science* **165**:131–137.

Haffer, J., 1974, Avian speciation in tropical South America, *Publ. Nuttall Ornithol. Club* **14**:1–390.

Hull, D. L., 1976, Are species really individuals? *Syst. Zool.* **25**:174–191.

Hull, D. L., 1977, The ontological status of species as evolutionary units, in: *Foundational Problems in the Special Sciences* (R. Butts and J. Hintikka, eds.), D. Reidel, Dordrecht-Holland, pp. 91–102.

Hull, D. L., 1978, A matter of individuality, *Phil. Sci.* **45**:335–360.

Hull, D. L., 1980, Individuality and selection, *Annu. Rev. Ecol. Syst.* **11**:311–332.

Johnson, N. K., 1982, Retain subspecies—at least for the time being, *Auk* **99**:605–606.

Keast, A., 1961, Bird speciation on the Australian continent, *Bull. Mus. Comp. Zool.* **123**(8):303–495.

Kleinschmidt, O., 1900, Arten oder Formenkreise? *J. Ornithol.* **48**:134–139.

Lanyon, W. E., 1982, The subspecies concept: Then, now, and always, *Auk* **99**:603–604.

Levin, D. A., 1979, The nature of plant species, *Science* **204**:381–384.

Lewontin, R. C., 1970, The units of selection, *Annu. Rev. Ecol. Syst.* **1**:1–18.

Mayr, E., 1940, Speciation phenomena in birds, *Am. Natural.* **74**:249–278.

Mayr, E., 1942, *Systematics and the Origin of Species*, Columbia University Press, New York.

Mayr, E., 1948, The bearing of the new systematics of genetical problems: The nature of species, *Adv. Genet.* **11**:205–237.

Mayr, E., 1957a, Species concepts and definitions, in: *The Species Problem* (E. Mayr, ed.), Am. Assoc. Adv. Sci. Publ. No. 50, Washington, D.C., pp. 1–22.

Mayr, E., 1957b, Difficulties and importance of the biological species concept, in: *The Species Problem* (E. Mayr, ed.), Am. Assoc. Adv. Sci. Publ. No. 50, Washington, D.C., pp. 371–388.

Mayr, E., 1963, *Animal Species and Evolution*, Harvard University Press, Cambridge.
Mayr, E., 1969, *Principles of Systematic Zoology*, McGraw-Hill, New York.
Mayr, E., 1970, *Populations, Species and Evolution*, Harvard University Press, Cambridge.
Mayr, E., 1982, *The Growth of Biological Thought*, Harvard University Press, Cambridge.
Mayr, E., and Phelps, W. H., Jr., 1967, The origin of the bird fauna of the south Venezuelan highlands, *Bull. Am. Mus. Nat. Hist.* **136:**269–328.
Mayr, E., and Short, L. L., 1970, Species taxa of North American birds, *Publ. Nuttall Ornithol. Club* **9:**1–127.
Monroe, B. L., Jr., 1982, A modern concept of the subspecies, *Auk* **99:**608–609.
Mengel, R. M., 1964, The probable history of species formation in some northern wood warblers (Parnulidae), *Living Bird* **3:**9–43.
Nelson, G. J., 1978, From Candolle to Croizat: Comments on the history of biogeography, *J. Hist. Biol.* **11:**269–305.
Nelson, G. J., and Platnick, N. I., 1981, *Systematics and Biogeography: Cladistics and Vicariance*, Columbia University Press, New York.
O'Neill, J. P., 1982, The subspecies concept in the 1980's, *Auk* **99:**609–612.
Paterson, H. E. H., 1981, The continuing search for the unknown and unknowable: A critique of contemporary ideas on speciation, *S. Afr. J. Sci.* **77:**113–119.
Patterson, C., 1981, The development of the North American fish fauna—a problem of historical biogeography, in: *The Evolving Biosphere* (P. L. Forey, ed.), British Museum (Natural History), London, pp. 265–281.
Platnick, N. I., 1977, Review of *Concepts of Species* by C. N. Slobodchikoff, *Syst. Zool.* **26:**96–98.
Platnick, N. I., and Nelson, G. J., 1978, A method of analysis for historical biogeography, *Syst. Zool.* **27:**1–16.
Popper, K. R., 1959, *The Logic of Scientific Discovery*, Harper Torchbooks Edition (1968), Harper and Row, New York.
Popper, K. R., 1972, *Objective Knowledge*, Oxford University Press, London.
Raven, P. H., 1976, Systematics and plant population biology, *Syst. Botan.* **1:**284–316.
Rosen, D. E., 1978, Vicariant patterns and historical explanation in biogeography, *Syst. Zool.* **27:**159–188.
Rosen, D. E., 1979, Fishes from the uplands and intermontane basin of Guatemala: Revisionary studies and comparative geography, *Bull. Am. Mus. Nat. Hist.* **162:**267–376.
Scudder, G. G. E., 1974, Species concepts and speciation, *Can. J. Zool.* **52:**1121–1134.
Selander, R. K., 1971, Systematics and speciation in birds, in: *Avian Biology*, Volume 1 (D. S. Farner and J. R. King, eds.), Academic Press, New York, pp. 57–147.
Short, L. L., 1969, Taxonomic aspects of avian hybridization, *Auk* **86:**84–105.
Sibley, C. G., 1957, The evolutionary and taxonomic significance of sexual selection and hybridization in birds, *Condor* **59:**166–191.
Sokal, R. R., and Crovello, T. J., 1970, The biological species concept: A critical evaluation, *Am. Natural.* **104:**127–153.
Streseman, E., 1936, The formenkreis-theory, *Auk* **53:**150–158.
Templeton, A. R., 1980, Modes of speciation and inferences based on genetic distances, *Evolution* **34:**719–729.
Vuilleumier F., 1980, Reconstructing the course of speciation, in: *Proc. XVII Int. Ornithol. Cong.* Deutsch. Ornithol-Gesells., Berlin, pp. 1296–1301.
White, M. J. D., 1978, *Modes of Speciation*, W. H. Freeman & Co., San Francisco.
Wiley, E. O., 1978, The evolutionary species concept reconsidered, *Syst. Zool.* **27:**17–26.

Wiley, E. O., 1980, Is the evolutionary species fiction?—A consideration of classes, individuals and historical entities, *Syst. Zool.* **29**:76–80.

Wiley, E. O., 1981, *Phylogenetics: The Theory and Practice of Phylogenetic Systematics,* John Wiley and Sons, New York.

Wilson, E. O., and Brown, W. L., Jr., 1953, The subspecies concept and its taxonomic application, *Syst. Zool.* **2**:97–111.

Zusi, R. L., 1982, Infraspecific geographic variation and the subspecies concept, *Auk* **99**:606–608.

CHAPTER 7

BIRD CHROMOSOMES

GERALD F. SHIELDS

1. INTRODUCTION

The study of the chromosomes of birds is inherently problematic because a majority of species possesses large numbers of microchromosomes which are difficult to identify and count. Further, the development of methods of chromosome preparation and analysis specific to birds has lagged behind that for mammals. As a result, only about 5% of the 8900 extant species of birds have been karyotyped and most of these have not been done well. The study of the chromosomes of birds is presently in transition away from the use of conventional techniques toward the use of techniques which employ a variety of procedures each of which is designed to differentially stain chromosomes in a specific way. These differential procedures have been collectively referred to as chromosome banding techniques. I (Shields, 1982) have summarized our knowledge of conventionally stained material and in the present paper address the utility of the use of tissue culture and chromosome banding procedures together with some discussion of comparative results and suggestions for future study. The organization of the avian genome and its relation to chromosome form and function has recently been reviewed (Shields, 1983).

GERALD F. SHIELDS • Institute of Arctic Biology and Division of Life Sciences, University of Alaska, Fairbanks, Alaska 99701.

2. THE DIPLOID NUMBER PROBLEM

The difficulty of accurately determining diploid numbers of chromosomes is probably the single most annoying feature of avian karyology. Since the microchromosomes are small they can be overlapped by macrochromosomes or lost altogether when they are squashed or air dried on slides. It is customary for some authors to report the detail of size and centromere position of macrochromosomes but to either ignore the microchromosomes or to attempt to identify and count them while reporting a range of number variation even for single individuals. Reliable reports of diploid number differences between species of birds are thus rare to nonexistent. I have developed a variety of methods over the years each of which helps to alleviate the difficulties of determining diploid numbers.

2.1. Utility of Cell Culture Procedures

One can debate the merits of various procedures in terms of their appropriateness for a given situation. For example, the feather pulp procedure of Shoffner et al. (1967) might be the preferred procedure for sexing young chicks in the wild where cell culture conditions are not available and when it is important that the birds not be killed. Alternatively, the blood leukocyte culture technique of de Boer (1976) might be preferred to analyze the karyotypes of birds in zoos from which blood can be taken. In the final analysis, however, there is no substitute for procedures which routinely result in large numbers of excellent spreads of chromosomes on slides which can then be analyzed in detail. In this regard chromosomes prepared from feather pulp tissue squashed on slides will almost always produce inferior results.

I have preferred to use laboratory cultures of kidney fibroblasts which can be manipulated to provide a continuous cell line from which cells can be taken periodically. While procedures using bone marrow (Lu, 1969) or blood leukocytes provide relatively large numbers of cells for analysis, they are techniques which are terminated with a single harvest of cells and are thus limited in their utility.

I typically maintain cell lines of kidney fibroblasts for weeks during which portions of the cells can be extracted and analyzed while the remainder is allowed to proliferate for future use. This procedure tends to be time consuming and requires at least the essential features of a tissue culture laboratory, but it seems to be one of the few methods of insuring large numbers of cells for analysis on a continuous basis.

The intent of any cell culture procedure is to establish a cell line which can then be manipulated depending on the intent of the study.

Most fibroblast culture procedures employ trypsin as a cell dispersing enzyme. Individual cells then enter a log growth phase during which extracts of cells can be taken. I have discontinued the use of trypsin, since its digestive properties are generally non-specific, which results in a significant loss of cells when cultures are initially established. I now prefer either to incompletely disperse cells by manual chopping with scissors or to chop the cells and then disperse them with a neutral protease, such as dispase. Dispase is collagen-specific and thus does not result in the large cell loss associated with the use of trypsin. Cells dispersed in dispase reach log phase sooner and the total number of cells in culture is significantly increased.

Dispase, however, is not preferred in the removal of proliferating fibroblasts from culture bottles because cells in monolayer are largely resistant to it. Instead we use trypsin since here its proteolytic action can be controlled. Cultures in log growth phase are treated with colcimide (10 μg) for 1 hr, removed from their culture flasks, treated with 0.075% KCl at 40°C for 20 min and then gently fixed with cold (2–4°C) carnoy solution. This fixation step is very important. Fixative is added drop-wise with a Pasteur pipette and the pellet is dispersed in fixative by forcing bubbles of air onto the pellet of cells. If cells are drawn up into the pipette during fixation, large numbers can be lost through their adherence to the side of the pipette. The pellet in fixative is then stored at 2°C overnight, fixed twice the next day, and finally cells are dropped onto slides which have been dipped once in 70% EtOH and then in distilled water. The fibroblast cell suspension is dropped onto the tilted side which is then covered with a petri dish and forced to dry over a minute's time period. This complete fixation together with the dropping of cells on tilted slides and prolonged drying all insure well spread elongate chromosomes which are a prerequisite to thorough analysis and subsequent banding of chromosomes.

Figure 1 shows the complete karyotype of a female Dark-eyed Junco (*Junco hyemalis*), which has been prepared in this way and homogeneously stained in Giemsa. It is obvious that the chromosomes are elongate and that the centromere positions are clear. This elongation, due partially to the prolonged drying, allows a clear identification of the macrochromosomes. Three pairs of chromosomes of intermediate size (arrows) are clearly metacentric. I had not noticed these three metacentrics in earlier work using brief fixation and quick air drying procedures (Shields, 1973). Thus, the true fundamental number of chromosome arms for this individual should be increased by six. Note also that several of the microchromosomes, although not paired, are clearly elongate, not dotlike, and are thus easier to count. I have analyzed large numbers of fibroblasts from this same female using this procedure and

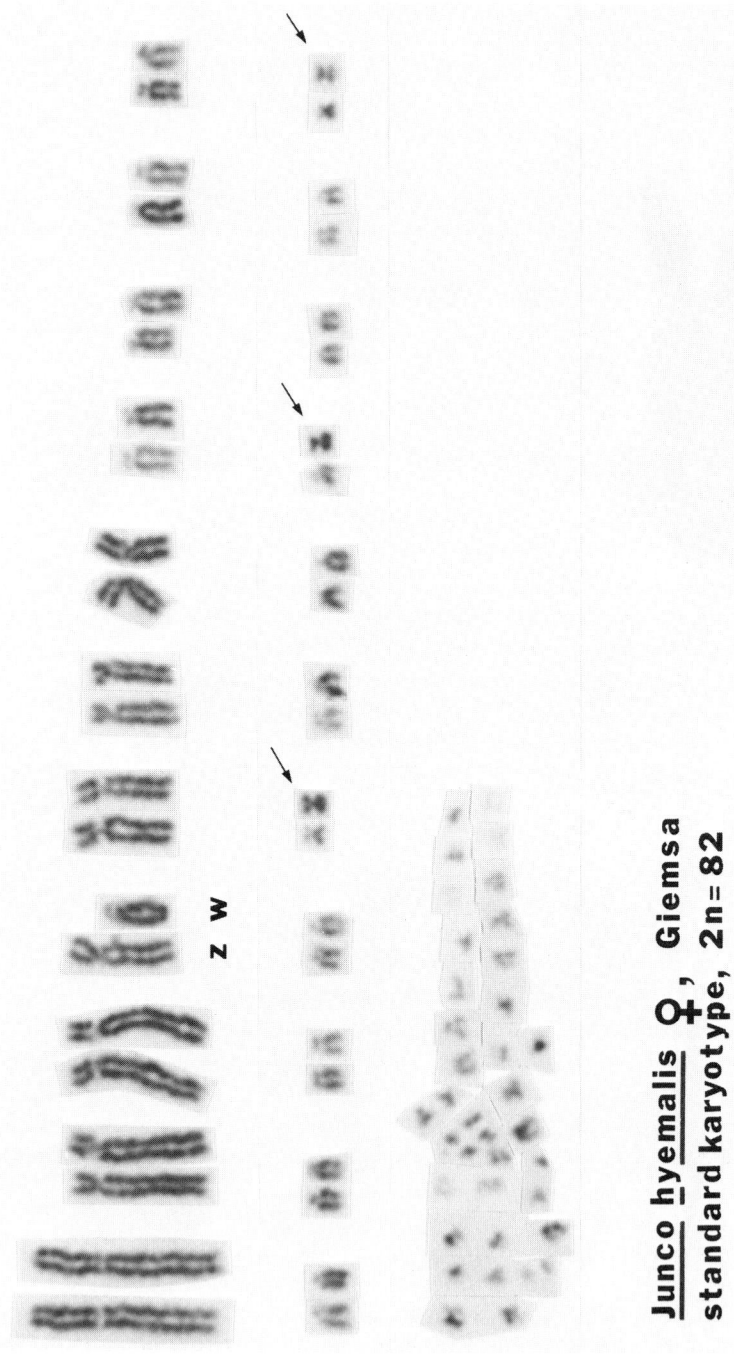

FIGURE 1. Karyotype of metaphase chromosomes of a female Dark-eyed Junco (*Junco hyemalis*) prepared from the primary culture of kidney fibroblasts. Note the metacentric chromosomes (arrows) of intermediate size, overlooked in earlier studies using brief drying times.

consistently count 82 chromosomes. An accurate diploid number can only be determined when large numbers of well-dried spreads are obtained.

2.2. Meiotic Procedures

Analysis of meiotic tissue has not been widespread in the study of avian chromosomes. Meiotic analyses were used to attempt to verify the chromosomal basis for rearrangements in the Dark-eyed Junco (Shields, 1976) and in the White-throated Sparrow (*Zonotrichia albicollis*; Thorneycroft, 1976). Such procedures provide an excellent tool for the determination of the diploid chromosome numbers of individual birds. Male birds are preferred in such studies since large numbers of metaphase I nuclei can be obtained from individuals which are actively undergoing spermatogenesis (Shields, 1976). Birds can be collected in breeding condition or induced into breeding condition by subjecting them to increased photoperiods. Testes are excised and placed in distilled water at 40°C for 20 min. The tunics are broken, and pieces of testes are dispersed in Carnoy fixative for 5 min, smeared onto albuminized slides, covered with coverslips, and squashed. The coverslips are then removed in the fixative and the slides are air dried, stained, and analyzed.

Figure 2 shows the entire metaphase I complement of a male Dark-eyed Junco which was treated in this way. Meiotic bivalents can be seen clearly. Hundreds of nuclei can be obtained from individual males and the total number of bivalents can readily be determined. It goes without saying that since homologous chromosomes are paired, the bivalent structures are larger than individual chromosomes at metaphase of mitosis and since their number is halved they are easier to count. I found no variation from the 41 bivalents counted for this male and it seems clear that his diploid number, like that of the female above, is 82. When used in unison, the mitotic fibroblast and meiotic procedures can provide unequivocal determinations of the diploid chromosome number of individual birds.

3. DIFFERENTIAL BANDING PROCEDURES

3.1. C-Banding

When metaphase chromosomes are treated with HCl and incubated in BaOH distinctive banding patterns can be seen along the chromosomes (Sumner, 1972). This C-banding procedure, as it is now known,

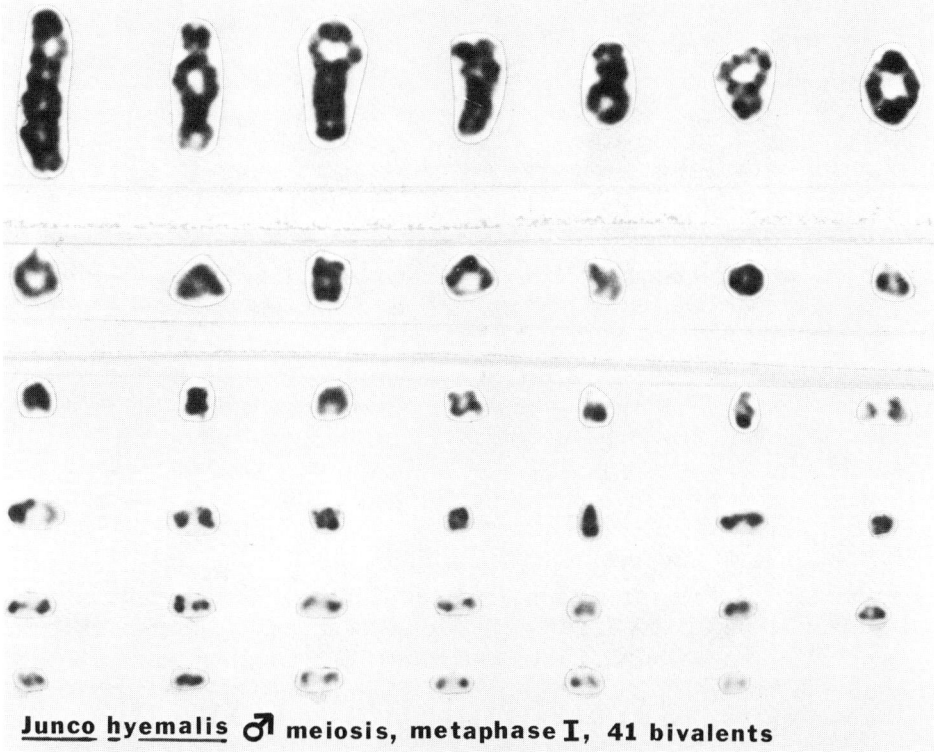

Junco hyemalis ♂ meiosis, metaphase I, 41 bivalents

FIGURE 2. Meiotic metaphase I karyotype of a male Dark-eyed Junco. Note that all the chromosomes are paired (as meiotic bivalents) and therefore their number is readily determined.

is specific for constitutive heterochromatin. It is clear that a differential rate of DNA loss in these preparations accompanies the treatment. The DNA in the heterochromatin is preferentially preserved (Hsu, 1979). It has been shown in mammals that constitutive heterochromatin can be totally centromeric, interstitial, terminal, or even to comprise total arms of chromosomes. Whenever highly repeated DNA sequences exist, C-bands appear. The appearance of C-bands is further independent of DNA base composition. In the mouse, DNA is AT-rich; in humans, a wide spectrum from AT- to GC-rich DNAs occur. In kangaroo rats, all satellites are GC-rich, yet each of these heterochromatin areas exhibits C-bands.

3.1.1. The W Chromosome

One of the most significant observations for the study of bird chromosomes is that the W chromosome of females tends to be partially or even totally heterochromatic and thus exhibits C-band positive regions.

A comparable situation exists in mammalian males where the Y chromosome is C-band positive. In the majority of birds the W chromosome in females is small, and can therefore be confused with any of the microchromosomes of similar size. It is not uncommon for authors to make a tentative identification of the W chromosome based on conventional staining of one female, or even to ignore its identification completely. At the present time, it is impossible to discuss morphologies and comparative aspects of W chromosomes of various avian species because the conventional data are so unreliable. However, based on several studies of other species, it appears that C-banding can now be used as a diagnostic test for the W chromosome.

Figure 3 shows complete C-banded karyotypes obtained from kidney fibroblasts of two cells of the same female Dark-eyed Junco. Several observations are of interest. The heavily stained W chromosome (arrow) can be seen clearly in both karyotypes. The utility of the C-band procedure is obvious in this regard. Using conventional staining procedures, the W chromosome is cryptic and often indistinguishable from a number of microchromosomes. When C-banding is used its presence is obvious.

There is, however, no consistent pattern for C-bands in W chromosomes of birds. In *Junco hyemalis* and Lesser-spotted Woodpeckers (*Dendrocopos minor*) W chromosomes appear entirely heterochromatic (Shields and Jarrell 1982). A similar situation exists in the Yellow-crowned Woodpecker (*Picoides mahrattensis;* Kaul and Ansari, 1978), the Great-horned Owl (*Bubo virginianus;* Biederman et al., 1979), and in the Parakeet (*Melopsittacus undulatus*), Pheasant (*Phasianus colchicus*), Mallard (*Anas platyrhynchos*), quail (*Colinus virginianus*), and chukar (*Alectoris graeca;* Stefos and Arrighi, 1971).

de Lucca (personal communication) reports a range of variation from 67–100% in amount of heterochromatin in W chromosomes of ten species of the Columbiformes. C-banding appears heavier in centromeric regions of W chromosomes of the pigeon (*Columba livia*), the Ring-necked Dove (*Streptopelia risoria*), and the domestic chicken (*Gallus domesticus;* Stock et al., 1974). This is similar to the situation in the Little Cormorant (*Phalacrocorax niger;* Patnaik et al., 1981), where heavier C-bands occur at the centromere region but become lighter in distal regions. The W chromosome of the variable Oystercatcher (*Hae-*

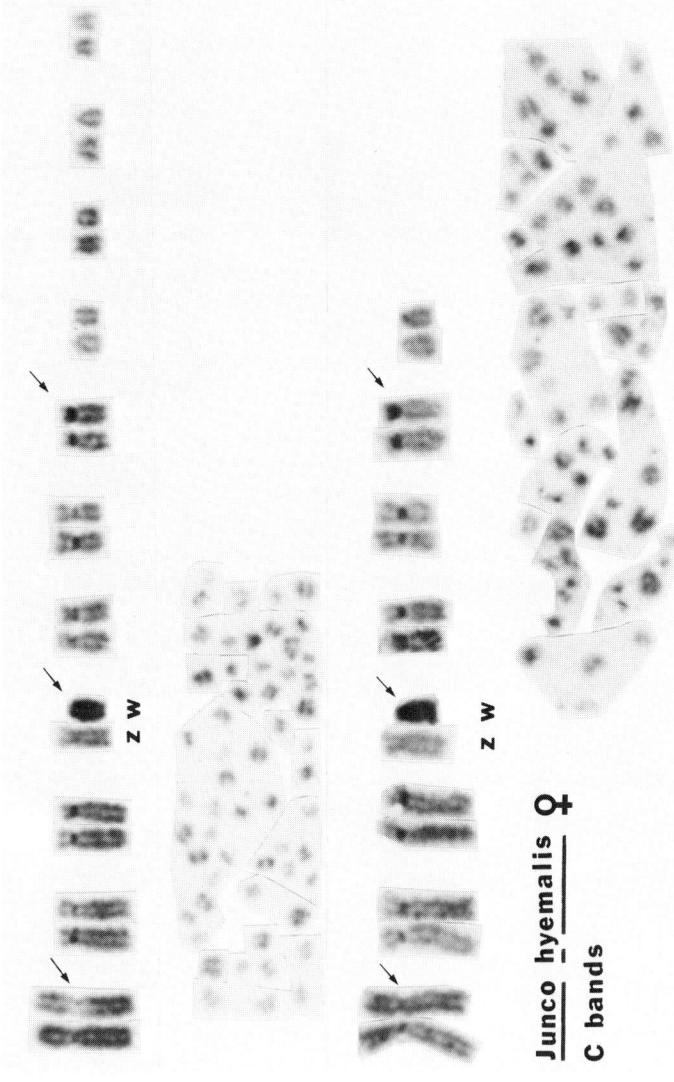

FIGURE 3. Mitotic metaphase karyotypes from two different nuclei of one female Dark-eyed Junco treated for C-bands. Arrows to the left and right indicate homologous sets of chromosomes showing differences in centromeric heterochromatin. Arrows in the middle indicate the W chromosome.

matopus unicolor) exhibits C-bands only in non-centromeric regions (Baker, 1982). All of this information can be summarized by saying that the W chromosome of female birds tends to be the most C-band positive chromosome of the complement. However, amounts and types of heterochromatin most certainly differ from one species to another. It is also important to point out that extreme caution should be used in the interpretation of the results of C-banding. The procedure itself involves a differential rate of DNA loss, while heterochromatin is preserved. Preparations should be exposed to HCl and BaOH for various time periods so that the range of C-band intensities can be determined in the same individual before comparisons with other individuals and species are made.

Stock et al. (1974) observed different amounts of C-band heterochromatin in telomere regions of the Z chromosomes of Gallus domesticus. Carlenius et al. (1981) verified this observation using the C-banding procedure of Sumner (1972).

3.1.2. C-Bands in Autosomes

Stefos and Arrighi (1971) observed heavy C-bands in microchromosomes of a variety of species. Additionally, they reported a lack of C-band heterochromatin in centromeric regions of macrochromosomes. However, latter studies by Stock et al. (1974) demonstrated the presence of C-band heterochromatin in centromeric regions of macrochromosomes of several of the species used in the earlier study.

It is now clear that the C-band heterochromatin variability observed in mammals will also be observed in birds when more species are studied. There appears to be no clear cut pattern of C-banding in autosomes. Figure 3 shows centromeric heterochromatin banding in macrochromosomes, but none in chromosomes of intermediate size. In addition, differences can be observed between centromeric regions of the same homologs (arrows). Note also that the Z chromosome appears devoid of C-band heterochromatin. It would be very informative to study patterns of C-band heteromorphism, W chromosome morphologies, and DNA base-pair compositions in a variety of birds representing various orders.

3.2. G-Bands

When air dried chromosomes on slides are treated with buffered trypsin and then stained in Giemsa, sets of crossbands result (Sumner et al., 1971; Patil et al., 1971; Drets and Shaw, 1971). Dark G-bands indicate high AT content and negative G-bands indicate high GC content. Thus, the earlier mentioned C-bands indicate repetition of DNA irrespective of base composition, while G-bands indicate base composition irrespective of sequence repetition. The G-band procedure thus allows for a differential banding sequence from one end of the chromosome to the other, depending on AT and GC base contents.

These techniques were introduced by mammalian cytogeneticists, and they have revolutionized the field of cytogenetics. Since there are so many bird chromosomes of similar size and centromere location, the G-banding procedures provided great hope not only for detecting chromosomal rearrangements but also for pairing homologous chromosomes correctly and establishing accurate diploid chromosome numbers.

I was ecstatic when I heard Maximo Drets talk about the Giemsa banding procedures at the American Society of Geneticists meeting in Rochester, New York in 1971. I returned to my laboratory and proceeded to fail at every attempt I made to induce G-bands in bird chromosomes. At that time I grew my fibroblasts on slides in petri dishes. Slides were air dried and stained. It was not until I removed the cells from the surface on which they grew and dropped them onto new slides that I finally obtained good G-bands.

Figure 4 shows the complete G-banded karyotype of a male White-crowned Sparrow (*Zonotrichia leucophrys*). Figure 5 shows the same spread arranged in a karyotype. Several features of this preparation are noteworthy. Exposure to trypsin tends to enlarge or expand the microchromosomes so that they can be recognized and counted more easily. When used in conjunction with mitotic fibroblast and meiotic analyses, G-banded preparations can again give an accurate diploid chromosome number. It can also be seen that homologous chromosomes can be paired more accurately when G-bands are used.

This ability to pair chromosomes with their presumptive homologs is of obvious value particularly when microchromosomes are involved. It remains difficult, however, to homologously pair chromosomes of the entire complement since many of the chromosomes of intermediate size have only single bands, while some microchromosomes have none. This problem can be partially alleviated by the use of the reverse R-banding procedure (discussed below).

My students and I are currently employing the use of G-bands and R-bands to help us determine the chromosomal basis for rearrangements in Juncos (Shields, 1973) and the White-throated Sparrow (Thorneycroft, 1976). These studies, however, will be reported elsewhere. Figure 6 shows the G-banded chromosomes of a partial karyotype of a female

Dark-eyed Junco that possesses the standard chromosome 2 and the homozygously inverted set number 5. Note that the W and Z sex chromosomes can be identified because of their unique banding sequences.

3.2.1. Utility of G-Banding

My students and I have been interested in the seemingly high frequency of chromosomal rearrangements that occur in various species of emberizid finches (Shields, 1982). In our preliminary analyses of banded karyotypes, we have observed polymorphic chromosomes in nearly every species studied. An example of such a case is shown in Fig. 7, which shows the complete karyotype of a male Fox Sparrow (*Passerella iliaca*). The karyotype of this bird is of interest since one set of chromosomes (arrow) appears dimorphic, and the short arm of the chromosome on the left is obviously longer than that of the chromosome on the right. When we G-banded chromosomes of this same individual, a diagnostic terminal band appeared at the tip of the larger chromosome. The partial karyotypes of two different cells of this individual are shown in Fig. 8. The photographs are slightly overexposed to bring out the presence of the terminal band. This band must represent an accumulation of AT-rich DNA at the terminal end of the chromosome. Whether this polymorphism is caused by addition or deletion of the terminal band is unknown. It is shown here as an example of the detail that can be obtained when several staining procedures are employed.

3.2.2. G-Banding Comparative Studies

Takagi and Sasaki (1974) compared G-banding patterns of the three largest macrochromosomes of a number of species of birds belonging to several diverse orders. They concluded that G-bands of the respective chromosomes in each species were identical. Further, they claimed that the standard G-band sequence in the three largest chromosomes of these birds is homologous to those of the three largest chromosomes in the turtle (*Geoclemys reevesii*). This study, more than any other, implied that chromosome evolution in birds was truly conservative, even to the point that homologies could be identified between birds and reptiles.

Stock and Mengden (1975) studied G-banded karyotypes of the domestic chicken, the Ring-necked Dove, the domestic pigeon, and the musophagid (*Gallirex porphyreolophus*), and compared these to those of the boid snakes, *Liasis olivaceus*, *L. boelni*, *L. amethystinus*, *Acrantophis dumereli*, and *Sanzinia madigascar-iensis*, the box turtle (*Ter-*

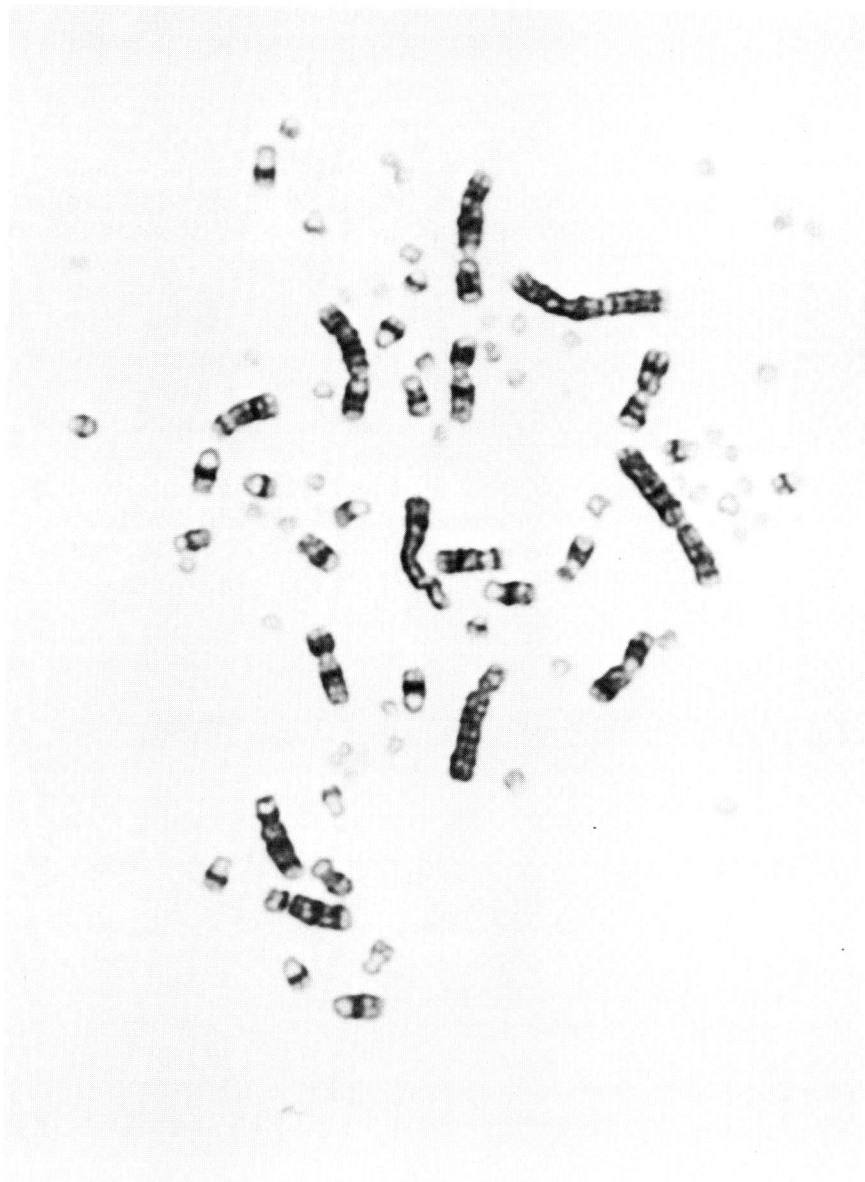

FIGURE 4. G-banded nucleus of a male White-crowned Sparrow (Zonotrichia leucophrys). Note the enlarged size of the microchromosomes.

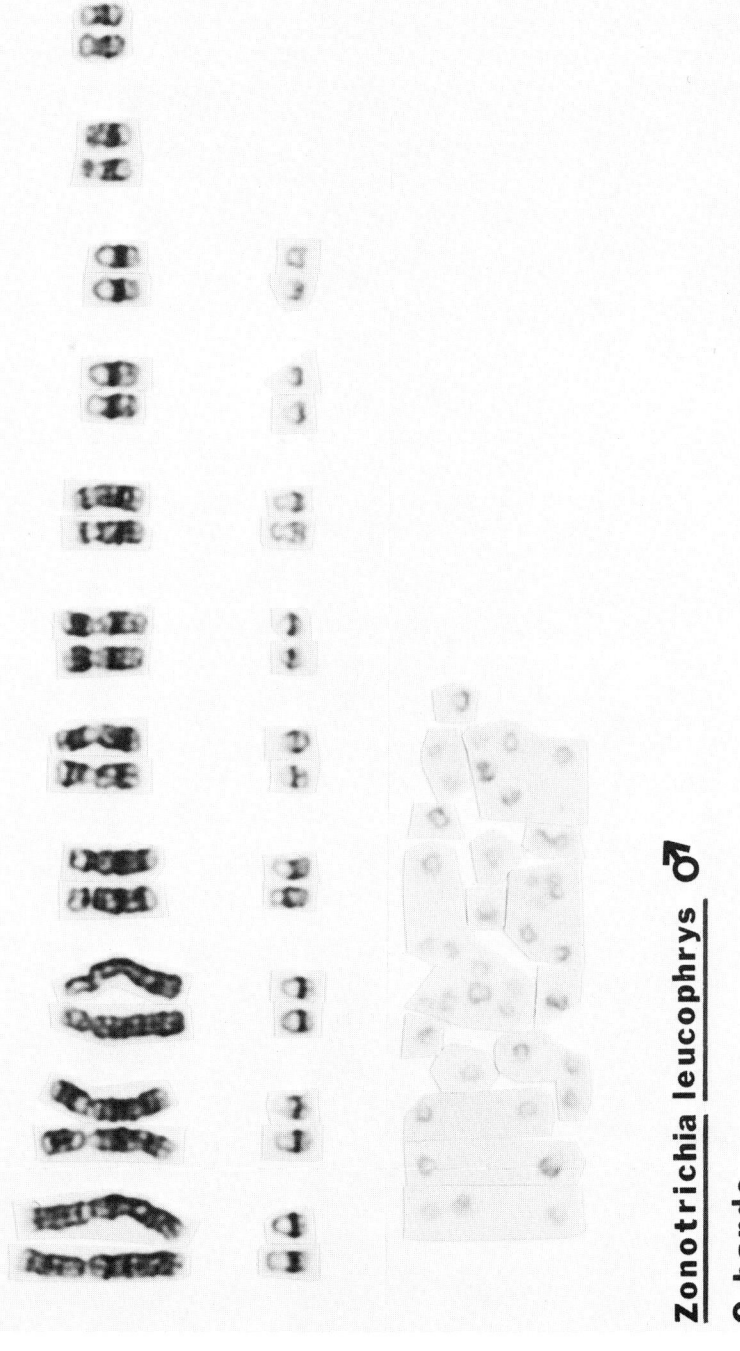

FIGURE 5. Complete G-banded karyotype of the nucleus shown in Fig. 4.

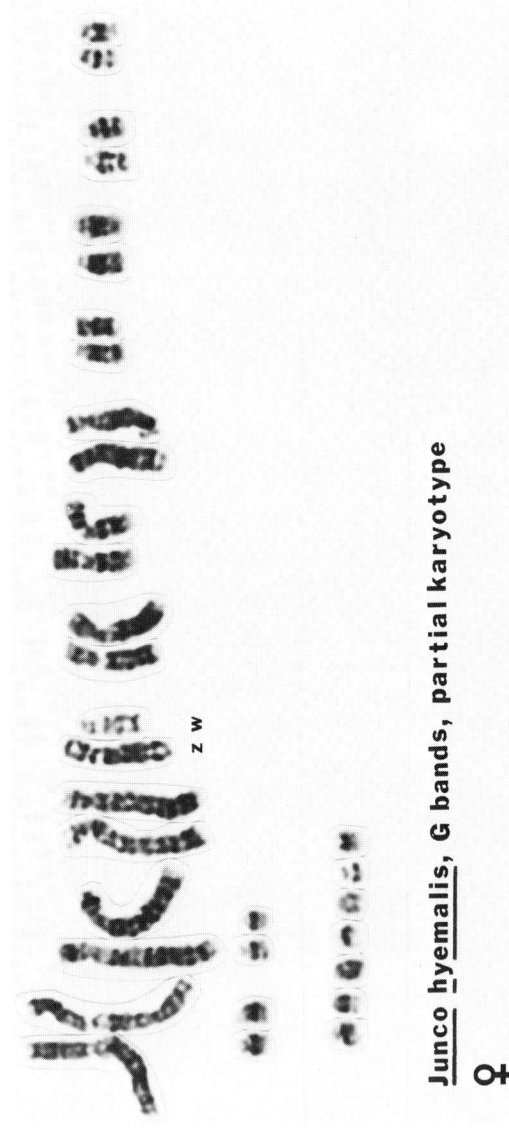

FIGURE 6. G-banded chromosomes of a partial karyotype of a female Dark-eyed Junco. Note the unique banding sequences of both W and Z sex chromosomes.

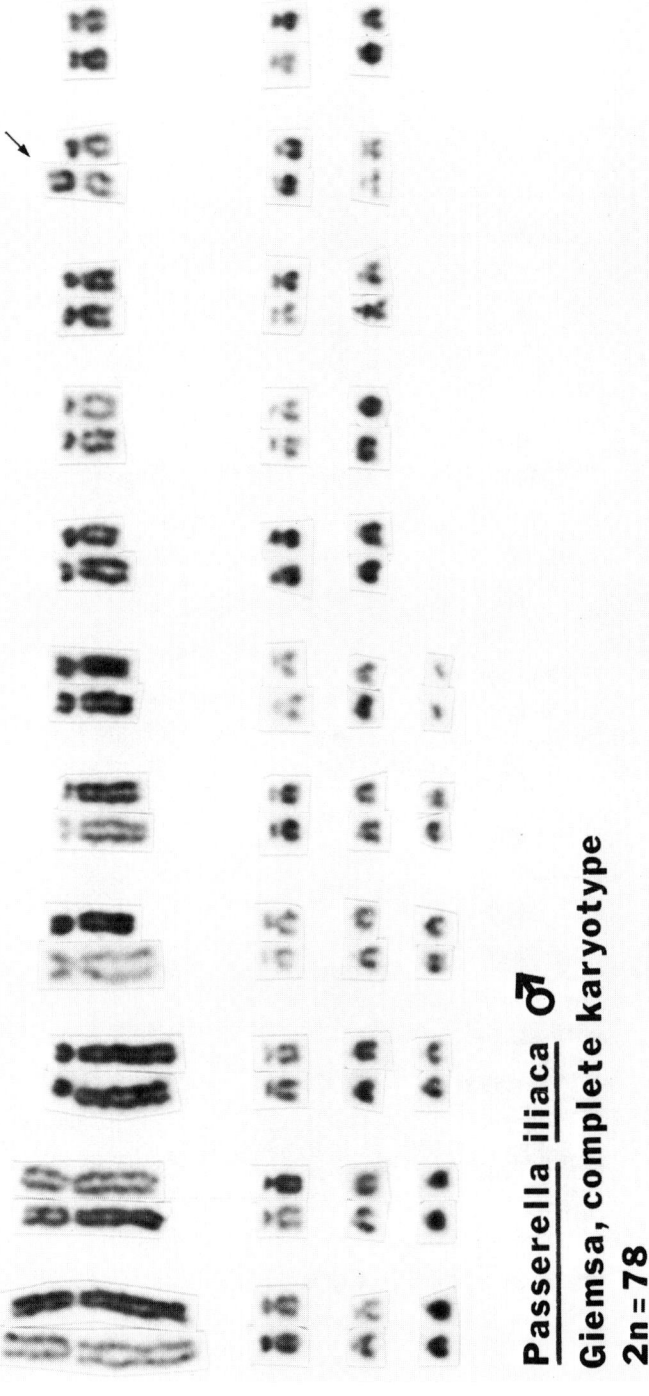

FIGURE 7. Complete metaphase karyotype of a male Fox Sparrow (*Passerella iliaca*) homogeneously stained with Giemsa. Note the heteromorphic chromosome pair (arrow).

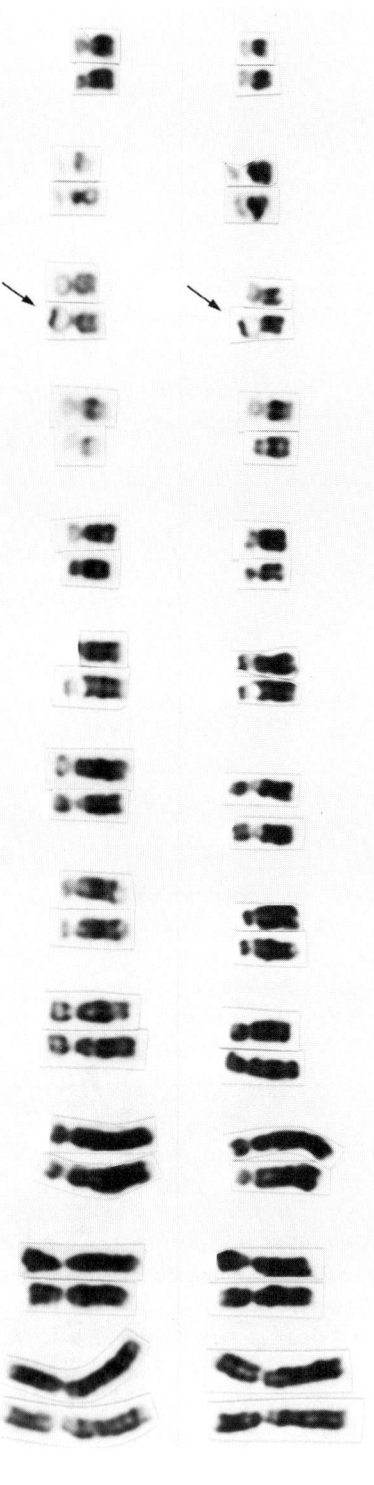

FIGURE 8. G-banded macrochromosomes taken from two nuclei of one male Fox Sparrow. Note the heavy terminal band (arrow) on the left-hand homologue of the heteromorphic chromosome pair.

repene carolina) and the African clawed toad (*Xenopus muelleri*), and concluded that the interpretation by Takagi and Sasaki (1974) of broad homology between diverse orders of birds, and to chelonian reptiles, was unwarranted. The crucial observation in both these studies is that differences in G-band patterns between species are obliterated when mid-metaphase chromosomes are analyzed. Chromosomes at midmetaphase are highly contracted, and thus bands that otherwise appear discrete become fused into blocks of bands, which then appear identical. When late prophase or early metaphase chromosomes are G-banded, much more detail can be observed. Comparison of G-banded karyotypes between species must, therefore, be done with extreme caution, and as an absolute prerequisite, one must determine the within-species range of G-band variability before species comparisons are made. Stock and Mengden (1975) describe considerable variability in G-banded patterns in macrochromosomes of the four bird species studied. Further they showed no real resemblance between bird chromosomes and those of turtle, snake, or amphibian.

Few studies comparing G-band patterns of chromosomes of closely related species of birds have been done. Ryttman *et al.* (1979) have found that G-banded karyotypes of the Lesser and Greater Black-backed Gulls, Herring Gulls, and Mew Gulls are identical. A single egg was used to describe G-bands of chromosomes of Great Black-backed Gulls and Mew Gulls, so nothing is known about intraspecific variability in these birds. Biederman (personal communication) has recently shown that G-bands on chromosomes of the Great Horned Owl, the Snowy Owl (*Nyctea scandiaca*), the Long-eared Owl (*Asio otus*), and Short-eared Owls (*A. flammeus*), are identical. G-band patterns appear unaltered in the 12 largest autosomes of *Tauraco livingstonii, T. erythrolophus, T. leucolophus, Musophaga rossae,* and *Corythaixoides concolor* (Van Tuinen, personal communication). Three different Z chromosomes are present, however. de Lucca (personal communication) has shown considerable variability in G-banded chromosomes of ten species of the Columbiformes, in fact no two G-band patterns appear identical.

It seems clear, however, that the rate of chromosome change in birds during evolution has not been constant. Large differences may exist between closely related taxa while virtually identical karyotypes may occur in taxa of remote ancestry.

3.3. R-Banding

Dutrillaux and Lejeune (1971) heated cytological preparations in hot phosphate buffer and stained in Giemsa. The amazing result was that crossbanding was opposite to the G-banding patterns. AT-rich areas are lightly stained while GC-rich areas are heavily stained. Aside from complementing G-band patterns, R-bands have a unique property. Terminal chromosome segments tend to be G-band negative (see Figs. 5 and 6; termini of first chromosomes), thus, it is difficult to determine actual chromosome lengths from G-band preparations. When used together, G- and R-banding can only lead to a better understanding of the complementary patterns. R-banding has not been widely used in avian cytogenetic studies. Carlenius et al. (1981) describe G- and R-band complementarity in the domestic fowl.

3.4. NOR-Banding

When chromosome preparations are treated with borate buffer and aqueous silver nitrate, nucleolus organizer regions (NORs) are stained with silver (Bloom and Goodpasture, 1976). Achromatic NORs represent sites of ribosomal cistrons. Each is thus a diagnostic chromosome marker. When NORs are terminal, it is difficult to differentiate them by conventional staining, and they can be observed only when bordered by euchromatic segments. Silver staining, however, is specific for all NORs whether terminal or interstitial.

It is known that NORs on different chromosomes may synapse and undergo genetic crossing over. Thus, ribosomal cistrons can be transferred from one chromosome to another. We consistently see figures at metaphase of mitosis which suggest that NORs of microchromosomes are fused. Silver staining has identified these fused areas as NORs. Like R-banding, this procedure has not been widely exploited by avian cytogeneticists.

3.5. Sequential Banding

It is often useful to subject a chromosome preparation to more than one banding procedure. Such techniques are called sequential banding (Jarrell, personal communication). Information specific to two different procedures can thus be obtained from one preparation. All sequential procedures involve an intermediate destaining. Figure 9 shows the macrochromosomes of a male Tree Sparrow (Spizella arborea). The top row is homogeneously stained with Giemsa while the bottom row is sequentially G-banded after exposure to urea. The seventh chromosome is polymorphic, one homologue is metacentric while the other is acrocentric. Urea G-banding suggests an intrachromosomal rearrangement, probably a pericentric inversion. Sequential staining has not been widely employed by avian cytogeneticists.

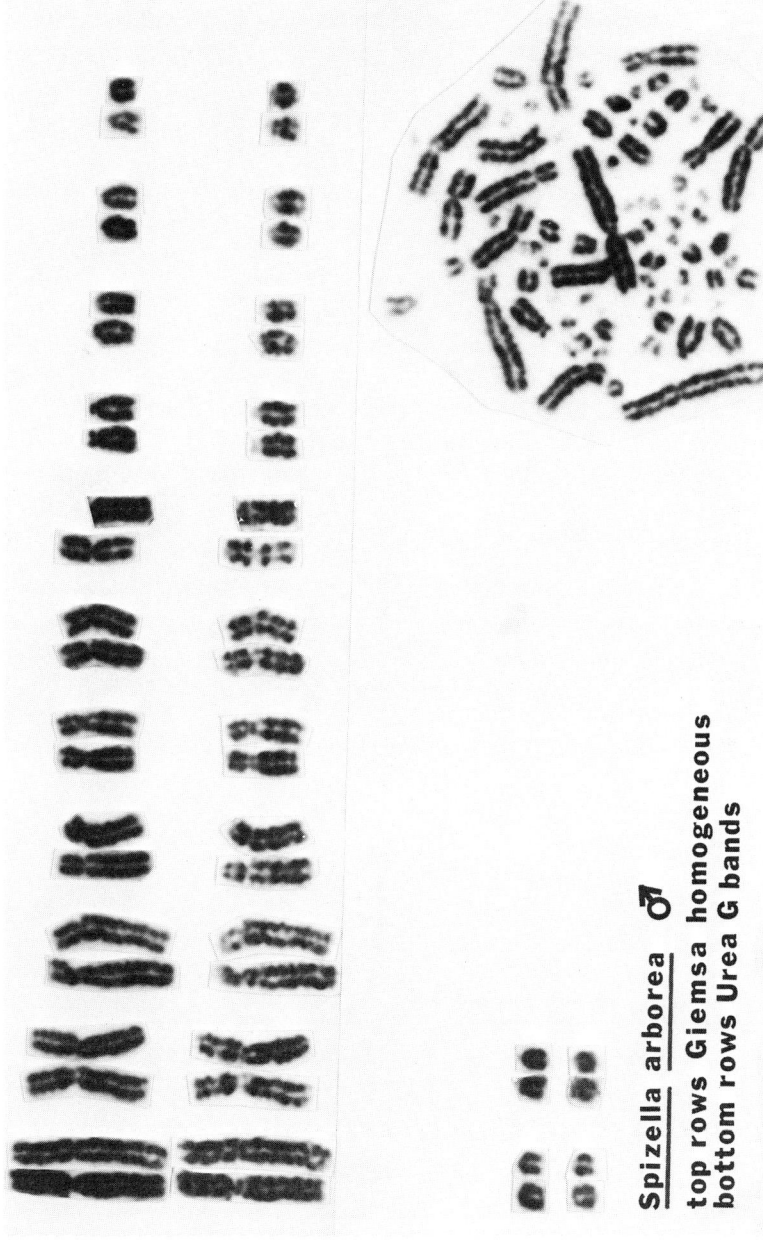

FIGURE 9. Sequential staining of the macrochromosomes of a male Tree Sparrow (Spizella arborea). Top row is stained homogeneously with Giemsa. Bottom row is G-banded after urea pretreatment. Inset shows entire complement.

4. CONCLUDING REMARKS

Procedures discussed here have generally proven most useful for comparative cytogenetics, and they can be used successfully with birds. Data from such procedures are not merely redundant observations on one cytogenetic theme. Each procedure is specific for the demonstration of a particular chromatin component, and with continued research a deeper understanding of the interplay between DNA on the one hand and chromosome form and function on the other will result. Already, we have detailed DNA sequence data for a relatively large number of avian genes including those of the hen oviduct (Breathnach et al., 1978; Cochet et al., 1979; Gannon et al., 1979; Lai et al., 1979). The details of these DNA sequence studies parallels that for mammals. Yet in most cases we are ignorant of the chromosomal localization of the genes that have been sequenced in birds. An increased research activity in avian cytogenetics, including the use of current banding and localization procedures, can only complement the more detailed molecular studies and result in a more thorough understanding of this largely neglected aspect of birds.

ACKNOWLEDGMENTS. I would like to thank B. Biederman and Peter Van Tuinen who kindly allowed me to cite their unpublished data. I thank my students Declan Troy and Gordon Jarrell for their cooperation. I dedicate this paper to my professors: James J. Manion, Jared Verner, James D. Rising, and Klaus H. Rothfels, who have always encouraged my studies in this area. Finally, I thank the Institute of Arctic Biology, University of Alaska for support.

REFERENCES

Baker, A. J., Parslow, M., and Chambers, D., 1981, Karyological studies of a female Variable Oyster Catcher (*Haematopus unicolor*), Can. J. Genet. Cytol. **23**:611–619.

Biederman, B. M., Florence D., and Lin, C. C., 1979, Cytogenetic analysis of Great Horned Owls (*Bubo virginianus*), Cytogenet. Cell Genet. **28**:79–86.

Bloom, S. E., and Goodpasture, C., 1976, An improved technique for selective silver staining of nucleolar organizer regions in human chromosomes, Hum. Genet. **34**:199–206.

Breathnach, R., Benoist, C., O'Hare, K., Gannon, F., and Chambon, P., 1978, Ovalbumin gene: Evidence for a leader sequence in mRNA and DNA sequences at the exon-intron boundaries, Proc. Natl. Acad. Sci. USA **75**:4853–4857.

Carlenius, C., Ryttman, H., Tegelstrom, H., and Jansson, H., 1981, R-, G-, and C-banded chromosomes in the Domestic Fowl (*Gallus domesticus*), Hereditas **94**:61–66.

Cochet, M., Gannon, F., Hen, R., Maroteaux, L., Perrin, F., and Chambon, P., 1979, Organization and sequence studies of the seventeen-piece chicken conalbumin gene, Nature **282**:567–574.

de Boer, L. E. M., 1976, The somatic chromosome complements of 16 species of Falconiformes (Aves) and the karyological relationships of the order, *Genetica* **46**:77–113.
Drets, M. E., and Shaw, M. W., 1971, Specific banding patterns of human chromosomes, *Proc. Natl. Acad. Sci. USA* **68**:2073–2077.
Dutrillaux, B., and Lejeune, J., 1971, Sur une nouvelle technique d'analyse du caryotype humain, *C.R. Acad. Sci. Paris* **272**:2638–2640.
Gannon, F., O'Hare, K., Perrin, F., LePennic, J. P., Benoist, C., Cochet, M., Breathnach, R., Royal, A., Garapin, A., Cami, B., and Chambon, P., 1979, Organization and sequences at the 5^1 end of a cloned complete ovalbumin gene, *Nature* **278**:428–434.
Hsu, T. C., 1979, *Human and Mammalian Cytogenetics: An Historical Perspective*, Springer-Verlag, New York.
Kaul, D., and Ansari, H. A., 1978, Chromosome studies in three species of Piciformes (Aves), *Genetica* **48**:193–196.
Lai, E. C., Stein, J. P., Catterall, J. F., Woo, S. L. C., Mace, M. L., Means, A. R., and O'Malley, B. W., 1979, Molecular structure and flanking nucleotide sequences of the natural chicken ovomucoid gene, *Cell* **18**:829–842.
Lu, M. R., 1969, A widely applicable technique for direct processing of bone marrow for chromosomes of vertebrates, *Stain Tech.* **44**:155–158.
Patil, S. R., Merrick, S., and Lubs, H. A., 1971, Identification of each human chromosome with a modified Giemsa stain, *Science* **173**:821–822.
Patnaik, A. K., Samanta, M., and Prasad, R., 1981, Chromosome complement and banding patterns in a pelecaniform bird, *Phalacrocorax niger*, *J. Hered.* **72**:447–449.
Ryttman, H., Tegelstrom, H., and Jansson, H., 1979, G- and C-banding in four *Larus* species (Aves), *Hereditas* **91**:143–148.
Shields, G. F., 1973, Chromosomal polymorphism common to several species of *Junco* (Aves), *Can. J. Genet. Cytol.* **15**:461–471.
Shields, G. F., 1976, Meiotic evidence for pericentric inversion polymorphism in *Junco* (Aves), *Can. J. Genet. Cytol.* **18**:747–751.
Shields, G. F., 1982, Comparative avian cytogenetics: A review, *Condor* **84**:45–58.
Shields, G. F., 1983, Organization of the avian genome, in: *Perspectives in Ornithology*, Volume 1 (A. H. Brush and G. A. Clark, eds.), Cambridge University Press, pp. 271–290.
Shields, G. F., and Jarrell, G. H., 1982, Enlarged sex chromosomes of Woodpeckers (Piciformes), *Auk* **99**:767–771.
Shoffner, R. N., Krishan, A., Haiden, G. J., Bammi, R. K., and Otis, J. S., 1967, Avian chromosome methodology, *Poult. Sci.* **46**:333–344.
Stock, A. D., Arrighi, F. E., and Stefos, K., 1974, Chromosome homology in birds: Banding patterns of the chromosomes of the domestic chicken, ring-necked dove, and domestic pigeon, *Cytogenet. Cell Genet.* **13**:410–418.
Stock, A. D., and Mengden, G. A., 1975, Chromosome banding pattern conservatism in birds and nonhomology of chromosome banding patterns between birds, turtles, snakes and amphibians, *Chromosoma* (Berlin) **50**:69–77.
Sumner, A. T., 1972, A simple technique for demonstrating centromeric heterochromatin, *Exp. Cell Res.* **75**:304–306.
Sumner, A. T., Evans, H. J., and Buckland, R. A., 1971, A new technique for distinguishing between human chromosomes, *Nature New Biol.* **232**:31–32.
Stefos, K., and Arrighi, F. E., 1971, Heterochromatic nature of the W chromosome of birds, *Exp. Cell Res.* **68**:228–231.
Takagi, N., and Sasaki, M., 1974, A phylogenetic study of bird karyotypes, *Chromosoma* (Berlin) **46**:91–120.
Thorneycroft, H. B., 1976, A cytogenetic study of the White-throated Sparrow, *Zonotrichia albicollis* (Gmelin), *Evolution* **29**:611–621.

CHAPTER 8

GENETIC STRUCTURE AND AVIAN SYSTEMATICS

KENDALL W. CORBIN

1. INTRODUCTION

Systematics has as its overall objective the ordering of taxa of organisms into a hierarchical scheme. To achieve a natural classification, one that reflects the evolutionary history and relationships of the organisms involved, the systematist should ideally have methods that enable the attainment of the following five intermediate objectives of classification: (1) the variation within species must be rigorously described so that individuals can be assigned unambiguously to species, (2) the evolutionary relationships among the species considered must be reconstructed, (3) estimates should be obtained for rates of evolutionary change along genetic lineages through geological time, (4) the structure within species should be characterized, and (5) the boundaries of taxa above and below that of the species must be delineated objectively.

This chapter will discuss how the analysis of genetic structure of natural populations leads to the achievement of these five intermediate objectives, emphasizing points (4) and (5). In particular, methods will be presented for the delineation of avian genera, species, and subspecies through the comparison of structural proteins and enzymes by means of electrophoresis and statistical analysis of the genetic data.

KENDALL W. CORBIN • Bell Museum of Natural History and Department of Ecology and Behavioral Biology, University of Minnesota, Minneapolis, Minnesota 55455.

Approximately 30 studies have used electrophoretic techniques to examine genetic variation within natural avian populations. For the purposes of this review and synthesis, the number of individuals and the number of gene loci examined should be relatively large, ideally 20 and 30 respectively, if the advice of Nei and Roychoudhury (1974) were followed. We shall not adhere strictly to this ideal, but for the purpose of the analyses presented here, the data base has been restricted to the following taxa and studies: starlings of the genus *Aplonis* (Corbin et al., 1974), the Northern Oriole (*Icterus galbula*; Corbin et al., 1979), wood warblers of the family Emberizidae* (Barrowclough and Corbin, 1978), the Yellow-rumped Warbler (*Dendroica coronata*; Barrowclough, 1980a,b), thrushes of the family Muscicapidae and their allies (Avise et al., 1980a), the White-crowned Sparrow (*Zonotrichia leucophrys*; Baker, 1975; Corbin, 1981, unpublished data), the Rufous-collared Sparrow (*Zonotrichia capensis*; Handford and Nottebohm, 1976; Corbin, unpublished data), representative species of the order Procellariiformes (Barrowclough et al., 1981, unpublished data), sparrows of the family Emberizidae (Avise et al., 1980b; Zink, 1982), and species of the family Phasianidae (Gutiérrez et al., 1983).

As a point of reference for the discussions that follow, I have tried here to summarize the basic concepts and describe the techniques of analysis in such a way as to be understandable to the novice. These summaries will include a brief description of the electrophoresis of proteins and a discussion of the kinds of genetic data one obtains from electrophoretic surveys. Also, there is a review of how one analyzes such data with respect to the calculation of Hardy–Weinberg equilibria, the estimation of genetic heterozygosity, the estimation of genetic distance between pairs of taxa, and the analysis of variance of allelic frequencies by means of F statistics.

2. METHODS OF DATA ACQUISITION AND ANALYSIS

For studies in biochemical systematics, three major kinds of techniques are now in use. These are electrophoresis, DNA–DNA or DNA–RNA hybridization, and chromosome banding. All three approaches are extensions of the classical, comparative method to the molecular level, with one overall distinction. Unlike data that characterize and

---

*The classification used here follows the changes that are incorporated into the Thirty-fourth Supplement to the American Ornithologists' Union Check-list of North American Birds.

reduce morphological variation, at the level of organisms and organ systems, to numerical values, data on proteins and nucleic acids provide direct measures of the variation in the underlying genetic structure of individuals and populations.

2.1. Electrophoretic Studies

Electrophoresis is the migration of a charged particle in an electric field. Proteins are charged particles of particular interest in evolutionary studies because they are the primary products of gene action and are direct translations of the underlying sequence of nucleotide base-pair codons, of which genes are comprised. The charge on a protein is determined both by the kinds of amino acids it possesses and by the pH of the medium in which it exists. In electrophoresis, the charge on proteins is controlled by altering the pH of the buffers used by the investigator who can thereby increase the rate of migration of proteins in an electric field of a given power. The power of the electric field is also controlled by varying voltage and amperage.

For comparative studies it is essential that proteins of different individuals be compared under nearly identical conditions of electrophoresis. Only in this way can one have confidence that the protein products of different genomes are identical or unique. Because many kinds of amino acid substitutions do not change the charge on a protein, theoretically only about 24% of all substitutions can be detected by electrophoresis. But amino acid substitutions may also affect the shape and size of a protein and such changes may be detected using certain electrophoretic techniques. This fact was well known, even in the 1960s, to those who were involved in the analysis of protein structure (Henning and Yanofsky, 1963). To maximize the probability of detecting amino acid differences among the proteins being analyzed, most electrophoretic systems utilize thin slabs of either hydrolyzed potato starch or acrylamide gel. One or the other of these support media was used in each of the studies cited here, making it possible to compare enzyme extracts from 10–30 individuals simultaneously.

By using such electrophoretic techniques it is possible to determine the allele composition of each individual for each type of protein extracted from its tissues. Although higher organisms synthesize thousands of different kinds of proteins, only 20–60 of these types are readily utilized in avian studies. But even this number provides an enormous amount of genetic information. The electrophoretic analysis of 50 unique proteins, i.e., coded by 50 gene loci, from 20 individuals of a local population makes it possible to identify and characterize 2000 alleles (40 genomes × 50 gene products = 2000 alleles examined). This wealth of genetic information can be reduced in a variety of ways.

2.2. Hardy–Weinberg Equilibria

The enumeration of genotypes provides genotypic frequency distributions which are used to calculate the observed frequencies of alleles and the expected genotype frequencies distributed according to Hardy–Weinberg equilibria. The expected frequency of each genotype is simply the binomial expansion (for a locus having only two alleles) or polynomial expansion (three or more alleles) of the observed allelic frequencies. A Hardy–Weinberg equilibrium distribution is the expected number of individuals of each genotype and is estimated by multiplying each expected genotype frequency by the number of individuals in the population sample. The test of whether a population is in Hardy–Weinberg equilibrium at a locus is evaluated by calculating chi-squared values for each genotype and then summing over genotypes to obtain a total χ^2 value, which measures the deviation of observed numbers of individuals of each genotype from the number expected. The number of degrees of freedom for this statistic equals $g - a - 1$, where g is the number of possible genotypes given the alleles observed, and a is the number of independent alleles, which is equal to the total number of alleles observed at a locus minus 1.

It is understood that observed genotypic frequencies often deviate significantly from expected values because equilibria are perturbed by: (1) gene flow via migration of individuals among populations having different allelic frequencies, (2) mutations from one allele to another, which are expressed as amino acid substitutions in proteins, (3) selection for or against an allele, (4) genetic drift in small populations, and (5) non-random mating within local populations. Sampling errors of several kinds may also contribute significantly to the deviation from expected equilibrium values. A particularly serious kind of sampling error causes what is known as the Wahlund effect (Wahlund, 1928). This error results from combining, within a single sample, individuals from genetically distinct populations.

2.3. Genic Heterozygosity

Estimates of heterozygosity can be made in several ways, and may pertain to variation within individuals, loci, populations, or samples of individuals representing each taxonomic level. Heterozygosity values may be estimated as observed frequencies of heterozygotes based on the distribution of genotypes within individuals or population samples, they may be based on the expected frequencies of heterozygotes as calculated from Hardy–Weinberg equilibria, or they may be calculated from allelic frequency data. To help distinguish one heterozygosity parameter from another, subscripts coupled with h are used here to denote the heterozygosity of individuals, h_i, or of loci, h_l. When averaged over individuals or loci, observed heterozygosity will be denoted as $\bar{H}_o$ and calculated heterozygosity as $\bar{H}_c$. These distinctions normally have not been made, which has resulted in some confusion in the literature. Not only do these different parameters pertain to alternative methods of estimating genetic variability, but there are also significant differences in the amount of variance associated with each parameter.

The observed heterozygosity of individuals, h_i, is simply the ratio of the number of an individual's heterozygous loci to the total number of loci assayed for that individual. Obviously, the larger the number of loci assayed, the more rigorous this estimate will be. Averaging h_i over individuals gives the average observed heterozygosity $\bar{H}_o$ (Corbin, 1981). The variance associated with this estimate of $\bar{H}_o$ is easily reduced by increasing the number of individuals sampled. The other methods for estimating heterozygosity all require that additional loci be assayed in order to reduce the variance, but usually this is both impractical and ineffective.

A second method of estimating $\bar{H}_o$ is to calculate the observed frequency of heterozygotes per locus, h_i, and then average these values over loci (Corbin et al., 1974, 1979; Barrowclough and Corbin, 1978). Given the same data set as that used to calculate the h_is above, $\bar{H}_o$ would be identical, but the variance associated with $\bar{H}_o$ estimated by this second method will be larger, and significantly so in many cases. This is a direct consequence of the high diversity in the number of alleles per locus. Among avian species, the number of polymorphic loci, expressed as a percentage of the total number of loci examined, varies from a high value of 64.3% for steamer ducks of the genus *Tachyeres* to the lower limit of 0.0% for the Lesser Prairie Chicken (*Tympanuchus pallidicinctus*; Table I).

A third method of estimating $\bar{H}_o$, if the distribution of genotypes corresponds to Hardy–Weinberg expectations, or $\bar{H}_c$ if it does not, is a modification of the second method. In this case, the expected frequencies of heterozygotes, based on the calculated equilibrium distributions, are used as estimates of h_l for each locus. These are then averaged over loci to obtain the average heterozygosity (Avise et al., 1980a). Again, however, the variance is large due to the inclusion of loci that are monomorphic along with those that are polymorphic to various degrees.

A fourth method estimates the heterozygosity per locus, h_l, as

TABLE I
Percent Polymorphism and Average Calculated Heterozygosity, $\bar{H}_c$ for Species of Birds[a]

Family and species[b]	Number of genomes	Number of loci	Percent polymorphism	$\bar{H}_c \pm$ SD	Reference
Procellariidae					
Thalassoica antarctica	16	16	0.250	0.114 ± 0.240	Barrowclough, Corbin, and Zink (unpublished)
Daption capense	16	16	0.313	0.143 ± 0.239	
Pagodroma nivea	22	16	0.438	0.175 ± 0.231	
Pachyptila desolata	16	16	0.313	0.134 ± 0.225	
Hydrobatidae					
Oceanites oceanicus	24	16	0.250	0.096 ± 0.180	Barrowclough, Corbin, and Zink (unpublished)
Anatidae					
Tachyeres patachonicus	68	14	0.714[c]	0.307 ± 0.017	Corbin, Humphrey, and Livezey (unpublished)[d]
T. leucocephala	12	14	0.643	0.164 ± 0.154	
Phasianidae[e]					
Tetraoninae					
Tympanuchus pallidicinctus	26	27	0.000	0.0	Gutiérrez et al. (1983)
Phasianinae[e]					
Phasianus colchicus	26	27	0.111	0.031 ± 0.100	Gutiérrez et al. (1983)
Coturnix coturnix	60	27	0.197[c]	0.075 ± 0.031	
Alectoris chukar	24	27	0.148	0.052 ± 0.138	
Lophortyx californicus	72	27	0.185[c]	0.025 ± 0.008	
L. gambelii	44	27	0.185	0.031 ± 0.098	
Callipepla squamata	58	27	0.130[c]	0.037 ± 0.023	
Colinus virginianus	30	27	0.148[c]	0.034 ± 0.004	
Oreortyx pictus	32	27	0.111	0.024 ± 0.092	

Cyrtonyx montuzumae	62		0.204[c]	0.039 ± 0.009	
Mimidae					
Dumetella carolinensis	16	27	0.208	0.028 ± 0.059	Avise et al. (1980a)
Muscicapidae[e]					
Turdinae					
Catharus ustulatus	34	27	0.296	0.065 ± 0.140	Avise et al. (1980a)
C. guttatus	26	27	0.259	0.045 ± 0.089	
C. fuscescens	10	27	0.185	0.048 ± 0.107	
Hylocichla mustelina	10	27	0.148	0.060 ± 0.150	
Turdus migratorius	10	26	0.077	0.031 ± 0.108	
Sialia sialis	14	27	0.148	0.045 ± 0.121	
Sylviinae[e]					
Regulus calendula	20	23	0.174	0.037 ± 0.097	Avise et al. (1980a)
Sturnidae					
Aplonis metallica metallica	372	18	0.118[c]	0.054 ± 0.157	Corbin et al. (1974)
A. metallica nitida	278	18	0.111[c]	0.046 ± 0.136	
A. metallica purpuriceps	58	18	0.111	0.027 ± 0.100	
A. cantaroides	188	18	0.125[c]	0.016 ± 0.061	
Emberizidae[e]					
Parulinae					
Vermivora peregrina	26	30	0.333	0.158 ± 0.243	Barrowclough and Corbin (1978)
V. celata	16	31	0.355	0.130 ± 0.203	
V. ruficapilla	44	31	0.335	0.123 ± 0.186	
Mniotilta varia	16	26	0.154	0.064 ± 0.157	
Dendroica magnolia	28	31	0.323	0.117 ± 0.208	
D. coronata	70	31	0.419	0.121 ± 0.175	
D. palmarum	24	31	0.323	0.134 ± 0.214	

(Continued)

TABLE I (Continued)

Family and species[b]	Number of genomes	Number of loci	Percent polymorphism	$\bar{H}_c \pm$ SD	Reference
D. striata	12	31	0.032	0.016 ± 0.090	
Seiurus auricapillus	20	31	0.290	0.126 ± 0.209	
S. noveboracensis	48	31	0.419	0.147 ± 0.219	
Oporornis philadelphia	16	27	0.222	0.081 ± 0.163	
Geothlypis trichas	32	30	0.233	0.084 ± 0.169	
Setophaga ruticilla	24	30	0.200	0.069 ± 0.159	
Dendroica coronata coronata	96	32	0.156[c]	0.031 ± 0.003	Barrowclough (1980a)
D. coronata auduboni	114	32	0.167[c]	0.034 ± 0.003	
Icterinae[e]					
Icterus galbula galbula	246	19	0.105	0.071 ± 0.005	Corbin et al. (1979)
I. galbula bullockii	132	19	0.105	0.073 ± 0.001	
Emberizinae[e]					
Zonotrichia leucophrys nuttalli	298	19	0.316	0.098 ± 0.158	Baker (1975)
Z. leucophrys oriantha	156	19	0.158	0.065 ± 0.100	
Z. leucophrys nuttalli	128	46	0.246[c]	0.085 ± 0.013	Corbin (1981; unpublished)
Z. leucophrys pugetensis	82	46	0.219[c]	0.096 ± 0.005	
Z. capensis hypoleuca	390	14	0.428	0.099 ± 0.009	Handford and Nottebohm (1976)
Z. capensis carabayae	126	45	0.489[c]	0.096 ± 0.159	Corbin (unpublished)
Z. melodia	14	20	0.050	0.007 ± 0.029	Avise et al. (1980b)
Z. georgiana	20	21	0.190	0.031 ± 0.071	
Z. albicollis	20	21	0.333	0.062 ± 0.109	
Junco hyemalis	20	21	0.143	0.045 ± 0.138	
Ammodramus sandwichensis	20	20	0.250	0.042 ± 0.089	

Species	N	% Polymorphism	$\bar{H}_c$	Reference
A. savannarum	20	0.200	0.051 ± 0.132	Avise et al. (1980b)
Spizella passerina	22	0.143	0.030 ± 0.083	
S. pusilla	21	0.143	0.054 ± 0.150	
Amphispiza bilineata	14	0.095	0.028 ± 0.101	
Pipilo fuscus	10	0.048	0.009 ± 0.039	
Zonotrichia leucophrys	38	0.231	0.032 ± 0.081	Zink (1982)
Z. albicollis	24	0.231	0.034 ± 0.075	
Z. atricapilla	30	0.154	0.041 ± 0.117	
Z. querula	36	0.128	0.019 ± 0.058	
Melospiza melodia	28	0.179	0.038 ± 0.097	
M. lincolnii	16	0.179	0.054 ± 0.128	
M. georgiana	32	0.231	0.040 ± 0.087	
Passerella iliaca	114	0.308	0.032 ± 0.069	
Junco hyemalis	96	0.179	0.029 ± 0.095	

[a] Percent polymorphism is the ratio of polymorphic loci to the total number of loci assayed. $\bar{H}_c$ is calculated as $\Sigma h/N$, where $h = 1-\Sigma x_i^2$, x_i is the frequency of the ith allele at locus x, and N is the number of loci assayed.
[b] Species were omitted from this summary if fewer than 5 individuals were examined; generic names in accord with those used by authority cited.
[c] These values of % polymorphism are the averages of two or more populations.
[d] Only liver enzymes have been examined in this study.
[e] Classification according to the Thirty-fourth Supplement to the A.O.U. Check-list of North American Birds.

$1 - \Sigma x^2_i$ (Nei, 1975), where x_i is the frequency of the ith allele at locus x. As in methods two and three, the values of h_l are averaged over loci, but now the average calculated heterozygosity, $\bar{H}_c$, is obtained. Previously, Corbin (1978) suggested that this approach to the estimation of average heterozygosity provides one of the few parameters available for comparing levels of genetic variability both across species and across studies by different laboratories. This is because the calculation of $\bar{H}_c$ does not depend on the electrophoretic analysis of identical or even homologous protein systems. However, in many cases it will be inappropriate to use this parameter because the method assumes that all loci of all populations included in the analysis are in Hardy–Weinberg equilibrium. Indeed, for all studies of avian populations in which this author has been involved, these conditions are not met. Nevertheless, it continues to be used, as a method of last resort, in the comparison of results from different studies (Table I), because it has been common practice to publish allelic frequencies rather than observed genotypic frequencies. As in methods two and three, the variance associated with $\bar{H}_c$ is large and can be reduced only by increasing the number of loci assayed. Repetitive sampling of local populations would provide independent estimates of $\bar{H}_c$, which could be averaged and the variance thereby reduced, but many avian populations could not withstand such an impact.

2.4. Genetic Distance between Taxa

The allelic frequencies of any two taxa can be compared and reduced to a single value that measures the genetic similarity or distance between these taxa. Two commonly used techniques have been devised for this purpose. The method of Nei (1972, 1978) estimates either the normalized genetic identity, I, of alleles shared by a pair of taxa, or their standard genetic distance, D. The parameter D is a nonlinear function of I, being calculated as the negative natural logarithm of I (Nei, 1972). The value D thus ranges between 0.0, for the comparison of two taxa that possess identical allelic frequencies at each and every locus, to a practical limit of 223 ($= -\ln 10^{-99}$). At the lower end of this range, i.e., for values of D between 0.0–4.6, which correspond to values of I between 1.0–.01, D may approximate the number of nucleotide base-pair substitutions per gene locus (assuming an average protein length of 100 amino acids = 300 nucleotide base-pairs) that have accumulated during geological time since the divergence of these taxa from their common ancestor. Although it requires additional explanation, which is given below in another section, Fig. 1 shows this relationship between I and D.

An alternative genetic distance metric is Rogers' index of similarity, S, which estimates the mean geometric distance between allelic frequencies, and reduces this information for all loci to a single, comparative value (Rogers, 1972).

Both Nei's and Rogers' indices have been widely used as the input for computer programs that reconstruct evolutionary relationships, e.g., the algorithms of Sneath and Sokal (1973) used to cluster taxa into hierarchical levels according to shared characteristics (genetic similarity) and the Wagner tree algorithm of Farris (1972, 1973) which fits genetic distance metrics to most-parsimonious branching networks (phylogenetic trees). Nei's distance does not, however, satisfy the conditions necessary for triangle inequality (Barrowclough and Corbin, 1978). This means that in the comparison of any three taxa, the estimated genetic distance between a given pair may not be less than the summed distances between the other two pairs. This condition results in negative branch lengths within a phylogeny reconstructed in this way, an artifact of the method. Rogers' index does satisfy the triangle inequality and has been used successfully to reconstruct evolutionary relationships among the wood warblers of the family Emberizidae (Barrowclough and Corbin, 1978), sparrows of the family Emberizidae (Zink, 1982), and species of the family Phasianidae (Gutiérrez et al., 1983).

2.5. Analysis of Genetic Variance

The analysis of the genetic structure of natural populations requires some means of determining the genic homogeneity of samples, both within and among populations. Genetic structure refers to several characteristics of populations that can be measured with data on population size and allelic frequencies or genotypic frequencies. It refers to the composition of a population with regard to its alleles and genotypes. In addition, the term refers to the relationships among genotypes within and between populations and the various subdivisions of populations. The degree of deviation from expected genotypic frequencies as a result of non-random mating and the amount of variance associated with such deviations are aspects of genetic structure. The subdivision of populations within a taxon is another aspect of genetic structure which is of particular interest to the systematist.

One way to analyze genetic structure is to determine how well a population conforms to the genotypic distributions expected on the basis of Hardy–Weinberg considerations. But this approach does not allow the simultaneous comparison of genotypic distributions at more than one locus or population. Workman and Niswander (1970) devised a test parameter, known as the genic contingency chi-squared value,

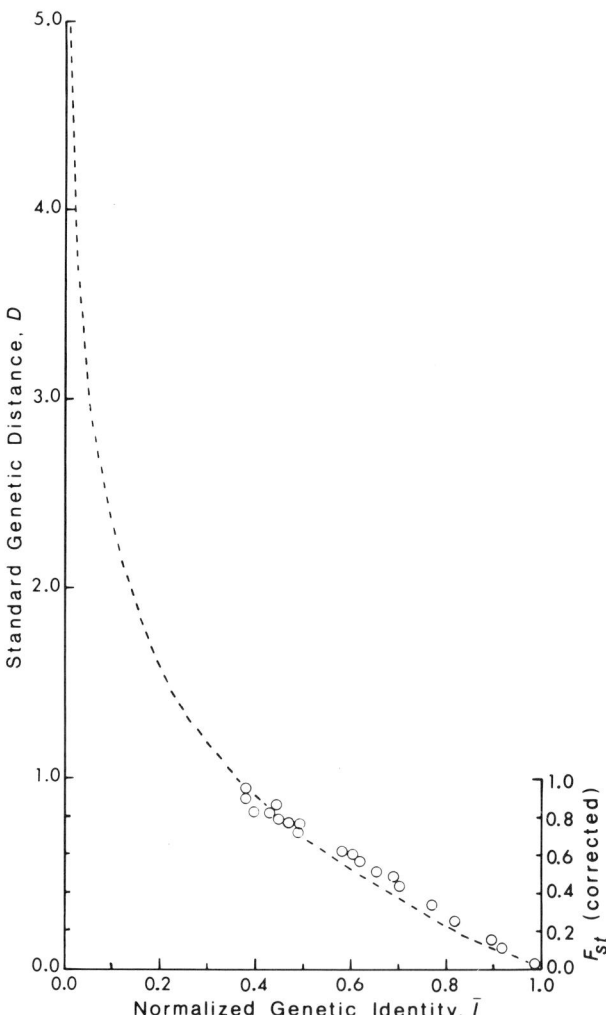

FIGURE 1. The dashed curve shows the relationship between Nei's (1972) normalized genetic identity, I, and Nei's standard genetic distance, D, which equals $-\ln I$. The open circles plot the Fixation Index, F_{st} (corrected for sample size bias), as a function of I, with which it is correlated at levels of r greater than 0.99 (see Fig. 5–8). These estimates of F_{st} vs. Nei's I are based on data for species of the family Phasianidae (Gutiérrez et al., 1983). Because F_{st} cannot be greater than 1.0, it is obvious that for these data the relationship between F_{st} and I will disintegrate at values of I below 0.3.

that does allow the simultaneous comparison of the allelic frequencies of any number of populations, but for only one locus at a time.

Interestingly, Wright's method of path analysis had solved this problem long before there were electrophoretic data to analyze (Wright, 1921). The resulting coefficients subsequently were called F statistics (Wright, 1951, 1965) to designate the different kinds of *fixation* indices. Some of these were measures of inbreeding at neutral gene loci while others related to genotypes undergoing assortative mating or differential selection (Wright, 1965). Unfortunately, Wright's method of path analysis was seldom employed by others outside the field of animal husbandry, probably due to the complexity of the technique. However, with the advent of powerful computers having large memory banks and the rigorous mathematical formulation of the various F parameters (Wright, 1965), this approach to the analysis of genetic variation in natural populations rapidly gained the favor of those who had electrophoretic data to analyze.

As developed by Wright, F statistics are correlations of several kinds: between homologous alleles of gametes uniting under different mating conditions, between alleles of non-combining gametes within individuals, and between alleles of gametes produced by siblings or individuals having other degrees of relationship to one another. For our purposes, the three F parameters proposed by Wright (1951) to describe the hierarchical nature of the subdivisions in natural populations, are of greatest interest. These are: (1) F_{st}, which is the correlation between alleles of randomly sampled gametes, within two or more subdivisions of a population or taxon, to the allelic distribution of the entire population, (2) F_{it}, which is the correlation between alleles of gametes that unite to produce individuals relative to the distribution of alleles in the total population, and (3) F_{is}, which is the average, over all subdivisions, of the correlations between alleles of uniting gametes within subdivisions relative to allelic distributions within those subdivisions. These three F statistics are interrelated according to the formula $F_{st} = (F_{it} - F_{is})/(1 - F_{is})$. Also, they are related directly to the degree of panmixia within and among subdivisions of a population by the formula $P_{it} = P_{is} \cdot P_{st}$ (Wright, 1965), where P, the panmictic index, equals $1 - F$. More recently, Wright (1978) has devised ways to account for sampling error in the calculation of F_{st}, which thus results in the estimation of three different values of F_{st}. These I shall designate here as F_{st} (uncorrected), F_{st} (corrected) and F_{st} (Wright), where the first does not correct for sampling error and the other two do so, but in different ways. Variance associated with the F_{st} parameter can be estimated by using the second method, in which case F_{st} values are first calculated for each locus and then averaged over loci. Of the three F_{st} values, F_{st} (corrected) is perhaps of greatest interest in avian systematics.

Some final comments are appropriate to help the reader understand and interpret the meaning of F_{st} values. As mentioned, Wright devised this method of analysis as a series of correlations between allelic distributions within individuals, subdivisions of populations, and the total gene pool. Cockerham (1969, 1973) showed that the same results can be achieved through an analysis of variance of allelic frequency distributions. If one corrects two of his working equations, it can be shown that the θ of Cockerham (1969, 1973) is identical to F_{st} (uncorrected) of Wright. Thus, one should view F_{st} as the probability that allelic frequency distributions, drawn at random for one or more loci from two or more populations, are unique. If they are identical, F_{st} will equal 0. If no alleles are shared, i.e., populations are fixed for alternative alleles, then $F_{st} = 1$. If, and only if, loci being compared possess only two alleles, and if sample sizes are identical for all populations, then F_{st} is directly related to the chi-squared test parameter as: $\chi^2 = 2NF_{st}$, where N is the total of all samples (Corbin, Cooke, and Weisberg, unpublished). Furthermore, Figs. 1 and 5–9 show that F_{st} (corrected) is nearly perfectly correlated with Nei's indices, corrected for sampling error (Nei, 1978). Figure 1 also shows the linear relationship between F_{st} and I, in contrast to the curvilinear relationship between I and D.

3. GENETIC DATA AND THEIR ROLE IN SYSTEMATICS

To reach the overall objective of systematics, i.e., a natural system of classification, it has been necessary to describe the variation found in natural populations, and then analyze and organize that variation into a hierarchic system. For these purposes only heritable variation is relevant.

The origin of heritable differences is, of course, mutation within local populations. Through time, new traits may spread to larger and larger units of genetic structure, from the local population to other portions of a deme, from one deme to others, and eventually perhaps to all the demes that comprise the species. During the process of speciation the trait is spread still farther, to higher taxa.

The processes by which variation, manifest as new genetic traits, is transferred from the local population to higher taxa is of considerable interest to the evolutionary biologist. Beginning in about the early 1900s and continuing through the 1960s, the prevailing viewpoint was that new traits increased in frequency, eventually to become fixed, as a result of natural selection favoring the new allele or genotype, while selecting against the old. With the advent of techniques that made it possible to characterize proteins and the genes controlling their synthesis, a new school of thought arose, the neutralist or non-Darwinian. This alternative viewpoint was first promulgated by Kimura (1968, 1969) and by King and Jukes (1969). To the extremists within the neutralist camp, nearly all variants were selectively equivalent (Kimura, 1968). Later it was conceded by some that although the majority of new mutations might be selectively equivalent, they were nevertheless selectively inferior to one or two predominant alleles found in natural populations. These more common alleles were maintained at high frequencies by natural selection for them, whereas many other alleles were maintained at much lower frequencies through the balancing processes of selection against an allele and recurrent mutation to it. This model has become known as the mutation–equilibrium hypothesis (Ohta, 1974), some details of which are identical to those proposed earlier by Corbin and Uzzell (1970).

This issue of how new mutants, new variants, are spread among populations and transferred to higher taxa holds important implications to the work of systematists. Evolution and dispersal of traits via natural selection alone would lead to plateaus of character differentiation in space and time. These plateaus would be recognized as gaps in character distributions and readily used to delimit taxa. On the other hand, evolution via the continuous, if not constant, accumulation of selectively equivalent alleles would lead inexorably to a steady increase in the degree of genetic differentiation between taxa, in which case gaps might not exist, except as superimposed by differential mutation rates and the accidents of extinction.

Evidence unequivocally supporting one or the other of these alternative processes of evolution has been difficult to obtain. One way to examine this issue is to compare the available genetic data with respect to Mayr's (1954) postulate that allopatric speciation results in "genetic revolution." Such changes would involve major reorganizations of species' gene pools during speciation. Natural selection would favor the incorporation of many new alleles into the gene pools of the incipient species. The end result would be relatively small genetic differences among intraspecific comparisons, with a jump to much larger genetic differences among interspecific comparisons. There may be disagreement, however, over how many loci must be involved before the reorganization is considered to be a genetic revolution.

To examine this hypothesis, it is necessary both to estimate genetic distances before, during, and after speciation events and to measure the rates of genetic divergence. It is not sufficient merely to show that full species are genetically more distinct than are sibling species, which are in turn more distinct than semispecies or subspecies, as presented for Drosophila (Avise, 1976). This would be true, even in the absence of genetic revolutions, if genetic divergence results from continuous processes.

To overcome these problems, the estimators of genetic distance have been evaluated as shown in Figs. 2 and 3. In each, Nei's D, corrected for bias due to sample size (Nei, 1978), is plotted as a function of F_{st} (corrected). These values are then used to estimate the two regression lines shown within each figure. One line passing through the open and closed circles defines intraspecific comparisons, whereas the second line passing through the squares defines interspecific comparisons. Among intraspecific comparisons, the open circles indicate comparisons between local populations whereas the closed circles indicate comparisons between subspecies. Figure 2 is based on data for sparrows of the family Emberizidae (Corbin, 1981, unpublished data; Zink, 1982). Figure 3 is based on data for wood warblers of the family Emberizidae (Barrowclough and Corbin, 1978; Barrowclough, 1980a,b). It should be noted that these estimates of D and F_{st} are based on fairly large data sets involving between 31–46 gene loci and the product of genomes times loci ranging from 496–4104.

Within regressions, the relationship between Nei's D and F_{st} is highly significant, with r being significantly different from zero at levels of p much less than 0.001. This means that F_{st} is also an outstanding estimator of genetic distance. But, whereas D varies between 0.0 and infinity (only to 223 for practical purposes), F_{st} varies only between 0.0–1.0. Thus, the linearity between D and F_{st} should disintegrate at higher values of D.

To examine the genetic revolution hypothesis, it is necessary to test the slope of the regression lines, within Figs. 2 and 3, against one another, i.e., to test the similarity of the regression coefficient, b, for intraspecific comparisons against b for interspecific comparisons. For comparisons involving taxa of sparrows (Fig. 2), b (intraspecific) is significantly different from b (interspecific) at a level of p much less than 0.005 ($F_{1,27}$ = 23.143; Steel and Torrie, 1960). For comparisons involving warblers (Fig. 3) the same is true ($F_{1,13}$ = 42.269, p << 0.005).*

*Comparisons of F_{st} to Nei's I, rather than D, yield the same results. That is, the regression coefficient resulting from intraspecific comparisons is significantly different from those of interspecific comparisons. Therefore, although a linear relationship between F_{st} and D may not exist at higher values of D, the effect of this deviation is negligible within the range of the values plotted in Figs. 2 and 3.

What does this mean in regard to the genetic revolution hypothesis? It indicates that genetic revolutions have occurred. There have been reorganizations of gene pools during the process of speciation, with the result that allelic frequency distributions are significantly different following speciation events. If it were otherwise, if species were only slightly farther along in the process of genetic divergence, in comparison to subspecies or local populations, than F_{st} and D should increase along a continuum. There would not be breaks or discontinuities in the progression from low to high values. In terms of the regression lines of Figs. 2 and 3, the expectation, in the absence of a genetic revolution, would be a single regression line. Clearly, this is not the case. Rather, the abrupt change in slopes, as values pass from intraspecific to interspecific comparisons, indicates that major reorganizations of genetic material have occurred at a time corresponding to speciation events.

Perhaps the above merely serves to confirm what taxonomists have known all along, that character distributions are discrete with respect to the boundaries between species. If it were otherwise, such that the evolution of species proceeded only via the constant accumulation of genetic differences, without an acceleration in the rate of divergence during speciation, then the delineation of species might be as difficult as is the recognition of the boundaries between taxa at other levels of classification.

3.1. Relationship between Genetic Structure and Higher Taxa

Taxa at all levels possess genetic structure. This structure can be described rigorously by means of F statistics, used in conjunction with measures of genetic distance. The results of these structural analyses can then be used as a basis for taxonomic decisions. How this might be done is presented below.

Of all the problems facing the taxonomist, few present more difficulty than the objective delineation of the boundaries of taxa, both above and below the species level, but the task of assigning individuals to species may be fairly simple. The two principal criteria against which we judge the boundaries of species are reproductive isolation and shared characteristics within groups of individuals. In practice, the latter criterion can be tested by means of sophisticated, multivariate statistical methods, whereas the former is assumed but not tested, except when intermediate forms are discovered. It then remains to be determined whether the intermediates are rare hybrids, or part of a clinal continuum between the previously recognized forms. This situation demands careful field observation and extensive sampling of the natural populations. In contrast to this, there is no single biological criterion for testing

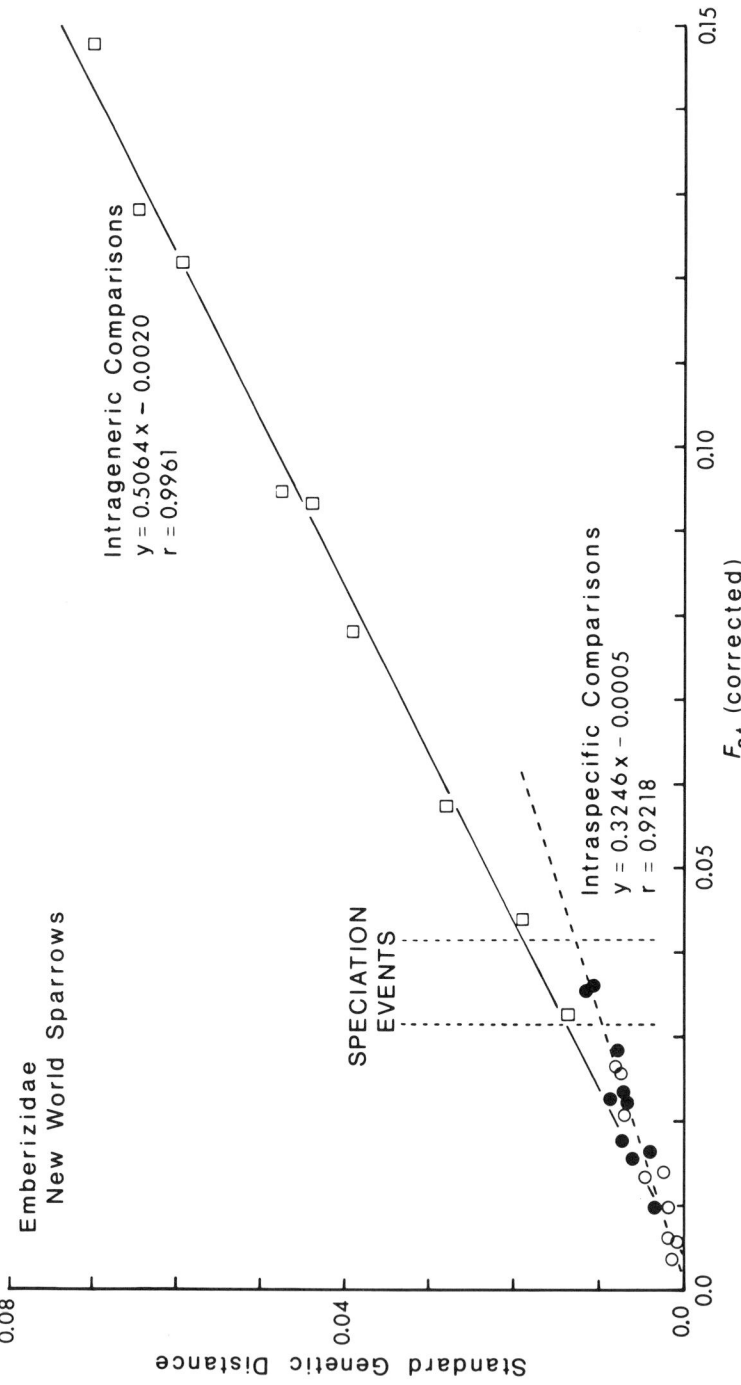

FIGURE 2. The hypothesis that allopatric speciation is accompanied by genetic revolution is examined here for sparrows of the family Emberizidae by comparing the slopes of the two regression lines. The regression coefficient 0.3246 of the dashed line, which passes through points (circles) that result from intraspecific comparisons, is significantly different from the regression coefficient 0.5064 obtained for interspecific comparisons (squares). Comparisons between local populations within *Zonotrichia leucophrys* are indicated by open circles (Corbin, unpublished). Comparisons between species of *Zonotrichia* or *Melospiza* are indicated by the open squares (based on allelic frequency data in Zink, 1982).

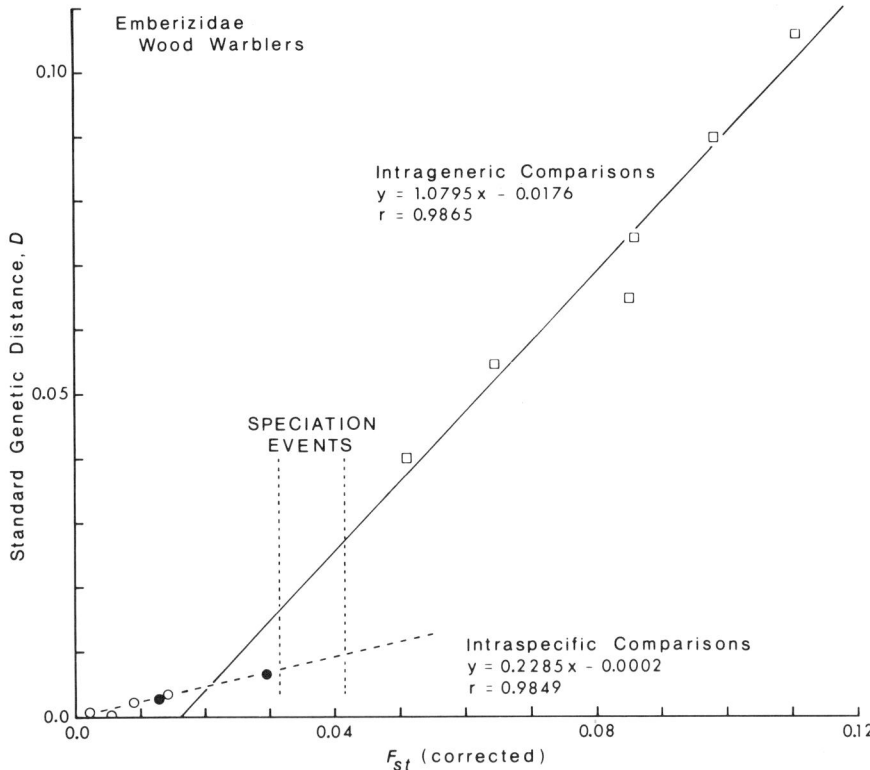

FIGURE 3. Examination of the genetic revolution hypothesis with respect to wood warblers of the family Emberizidae, subfamily Parulinae. As in Fig. 2, the regression coefficients are significantly different. Symbols have the same meaning as in Fig. 2, except that intraspecific comparisons involve local populations and subspecies of *Dendroica coronata* (based on allelic frequency data in Barrowclough, 1980a,b) and interspecific comparisons are made within *Dendroica*, *Vermivora*, or *Seiurus* (based on data in Barrowclough and Corbin, 1978).

whether two species belong in the same genus. In many respects the problem is little different from that faced by earlier taxonomists who, in the absence of the biological concept of species, had few means by which they could objectively measure whether two individuals belong to the same species.

Because population and genetic structures are attributes of biological species, it seems logical that these structural characteristics necessarily extend into the higher taxonomic levels if these categories have a biological reality. We merely need some means by which to recognize and measure this structure and then apply that knowledge to taxonomic practice.

The ability to obtain and rigorously analyze enormous amounts of genetic information would seem to provide the means for recognizing biologically meaningful higher categories. Nei's indices, either of genetic distance or similarity, and Wright's F statistics make it possible to distill this wealth of genetic information into single comparative numbers. It remains to examine the distribution of these parameters themselves, relative to firmly established boundaries between taxa, to discover the relationship, if any, between these parameters and taxonomic categories. The method used above to examine the genetic revolution hypothesis is equally suited to this kind of analysis.

A theoretical model against which to measure the relationship between taxonomic categories and F statistics or Nei's indices is presented in Fig. 4. The boundaries between the different kinds of comparisons, i.e., intrageneric, intergeneric (but intrafamilial), and interfamilial, would be established over time as sufficient data became available. Taxa of questionable relationship to others would not be used to establish the boundaries. That is, one would not include a monotypic genus in comparisons seeking to establish the limit of comparisons between genera because it might, in fact, belong within some other genus. The same precaution would be exercised for each of the other levels of comparison.

Within the various levels of comparison shown in Fig. 4, one expects to find clusters of closely related species, if there are discontinuities of allelic frequency distributions among genera. As illustrated, hypothetical species A, B, and C belong to one genus whereas species D belongs to another genus of the same family. Comparisons between A, B, and C all cluster together within the section labeled "intrageneric" comparisons. If they are indeed closely related, then each should have approximately the same degree of genetic differentiation in comparison to species D, as illustrated by the cluster within the intergeneric section of the diagram. Species E is genetically more similar to species D than to species A, B, or C, but E is unequivocally a member of another family. The "reality" of the genus composed of species A, B, and C is again demonstrated by their clustering together in comparison to species E. If now we suppose that yet another species, thought to be a member of the same genus as species A, B, and C, does not cluster with them at each level, this would be reason to question its assignment to that genus. Conversely, a species which is thought to belong to some other genus, but which clusters with species A, B, and C at each level, is probably misclassified.

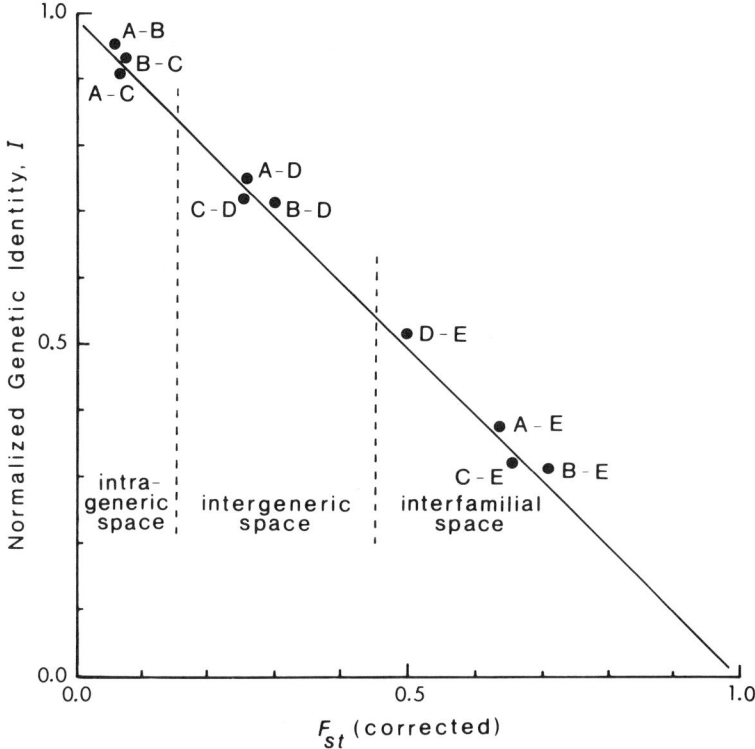

FIGURE 4. A model for the analysis of genetic structure at supraspecific levels of classification. Hypothetical species A, B, and C are closely related members of a genus. Comparisons of Wright's fixation index, F_{st}, or Nei's normalized genetic identity, I, involving these three species, cluster together within intrageneric space. Comparisons involving species A, B, and C to species D, which is placed in another genus *of the same family*, all cluster within intergeneric space. Species E is a member of a second family, but more closely related to species D than to species A, B, or C. The "reality" of the genus comprised of species A, B, and C is again demonstrated by their clustering together within interfamilial space.

The model proposed above is used to examine the relationships within the thrushes in Fig. 5 (based on the data in Avise et al., 1980a), the wood warblers in Fig. 6 (based on data in Barrowclough and Corbin, 1978), sparrows of the New World in Fig. 7 (based on data in Zink, 1982), and the family Phasianidae in Fig. 8 (based on data in Gutiérrez et al., 1983). Each point on these graphs, which relate F_{st} (corrected) to Nei's normalized genetic identity, represents the simultaneous pairwise comparison of the genetic data for all loci of all individuals examined for each species. The number of genomes sampled and the number of loci assayed varies for each species, as shown in Table I, and the product of these values ranges between 260, for the genus *Turdus* in Fig. 5, to a high of 4446, for the Fox Sparrow (*Passerella iliaca*) in Fig. 7. Both Nei's I and F_{st} values have been corrected for bias due to sample sizes.

In Fig. 5, the cluster labelled A, at the upper left corner, results from comparisons between three species of *Catharus*. The next lower cluster, B, results from comparisons between each of the three species

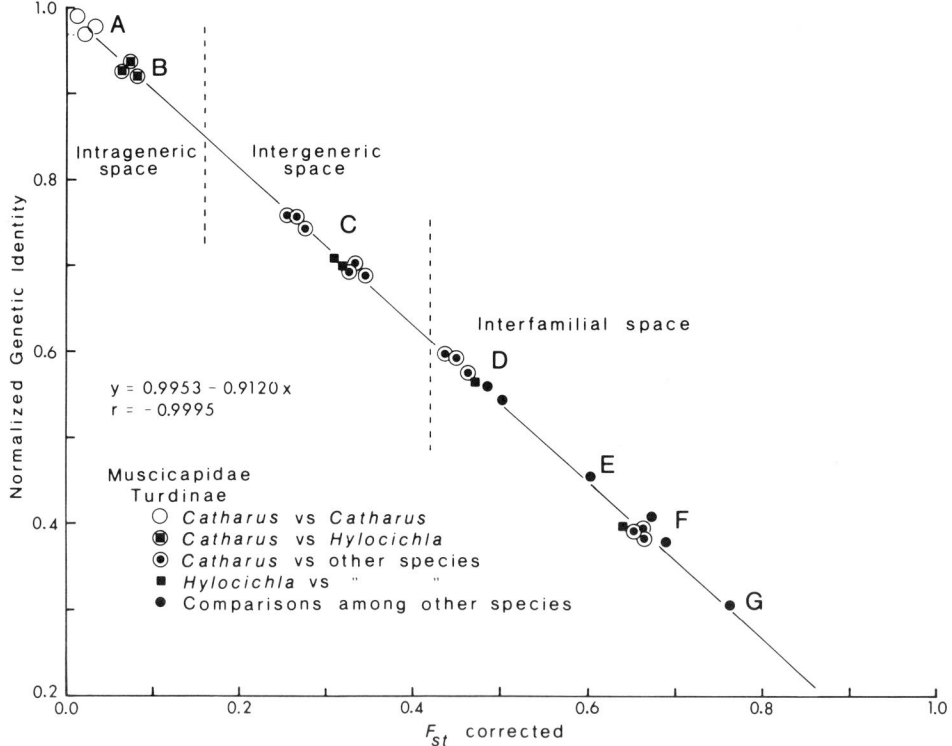

FIGURE 5. Analysis of genetic structure at supraspecific levels of classification as it relates to taxa of thrushes of the family Muscicapidae, subfamily Turdinae, and their allies. The division line between intra- and intergeneric space is based on information in Figs. 6–8. The intergeneric–interfamilial division line is placed so as to include all comparisons between turdids and the mimid (*Dumetella carolinensis*) within interfamilial space. Based on allelic frequency data in Avise et al. (1980a). Species are listed in Table I.

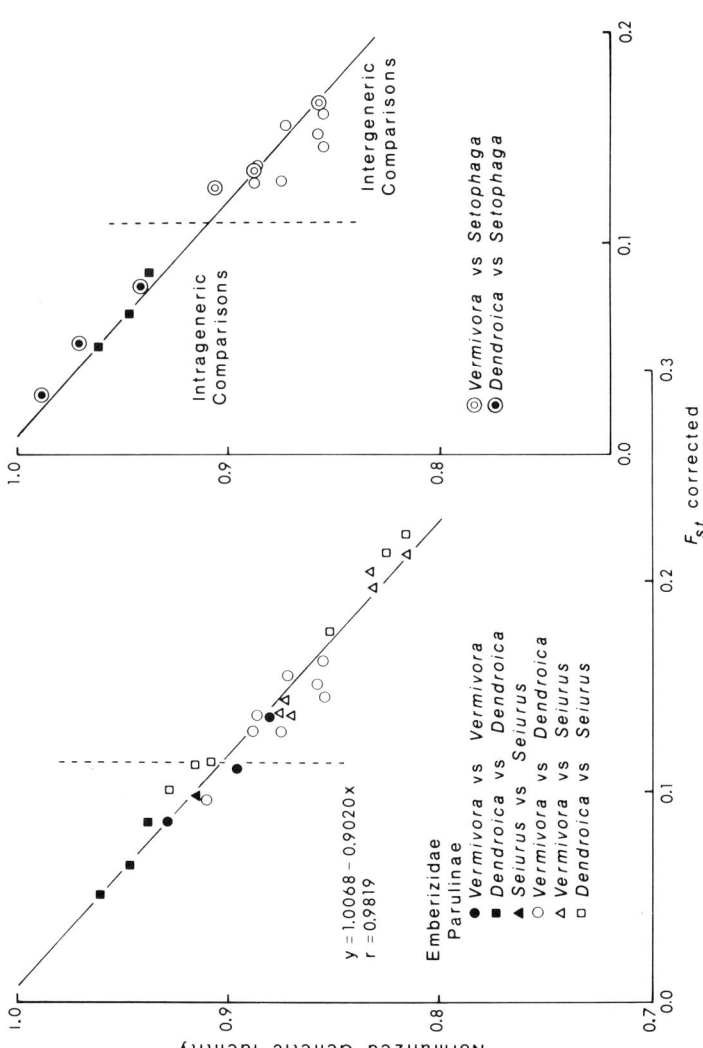

FIGURE 6. Analysis of genetic structure at supraspecific levels of classification as it relates to taxa of wood warblers of the family Emberizidae, subfamily Parulinae. As an aid to viewing, the comparisons involving Setophaga ruticilla to species of either Dendroica or Vermivora are placed separately on the right side. As points of reference, some of the comparisons on the left side are also shown on the right side. Based on allelic frequency data in Barrowclough and Corbin (1978). Species are listed in Table I.

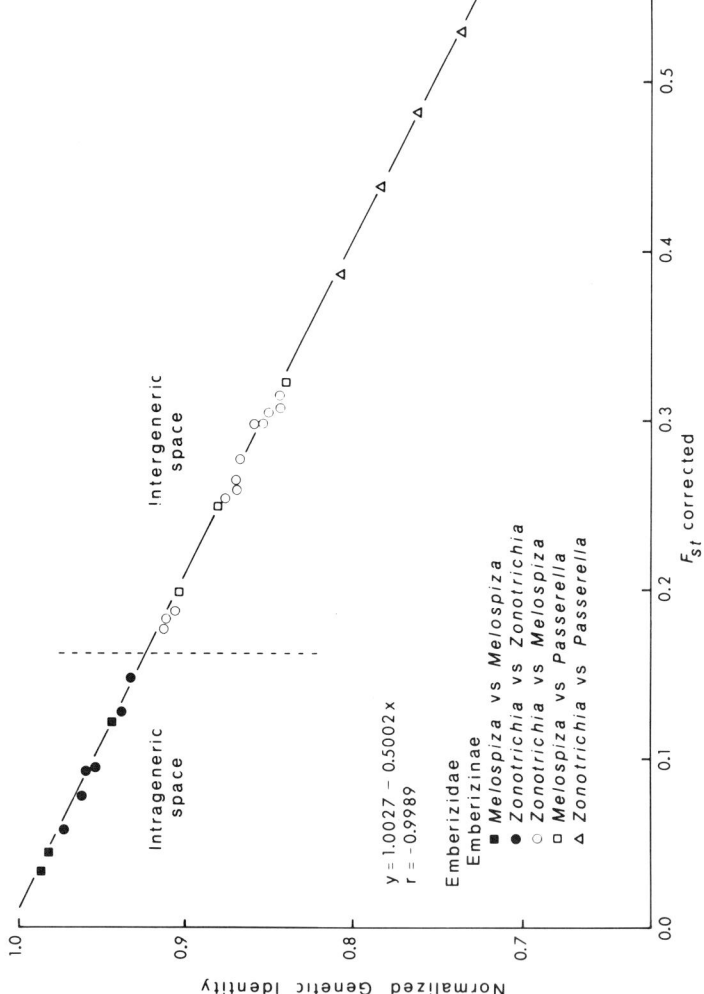

FIGURE 7. Analysis of genetic structure at supraspecific levels of classification as it relates to taxa of sparrows of the family Emberizidae, subfamily Emberizinae. Based on allelic frequency data in Zink (1982). Species are listed in Table I.

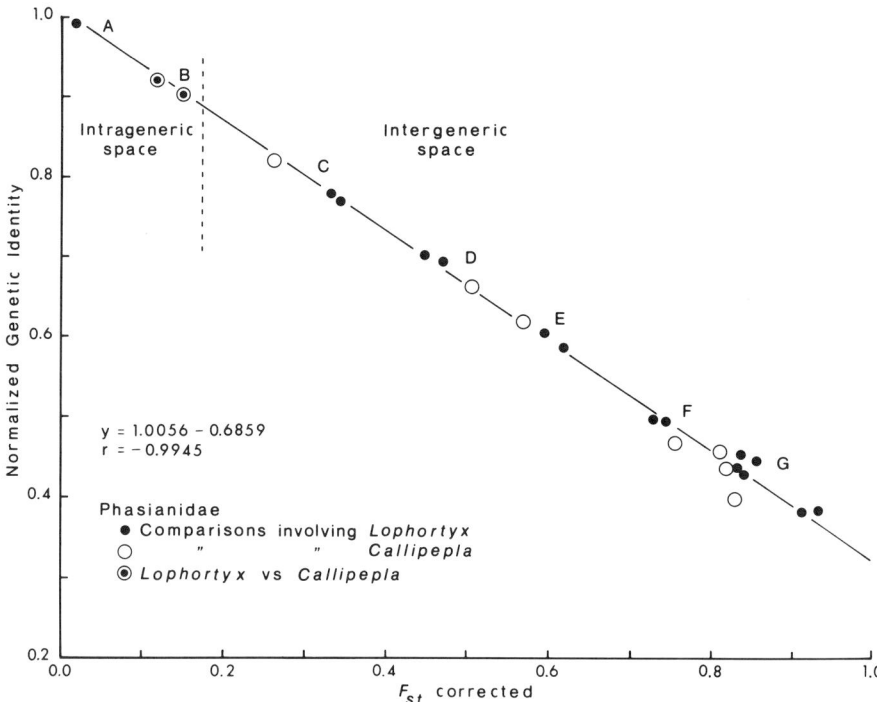

FIGURE 8. Analysis of genetic structure at supraspecific levels of classification as it relates to taxa of quails and their allies of the family Phasianidae. To simplify the diagram, only those comparisons involving species of Lophortyx and Callipepla, that is only 24 of 45 comparisons, are included. Based on allelic frequency data in Gutiérrez et al. (1983). Species are listed in Table I.

of *Catharus* and the Wood Thrush (*Hylocichla mustelina*). Cluster C results from comparisons of *Catharus* spp. and *Hylocichla* to the Eastern Bluebird (*Sialia sialis*) and the Robin (*Turdus migratorius*), all without question being intergeneric comparisons. With one exception, cluster D is composed of interfamilial comparisons between turdids and mimids, i.e., *Catharus* spp., *Hylocichla* or *Turdus* vs. the Gray Catbird (*Dumetella carolinensis*). The one exception is an intergeneric comparison between *Sialia* and *Turdus*. The remaining comparisons at E, F, and G are all interfamilial, at E between *Sialia* and *Dumetella*, at F between turdids and the sylviid *Regulus calendula*, the Ruby-crowned Kinglet, or between the mimid and sylviid, and at G between a turdid, *Sialia*, and the sylviid, *Regulus*.

The distribution of intrageneric comparisons in Figs. 6 and 7, which involve either parulines or fringillines, suggests that the genetic distinction between *Catharus* and *Hylocichla* falls well within the range of intrageneric comparisons. Furthermore, a close examination of clusters C and D of Fig. 5 reveals that intrafamilial comparisons involving *Hylocichla* and either *Turdus* or *Sialia* cluster tightly with those involving the *Catharus* spp. and these other two turdids. For these reasons, the intrageneric division line in Fig. 5 has been placed between clusters B and C, rather than between clusters A and B. These data therefore indicate that the monotypic genus *Hylocichla* should be merged with *Catharus*. This, then, is an example of a taxonomic decision, involving a category of classification higher than the species, that could be made objectively on the basis of biologically significant differences among other species within the order Passeriformes.

In Fig. 6, the left half shows an array of intrageneric and intergeneric comparisons for genera of warblers that seem firmly established, i.e., *Dendroica*, *Vermivora*, and *Seiurus*. Species within these genera are then compared to the American Redstart (*Setophaga ruticilla*). These are shown in the right-hand portion of Fig. 6, where it is obvious that *Setophaga* groups with species of *Dendroica* for both intra and intergeneric comparisons. If one accepts this result, it then appears that species of *Setophaga* are merely aberrant, with respect to details of coloration and principal mode of food gathering, in comparison to species of *Dendroica*, but that the majority of genes are either identical or very similar to those of *Dendroica*. This then suggests that the genus *Setophaga* should be synonomized with *Dendroica*. (As supporting evidence, *Setophaga* spp. are known to hybridize with *Dendroica* spp., e.g., Burleigh, 1944).

The comparisons among species of emberizines, shown in Fig. 7, do not present any surprises. Intrageneric comparisons involving species of either *Zonotrichia* or *Melospiza* all fall together within "intrageneric space," whereas comparisons between these genera and between them and the Fox Sparrow (*Passerella iliaca*) all fall within the "intergeneric space" of the diagram.

Figure 8 shows the relationships among comparisons of phasianids, and is notable with respect to the grouping of the Scaled Quail (*Callipepla squamata*). In comparisons involving *Callipepla*, it always clusters with two species of *Lophortyx*, the California Quail (*Lophortyx californicus*) and Gambel's Quail (*L. gambelii*).* The comparison of

---

*This analysis was completed before learning that the A.O.U. Check-list Committee had made the decision to synonomize *Lophortyx* with *Callipepla*, the generic name that takes precedence according to the International Rules of Zoological Nomenclature.

these two species of *Lophortyx* is shown at point A and comparisons between *Lophortyx* and *Callipepla* at cluster B, which falls within "intrageneric space," as judged from the comparisons in Figs. 5–7. (It is possible that intrageneric space may be larger in comparisons involving non-passerines, but much more information is needed to establish this relationship). The relationships between *Lophortyx* spp. and *Colinus, Oreortyx, Cyrtonyx, Alectoris, Tympanuchus,* and *Phasianus* are shown at clusters C through G, respectively, and indicated by the closed circles. Comparisons between each of these six genera and *Callipepla* are indicated by the open circles, which group with those involving *Lophortyx* in every case. As with the genera *Hylocichla* and *Setophaga*, this suggests that the recognition of a separate genus for *Callipepla squamata* forces the classification into an arrangement that is not natural, and into a distinction that is not biologically based.

The method of analysis presented here obviously needs refinement in a number of ways. Future studies might be designed to determine the limits of variation within many other firmly established genera. This will help to delineate intrageneric space and to determine whether this space is similar over taxa. At this time, the division line or zone between intra and intergeneric space appears to fall between F_{st} values (corrected) of 0.12–0.15, but this is based on meager data.

3.2. Genetic Structure at the Species Level

It has been argued above that a careful analysis of the genetic structure of natural populations leads to a biological basis of classification, not only at the species level but above it and below it as well.

Following the development of the biological concept of species (Mayr, 1963), the principal conceptual problems associated with the recognition of species were resolved. Thereafter, a species, as defined by Mayr (1940), has been recognized as groups of actually or potentially interbreeding natural populations that are reproductively isolated from other such groups. Although the evidence for major genetic reorganization during speciation was largely inferential, i.e., the fact that two organisms cannot interbreed or produce fertile progeny says little about the number of genetic differences between them, only that at least one inherited trait prevents the passage of genes to progeny, the taxonomist had little difficulty recognizing gaps in character distributions. Whether genetic reorganization extended to other traits that could be observed and measured was of little consequence to taxonomists. The presentation of evidence above suggesting that major genetic changes in "hidden" character systems do occur during speciation serves to confirm that significant gaps do exist between species in nature. These gaps involve many traits, not merely those of convenience to the taxonomist.

An implication of this is that electrophoretic data are well suited for the purpose of delineating species and ordering them into natural classifications. Few systematists have used such data to delineate species, but during the past decade it has been common practice to use allelic frequency information to reconstruct phylogenetic relationships. With the development of algorithms for this purpose (Farris, 1972; Sneath and Sokal, 1973), it was only a matter of time before phylogenies reconstructed in this way were shown to be congruent with those derived by using more classical techniques. For birds, analyses of this type have been carried out on wood warblers of the family Emberizidae (Barrowclough and Corbin, 1978), thrushes of the family Muscicapidae and their allies, the Ruby-crowned Kinglet (*Regulus calendula*), and the Gray Catbird (*Dumetella carolinensis*; Avise et al., 1980a), sparrows of the family Emberizidae (Avise et al., 1980b; Zink, 1982), and species of the family Phasianidae (Gutiérrez et al., 1983).

If these phylogenies were no more than similar to schemes of classification based upon the results of classical morphological analyses, they would serve the sole function of confirmation. They would not add significantly to our knowledge of avian relationships. However, they offer much more than mere confirmation of past efforts. On the one hand, they present an internal consistency that one normally does not find among the various cladistic arrangements derived on the basis of morphological data. That is, the various methods of cladistic analysis, each with its own set of assumptions concerning how best to reconstruct evolutionary relationships, produce nearly identical results when allelic frequency data are used. This was well demonstrated in the studies of Avise et al. (1980a) and Zink (1982). Often this is not true of cladistic analyses based on morphological data, and one is left to puzzle over which cladogram represents the natural arrangement.

Of equal or greater importance is the relationship between the underlying genetic blueprint and the data base used to reconstruct the evolutionary relationships. Regulatory genes must play an incredibly complex role in determining the size, shape, color, and behavior of organisms. As such, these attributes of organisms are secondary or tertiary expressions of gene action and subject to many kinds of confounding interactions that lead to convergent or parallel evolutionary changes. This complexity and lack of direct expression is then reflected in different cladistic arrangements, cladograms, that vary as a function of the particular sets of characters examined.

In contrast to this, proteins are a direct expression of genetic structure, unmodified so far as is known by the action of regulatory genes, and thereby less subject to the confounding effects of convergent or parallel evolution. They are not free from these influences, as shown by Corbin and Uzzell (1970) and Uzzell and Corbin (1971), only affected to a much lesser degree. The result is that different subsets of protein data give nearly identical results in cladistic analyses.

3.3. Genetic Structure within Species

There has been controversy for years concerning taxonomic problems at the subspecific level of classification. For instance, are subspecies biological entities or merely conveniences for pigeonholing organisms? If subspecies are to be recognized, on what basis should this be done? There are two published forums on this subject, one appearing nearly 30 years ago (*Systematic Zoology*, 1954, pp. 99–125), and the second more recently (*The Auk*, 1982, pp. 593–615).

To deny the existence of subspecies is to deny the biological reality of the genetic subunits of which species are composed. Admittedly there are no obvious limits to local populations or demes that make up subspecies. They nevertheless possess a genetic, and therefore biological, reality defined not in terms of physical structure, but in terms of differences in the frequencies of alleles at each and every locus.

The discussion in several sections of this chapter has dealt with methods for recognizing and describing genetic structure. In each case, the smallest unit of comparison was the distribution of the alleles of individuals at a single locus, and this was built upon, via averaging processes, to combine the information for all loci simultaneously. From this it is clear that the degree to which F_{st} (corrected) and Nei's indices measure genetic structure is a function of the size of the geographic area included in the comparison. The taxonomic problem, therefore, reduces to one of discovering homogeneous subsets of local populations that coincide with demes, which in turn may correspond to what taxonomists conceive to be subspecies. Corbin (1981) suggested that variation in the average observed heterozygosity, $\bar{H}_o$, of local populations, within and outside of zones of intergradation between subspecies, could be used to distinguish between subspecies that were genetically distinct. In that study, the measure of distinctness was F_{st}, and $\bar{H}_o$ decreased exponentially with distance from the midpoint of the zone of intergradation between *Zonotrichia leucophrys nuttalli* and *Z. l. pugetensis*.

An alternative approach to the analysis of genetic variation in zones of intergradation is shown in Fig. 9, which also is based on the allelic frequency data for the White-crowned Sparrow (*Z. leucophrys*; Corbin

1981, unpublished data). Again, F_{st} (corrected) is plotted against Nei's standardized genetic distance, but for comparisons between local populations that form a transect through all of the areas inhabited by Z. l. nuttalli, across the zone of intergradation and into areas inhabited by Z. l. pugetensis. The striking feature of Fig. 9 is that comparisons between populations separated by the greatest geographic distances, which

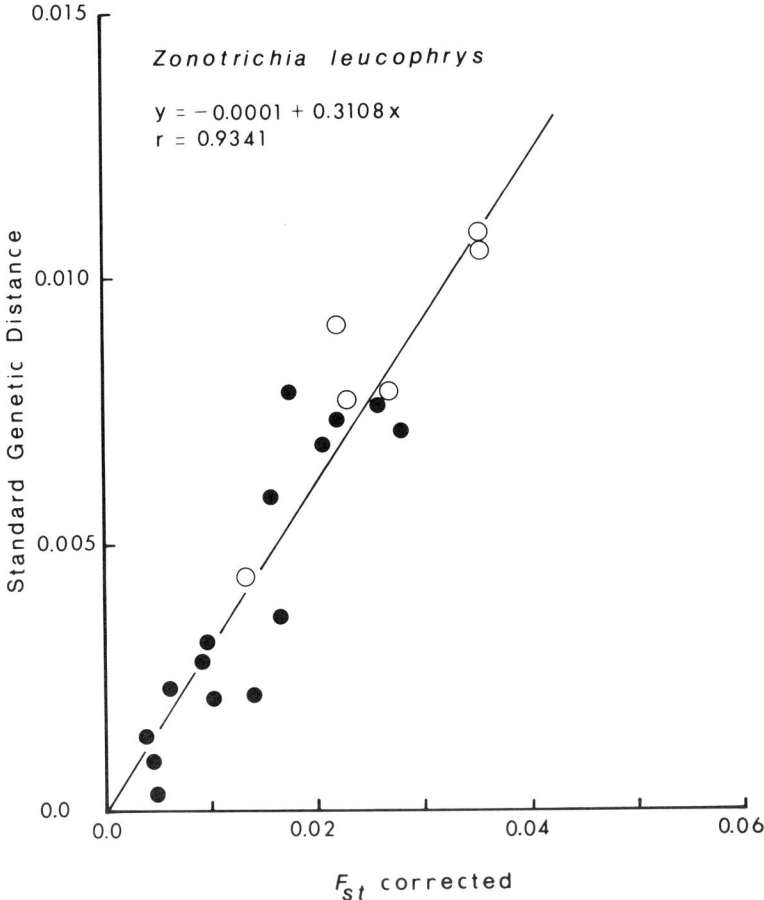

FIGURE 9. Analysis of genetic structure at the infraspecific level of classification as it relates to subspecies and a zone of intergradation between Zonotrichia leucophrys nuttalli and Z. l. pugetensis. The open circles indicate comparisons involving a local population at Manchester, California, which lies near the midpoint of the zone of intergradation, with each of the other local populations. Based on the unpublished data of Corbin.

involve both subspecies, do not yield the highest values. Rather, with one exception, all of the highest values, indicated by the open circles, involve comparisons between the population at Manchester, California, at the midpoint of the zone of intergradation, and each of the other populations. The one exception involves the comparison with a population of Z. l. pugetensis at Yachats, Oregon, which is apparently more similar to the population at Manchester than are the other populations.

The implication is that zones of intergradation between well-differentiated subspecies are even more distinct, in terms of allelic frequency distributions, than are the subspecies themselves. This result is counter intuitive, but it is understandable when one examines the allelic frequency distributions within zones of intergradation. Cline theory predicts that allelic frequencies within zones of intergradation will be intermediate to those in parental populations. (The shape of the cline is irrelevant here, only that frequencies of alleles are intermediate). In at least two studies of avian clines, this prediction has not been observed. In the contact zone between two subspecies of the Northern Oriole (Icterus galbula; Corbin et al., 1979) and between the two subspecies of the White-crowned Sparrow (Corbin, unpublished data) allelic frequencies vary dramatically, with several abrupt changes within the zone of intergradation to frequencies either higher or lower than those in the parental populations. At this time, there is no plausible explanation for these observations. Nevertheless, if future studies bear out these findings the systematist may be able to take advantage of this variability within zones of contact as an aid to the recognition of boundaries between subspecies that may indeed have a biological reality.

4. CONCLUSIONS

The analysis of genetic structure of avian populations, through the characterization of protein variation, provides a means whereby systematists can construct a natural classification of birds. Genetic structure exists at taxonomic levels above and below that of the species. At the subspecies level, genetic structure is expressed both as significant differences in allelic frequencies among demes and as increased genetic variability within zones of intergradation between subspecies. At the species level, gaps in allelic frequency distributions exist as a result of major genetic reorganizations that occur during speciation. These are expressed as significantly greater genetic distances between species, and these estimates of genetic distance can be used to reconstruct phylogenetic relationships. At the generic level, apparently natural clusters of species can also be recognized through analyses of variance of allelic frequencies to obtain Wrightian F statistics. When F_{st} is graphed as a function of Nei's genetic distance or identity, clusters form that correspond to intrageneric, intergeneric, and interfamilial expectations. More rigorous analyses, specifically designed to test these observations, are needed to define the boundaries of these clusters, but at this time such analyses seemingly provide a biological basis for recognizing taxonomic categories at several levels of classification.

ACKNOWLEDGMENTS. I extend my appreciation to Gloria M. Garcia, Susan I. Evarts, and Jane Stauffer who each put in long hours preparing the published data for computer analysis. For critical advice I am particularly grateful to Harrison B. Tordoff and Patricia J. Wilkie. Robert M. Zink kindly allowed me to use unpublished data in his manuscripts. Portions of this study were supported by a subcontract to NSF grant DEB 80-12403, two grants from the National Geographic Society, and grant No. 8599 from the Penrose Fund of the American Philosophical Society.

REFERENCES

Avise, J. C., 1976, Genetic differentiation during speciation, in: *Molecular Evolution* (F. J. Ayalla, ed.), Sinauer, Sunderlund, Massachusetts, pp. 106–122.
Avise, J. C., Patton, J. C., and Aquadro, C. F., 1980a, Evolutionary genetics of birds. I. Relationships among North American thrushes and allies, *Auk* **97**:135–147.
Avise, J. C., Patton, J. C., and Aquadro, C. F., 1980b, Evolutionary genetics of birds. II. Conservative protein evolution in North American sparrows and relatives, *Syst. Zool.* **29**:323–334.
Baker, M. C., 1975, Song dialects and genetic differences in white-crowned sparrows (*Zonotrichia leucophrys*), *Evolution* **29**:226–241.
Barrowclough, G. F., 1980a, Genetic differentiation in the *Dendroica coronata* complex, Ph.D. Dissertation, University of Minnesota.
Barrowclough, G. F., 1980b, Genetic and phenotypic differentiation in a wood warbler (Genus *Dendroica*) hybrid zone, *Auk* **97**:655–668.
Barrowclough, G. F., and Corbin, K. W., 1978, Genetic variation and differentiation in the Parulidae, *Auk* **95**:691–702.
Barrowclough, G. F., Corbin, K. W., and Zink, R. M., 1981, Genetic differentiation in the Procellariiformes, *Comp. Biochem. Physiol.* **69B**:629–632.
Burleigh, T. H., 1944, Description of a new hybrid warbler, *Auk* **61**:291–293.
Cockerham, C. C., 1969, Variance of gene frequencies, *Evolution* **23**:72–84.
Cockerham, C. C., 1973, Analyses of gene frequencies, *Genetics* **74**:679–700.
Corbin, K. W., 1978, Genetic diversity in avian populations, in: *Endangered Birds: Man-*

agement Techniques for Preserving Threatened Species (S. A. Temple, ed.), The University of Wisconsin Press, Madison, pp. 291–302.

Corbin, K. W., 1981, Genic heterozygosity in the White-crowned Sparrow: A potential index to boundaries between subspecies, Auk **98**:669–680.

Corbin, K. W., and Uzzell, T., 1970, Natural selection and mutation rates in mammals, Am. Natural. **104**:37–53.

Corbin, K. W., Sibley, C. G., Ferguson, A., Wilson, A. C., Brush, A. H., and Ahlquist, J. E., 1974, Genetic polymorphism in New Guinea starlings of the genus *Aplonis*, Condor **76**:307–318.

Corbin, K. W., Sibley, C. G., and Ferguson, A., 1979, Genic changes associated with the establishment of sympatry in orioles of the genus *Icterus*, Evolution **33**:624–633.

Farris, J. S., 1972, Estimating phylogenetic trees from distance matrices, Am. Natural. **106**:645–668.

Farris, J. S., 1973, A probability model for inferring evolutionary trees, Syst. Zool. **22**:250–256.

Gutiérrez, R. J., Zink, R. M., and Yang, S. Y., 1983, Genic variation, and systematic and biogeographic relationships of some galliform birds, Auk (in press).

Handford, P., and Nottebohm, F., 1976, Allozymic and morphological variation in population samples of rufous-collared sparrow, *Zonotrichia capensis*, in relation to vocal dialects, Evolution **30**:802–817.

Henning, U., and Yanofsky, C., 1963, An electrophoretic study of mutationally altered A proteins of the tryptophan synthetase of *Escherichia coli*, J. Mol. Biol. **6**:16–21.

Kimura, M., 1968, Evolutionary rate at the molecular level, Nature **217**:624–626.

Kimura, M., 1969, The rate of molecular evolution considered from the standpoint of population genetics, Proc. Natl. Acad. Sci. USA **63**:1181–1188.

King, J. L., and Jukes, T. H., 1969, Non-Darwinian evolution, Science **164**:788–798.

Mayr, E., 1940, Speciation phenomena in birds, Am. Natural. **74**:249–278.

Mayr, E., 1954, Change of genetic environment and evolution, in: *Evolution as a Process* (J. Huxley, A. C. Hardy, and E. B. Ford, eds.), Allen and Unwin, London, pp. 157–180.

Mayr, E., 1963, *Animal Species and Evolution*, Harvard University Press, Cambridge, Massachusetts.

Nei, M., 1972, Genetic distance between populations, Am. Natural. **106**:283–292.

Nei, M., 1975, *Molecular Population Genetics and Evolution*, North-Holland, Amsterdam.

Nei, M., 1978, Estimation of average heterozygosity and genetic distance from a small number of individuals, Genetics **89**:583–590.

Nei, M., and Roychoudhury, A. K., 1974, Genic variation within and between the three major races of man, Caucasoids, Negroids, and Mongoloids, Am. J. Hum. Genet. **26**:421–443.

Ohta, T., 1974, Mutational pressure as the main cause of molecular evolution and polymorphism, Nature **252**:351–354.

Rogers, J. S., 1972, Measures of genetic similarity and genetic distance, Univ. of Texas Publ. 7213:145–153.

Sneath, P. H., and Sokal, R. R., 1973, *Numerical Taxonomy*, W. H. Freeman and Company, San Francisco.

Steel, R. G. D., and Torrie, J. H., 1960, *Principles and Procedures of Statistics*, McGraw-Hill, New York, p. 320.

Uzzell, T., and Corbin, K. W., 1971, Fitting discrete probability distributions to evolutionary events, Science **172**:1089–1096.

Wahlund, S., 1928, Zuzammensetzung von Populationen und Korrelationserscheinungen vom Standpunkt der Vererbungslehre aus betrachtet, *Hereditas* **11**:65–106.

Workman, P. L., and Niswander, J. D., 1970, Population studies on Southwestern Indian tribes. II. Local genetic differentiation in the Papago, *Am. J. Hum. Genet.* **22**:24–49.

Wright, S., 1921, Systems of mating, *Genetics* **6**:111–178.

Wright, S., 1951, The genetical structure of populations, *Ann. Eugen.* **15**:323–354.

Wright, S., 1965, The interpretation of population structure by F-statistics with special regard to systems of mating, *Evolution* **19**:395–420.

Wright, S., 1978, *Evolution and the Genetics of Populations, Volume 4, Variability Within and Among Natural Populations*, The University of Chicago Press, Chicago.

Zink, R. M., 1982, Patterns of genic and morphologic variation among sparrows in the genera *Zonotrichia, Melospiza, Junco* and *Passerella, Auk* (in press).

CHAPTER 9

PHYLOGENY AND CLASSIFICATION OF BIRDS BASED ON THE DATA OF DNA–DNA HYBRIDIZATION

CHARLES G. SIBLEY and JON E. AHLQUIST

1. INTRODUCTION

The elements of phylogeny are *clustering*, *branching*, and *time*. Ideally, to reconstruct the phylogeny of a group of organisms it is necessary to define the monophyletic clusters of taxa, to determine the branching pattern of their divergences, and to place that pattern on the scale of absolute time. This definition is simple, but to reconstruct a phylogeny in which all three elements are correctly determined is far more difficult. To our knowledge it has never been done with a high degree of accuracy.

The technique of DNA–DNA hybridization yields data that can be used to delineate clusters and to determine the branching pattern in terms of *relative* time. To convert relative time to absolute time it is necessary to calibrate the DNA hybridization measurements against fossil evidence of divergence dates, or against dated geological events which caused phyletic dichotomies between tbe ancestors of living taxa. The validity of this procedure depends upon (1) the ability of the technique to measure the "genetic distance" between taxa, and (2) the ex-

---

CHARLES G. SIBLEY and JON E. AHLQUIST • Peabody Museum of Natural History and Department of Biology, Yale University, New Haven, Connecticut 06511.

istence of the same *average* rate of DNA evolution in all lineages. We will present evidence in support of these requirements. The accuracy of the absolute time dimension obviously depends upon the accuracy of the fossil and/or geological datings.

The derivation of a classification from a phylogeny may take various forms, depending upon the opinions of the classifier. We believe that a classification should reflect the cladistic pattern of the phylogeny, that taxonomic rank should be determined by the age of origin of taxa, and that "sister groups" should be of coordinate rank. We also believe that a classification derived from DNA hybridization data will express degrees of genetic divergence, although it may not reflect the degrees of morphological specialization ("grades") as judged by the human eye. We thus follow Hennig (1966) concerning the relationship between phylogeny and classification, although for somewhat different reasons.

Since 1974 we have produced and studied more than 17,000 DNA–DNA hybrids among avian taxa. The DNAs of over 1000 species of birds, representing all of the orders, and 163 of the 171 families recognized by Wetmore (1960), have been used in these comparisons. In this paper we describe the DNA hybridization technique and demonstrate its ability to improve our understanding of the systematic relationships and evolutionary history of birds.

2. THE DNA–DNA HYBRIDIZATION TECHNIQUE

2.1. DNA Structure and Properties

A moderately complete description (Sibley and Ahlquist, 1981a), and synoptic accounts (Sibley and Ahlquist, 1981b, c, 1982a–e; Sibley et al., 1982) of the DNA hybridization technique have been published. Our procedures are based on those of Britten and Kohne (1968), Kohne (1970), Kohne and Britten (1971), and Britten et al. (1974).

The DNA hybridization technique takes advantage of the double-stranded structure of the DNA molecule. The two strands of DNA are composed of linear sequences of four kinds of subunits, called nucleotides, which differ in the composition of their nitrogenous bases, namely, adenine (A), thymine (T), guanine (G), and cytosine (C). In double-stranded DNA, the bases occur as complementary pairs, an A in one strand pairs only with a T in the other, and a G pairs only with a C. A-T pairs are held together by two hydrogen bonds, G-C base pairs by three. Genetic information is encoded in the linear sequences of the bases. If native DNA is heated in solution to ca. 100°C, the hydrogen bonds between complementary base pairs will rupture ("melt") and the two strands will separate. Upon cooling the double-stranded molecules reform because complementary bases "recognize" one another and reassociate, i.e., hydrogen bonding is re-established. If the incubation temperature is maintained at or above ca. 50°C base pairing will occur only between long homologous sequences of bases because only long sequences of complementary base pairs have sufficient bonding strength to maintain stable duplexes at those temperatures, and only homologous sequences possess the necessary degree of complementarity. Thus, under appropriate conditions of temperature and salt concentration, the single strands of conspecific DNA will reassociate only with their homologous partners and the complementary matching of base pairs will be essentially perfect.

Similarly, if the single-stranded DNA molecules of two different species are combined under conditions favorable only for the reassociation of long complementary base sequences, "hybrid" double-stranded DNA molecules will form *only* between homologous base sequences. Such hybrid DNA molecules will contain mismatched base pairs because of the differences in their nucleotide sequences that have evolved as the result of the fixation of point mutations since the two species diverged from their most recent common ancestor. If such hybrid DNA molecules are then placed in a thermal gradient they will melt at a temperature lower than that required to melt conspecific double-stranded molecules. The reduction in the melting temperature of interspecific DNA–DNA hybrid molecules, compared with that of conspecific double-stranded molecules, is proportional to the percentage of mismatched bases between the two strands. Each 1°C reduction in melting temperature indicates that ca. 1% of the bases are noncomplementary (Bonner et al., 1973).

The extent of base pair mismatch between homologous DNA sequences is measured by determining (1) the extent of reassociation, i.e., the percentages of the two genomes that form stable hybrids at a given incubation temperature, e.g., 60°C, and (2) the thermal stability, i.e., melting temperature, of the reassociated double-stranded molecules.

At 60°C the rate of reassociation is maximal and the precision of base pairing is high. For these reasons, 60°C is used as the "standard" incubation temperature.

2.2. The Sequence Organization of the Genome in Relation to DNA–DNA Hybridization

The words "sequence," "nucleotide sequence," or "base sequence," designate a segment of DNA without reference to its size, function, or location. DNA sequences are arranged in tandem and they vary in the lengths of functional units ("genes") and in the number of identical, or similar, copies per genome. Under a given set of experimental conditions, the genomes of eukaryotes *appear* to be composed of two main frequency classes of nucleotide sequences: (1) those with a frequency of one copy per haploid genome are the "single-copy," "unique," or "nonrepeated" sequences, and (2) those occurring with a frequency greater than one copy per haploid genome are the "repeated," "redundant," or "multiple-copy" sequences.

The approximate number of copies of a sequence may be determined by measuring the rate of reassociation which is proportional to the number of copies present. The rate of reassociation is also affected by other factors, especially by the temperature and salt concentration. Therefore, under different sets of experimental conditions, the rate varies, hence the proportions of repeated and single-copy sequences also vary. Thus "the assignment of any given sequence to one or another category . . . is dependent upon the reaction conditions chosen by the experimenter" (McCarthy and Farquhar, 1972, p. 33). Therefore, whenever sequence frequencies are noted, the temperature and salt concentration must be specified.

Repeated sequences are produced by the "amplification" of single-copy sequences. A newly amplified sequence thus adds to the total *quantity* of DNA, but not to the *different* sequences or "sequence complexity" (Britten 1971) of the genome. However, like all nucleotide sequences, the originally identical repeated sequences gradually diverge, forming "families" of repeats which, eventually, will diverge enough from one another to behave like different sequences under certain experimental conditions. For example, the members of an old repeated sequence family may behave as single-copy sequences at 80°C, but as repeated sequences at 50°C. Thus, the definition of sequence frequency classes is based upon the kinetics of reassociation which are determined by the stringency of the experimental conditions. For additional information see Britten and Kohne (1968), Kohne (1970), Britten and Davidson (1971), Rice (1972), McCarthy and Farquhar (1972), Britten et al. (1974), and Straus (1976).

At the "standard criterion" of 60°C and 0.12 M sodium (Na^+) the single-copy sequences of birds make up 60–70% of the *volume* of nuclear DNA and the repeated sequences 30–40%. However, under these conditions, the single-copy fraction contains 95–98% of the *different* sequences, i.e., of the "sequence complexity," and only 2–5% of the different sequences are represented in the repeated sequence fraction. When DNA is prepared for use as the radiolabeled "tracer" for DNA hybridization comparisons, most of the copies of repeated sequences are removed, but *at least* one copy per genome of each of the families of repeated sequences will remain in the nonrepeated fraction. Thus, in "single-copy tracer" DNA, 100% of the sequences that are kinetically different under the experimental conditions are present. Figure 1 illustrates these points.

The average diploid avian genome contains ca. 3.8 pg of DNA (Bachman et al., 1972) which is equivalent to ca. 3.4×10^9 nucleotide pairs (NTP). The sequence complexity of single-copy tracer preparations is ca. 60% of this number, or 2×10^9 NTP.

There are ample reasons for removing most of the copies of repeated sequences from tracer preparations. Repetitive families may consist of from hundreds to millions of copies. Since the rate of reassociation is a function of copy number, the repeated sequences would contribute a disproportionate share to the "counts" of radioactivity. This would distort the data because, as Kohne et al. (1971, p. 490) pointed out:

> The fraction containing repeated sequences is diverse even within one species, and the differences observed between the repeated sequences of two species cannot be assumed to have occurred since the time of divergence of those species.
>
> Nonrepeated DNA sequences, on the other hand . . . [occur] only one time per haploid cell. Reassociated, nonrepeated DNA has a thermal stability that indicates that essentially perfect nucleotide pair matching is present. Nonrepeated DNA sequences held in common between species must be the descendants of the same common ancestor sequence that was present in the most recent common ancestor. Nonrepeated DNA is therefore highly suitable for determining the extent of nucleotide change since the divergence of two species.

Kohne (1970, p. 361) also noted that "Hybridization experiments using nonrepeated DNAs provide information concerning the rate of nucleotide substitution during evolution. Similar experiments using the repeated fraction provide a different type of information, namely information concerning the rate at which new DNA was incorporated into the genome by formation of new families of repeated sequences."

The removal of most of the copies of repeated sequences from the "tracer" DNA is therefore necessary to obtain a true measure of nucleotide sequence evolution, i.e., the accumulation of fixed point mutations since the two species being compared diverged from their most recent common ancestor.

Another question concerns the possibility that repeated sequences alone might yield an answer different from that obtained using single-copy tracer DNA. Since the repeated fraction contains only 2–5% of the different sequences at the standard criterion, it may be expected to be less informative than the single-copy fraction. In our experience (Sibley and Ahlquist, 1981a, pp. 320–321) we have found no discrepancy between the results using single-copy vs. repeated DNA tracers. However, Eden et al. (1978) reported a discrepancy in some avian comparisons that we believe was due to the misidentification of one of their

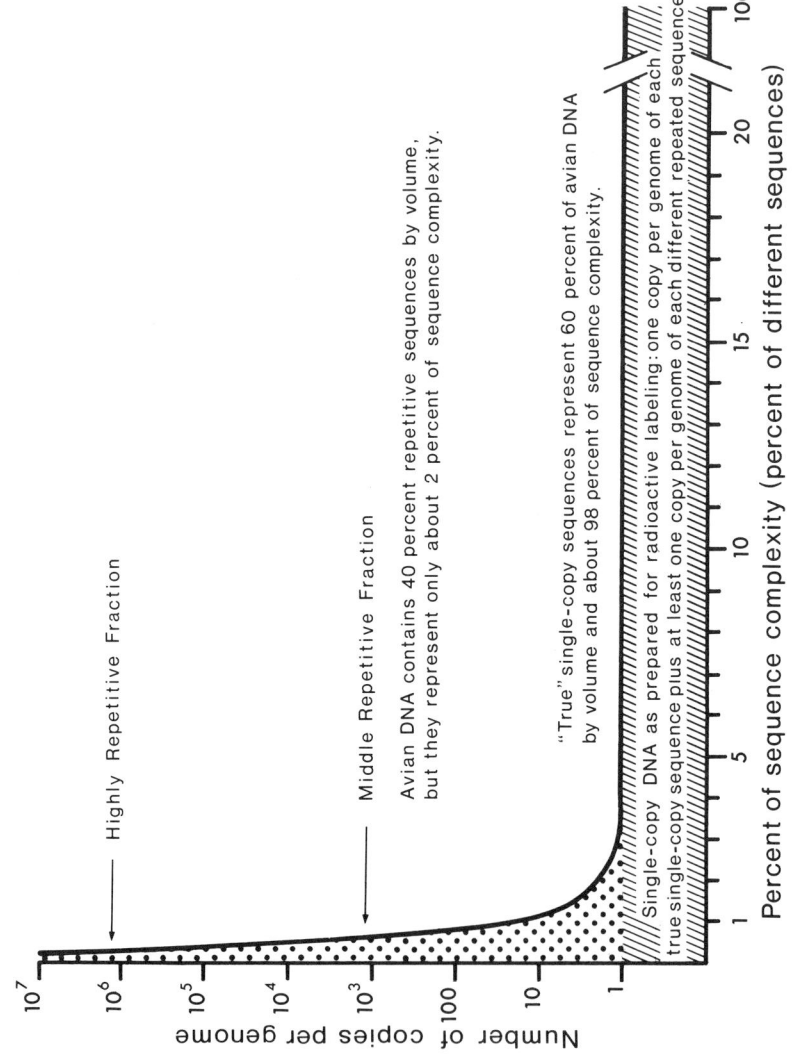

FIGURE 1. The sequence complexity of the avian genome.

preparations. We have compared the same species and found no anomalies. We conclude that single-copy tracers are preferable for studies of phylogeny and relationships. "Middle repetitive" sequences can be informative when comparing closely related taxa, including intraspecific comparisons (Mednikov, 1980). Gillespie (1977) found that comparisons of the repeated sequences of several taxa of primates agreed with traditional sources of evidence and that the rate of change of "newly evolved repeated sequences" was essentially the same as for single-copy sequences.

Although the definition of frequency classes depends upon the experimental conditions, it is possible to obtain consistent and comparable data by using the same conditions for all comparisons.

Another question relates to the possible effects on DNA–DNA hybridization data of such genomic components as intervening sequences ("introns"), moveable elements ("transposons"), etc. Because the long-chain native DNA molecules are sheared into ca. 500 nucleotide fragments, all sequences are free to find their homologous complements and to form duplexes during incubation, regardless of their original location in the genome. Thus, only homology is detected and the conditions of the experiments ensure that only homologous sequences can form double-stranded molecules.

There may be other questions that can be raised but it seems unlikely that the internal consistency we observe in DNA distance values could be obtained if there were serious flaws in the technique. We believe that the evidence from more than 17,000 avian DNA–DNA hybrids indicates that this technique measures the net divergence between the homologous nucleotide sequences of the species being compared. The *average* divergence between the sequences of DNA hybrids is obtained by calculating the $T_{50}H$ statistic, which is defined as the temperature at which 50% of the sequences are in the hybrid form. $T_{50}H$, and other aspects of data analysis, are discussed in Section 3.

2.3. Factors That Determine the Rate and Extent of Reassociation, and the Thermal Stability of DNA–DNA Hybrids

2.3.1. Temperature

The optimal temperature for the reassociation of conspecific DNA strands is ca. 60°C, at which the base pairs must be more than ca. 80% correctly matched to form a stable duplex. This precludes the formation of stable duplexes between nonhomologous sequences. Below 60°C, more mismatching is tolerated, the rate of reassociation increases, and at ca. 40°C, random reassociation occurs. Above 60°C, the rate of reassociation decreases and progressively more perfect base pairing is required for duplex formation. Also, as noted above, at higher temperatures less of the genome appears to consist of repeated sequences, and at lower temperatures more repeated sequences appear. Thus, the rate of reassociation and the sequence frequency composition of the genome vary with the stringency of the experimental conditions.

2.3.2. Salt Concentration

Monovalent cation concentration, usually as sodium (Na^+), has a major effect on the rate of reassociation. Below 0.01 M Na^+ no reassociation occurs. The relative rate at 60°C in 0.12 M Na^+ has been set at 1.0, and in 0.48 M Na^+ the relative rate is 5.65 (Britten et al. 1974, p. 364).

2.3.3. Concentration of DNA

Because random collisions between homologous single strands of DNA must occur for the formation of duplex molecules, the rate of reassociation is dependent upon the concentration of each different sequence in the reaction mixture. DNA reassociation follows second order kinetics and the rate can be controlled by varying the concentration. The product of the initial concentration and the length of the period of incubation has been designated "$C_o t$" which is expressed in moles of nucleotides per liter times the incubation period in sec. For example, a $C_o t$ of 1.0 is attained if single-stranded DNA at a concentration of 83 $\mu g/ml$ is incubated for 1 hr at 60°C in 0.12 M sodium phosphate buffer (Britten and Kohne, 1968).

The time course of DNA reassociation may be represented by a "$C_o t$ Plot," as in Fig. 2. The plot of $C_o t$ vs. % single-stranded DNA is used to determine sequence repetition frequencies. Most of the repeated sequences reassociate below a $C_o t$ of 200 but, to obtain a fraction containing a minimal number of repeated copies, we allow reassociation to proceed to a $C_o t$ of 1000 at 50°C when preparing single-copy tracer DNAs. Kohne (1970) and Britten et al. (1974) have discussed the reassociation of DNA fragments.

2.3.4. Fragment Size

The rate of reassociation is also affected by the length of the strands of DNA. To achieve reproducible rates, and to be able to remove the

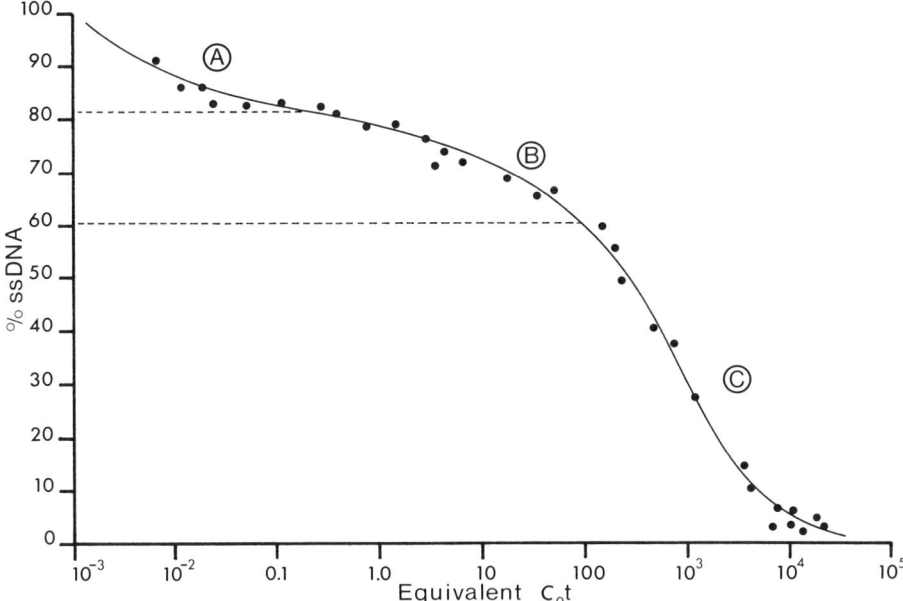

FIGURE 2. A C_ot Plot of the reassociation kinetics of whole, sheared (500 base fragments) Herring Gull (*Larus argentatus*) DNA. Values up to a C_ot of ca. 1.0 were determined spectrophotometrically in 0.12 M phosphate buffer. Higher orders of C_ot were obtained by incubating a known concentration of DNA for various lengths of time and separating single-stranded (ssDNA) from double-stranded DNA on hydroxyapatite columns. These reassociations were carried out in 0.48 M phosphate buffer at the equivalent C_ot values relative to 0.12 M, assuming a rate constant of 5.65. Ca. 18% of the Herring Gull genome consists of highly repeated sequences (A), ca. 22% is in a middle repetitive class (B), and the remaining 60% is composed of single-copy sequences (C).

excess copies of repeated sequences, the long-chain DNA is "sheared" to an average length of 500 NTP. Wetmur and Davidson (1968), Kohne (1970), and Britten et al. (1974) have discussed this subject. Although the latter expressed reservations about using high frequency sound ("sonication") to shear DNA into fragments we have found it to be a satisfactory procedure (Sibley and Ahlquist, 1981a, p. 307).

In our DNA hybrids we use 200 ng of single-copy tracer DNA, which is the amount in 180,000 genomes. When sheared into ca. 500 NTP fragments this results in ca. 3.7×10^{11} fragments of tracer in each combination of labeled and unlabeled DNA fragments. To increase the probability that hybrid double-stranded molecules will be formed, the ratio between tracer and unlabeled ("driver") fragments is 1:1000. This reduces the percentage of "selfed" hybrids, i.e., tracer x tracer duplexes, to ca. 1–2%. Ratios of 1:5000 or 1:10,000 are sometimes used.

2.3.5. Base Composition

Because there are two hydrogen bonds between A-T base pairs, and three between G-C base pairs, the A-T pairs will melt at a temperature lower than that required to melt G-C pairs. The melting curve of conspecific double-stranded DNA is a cumulative plot of the melting behavior of different regions having different A-T:G-C ratios. The A-T rich sequences begin to melt ca. 80°C, the maximum melting rate occurs near 86–87°C, and the G-C rich regions melt at still higher temperatures. Thus, the thermal transition for native DNA is 14–15°C.

Homologous, i.e., conspecific, DNA hybrids have melting curves like those of native DNAs, but the melting behavior of heterologous, i.e., interspecific, DNA–DNA hybrids is also affected by base pair mismatching. The differences (i.e., delta values) between the melting curve of a homologous hybrid (i.e., tracer x tracer) and those of heterologous hybrids are due to base pair mismatches.

The A+T:G+C ratio in the Domestic Fowl (G. gallus) is ca. 1:1.33 (Shapiro, 1976). A 1% change in this ratio causes a change of 0.45°C in the median melting temperature (T_m) of native DNA. The range of this ratio is small in birds and the effect of the variation is not a significant factor.

The thermal stability curve of a heterologous DNA–DNA hybrid is a plot of base sequence divergence between the two species whose DNAs compose the heteroduplex. The duplexes containing the largest percentages of mismatched bases melt at the lowest temperatures, those that are perfectly matched at the highest temperatures. Grula et al. (1982) showed that a thermal stability curve is also a plot of rates of nucleotide substitution. Thus the DNA–DNA hybrids that melt at low temperatures are composed of the "derived" or "apomorphous" sequences, and those that melt at the highest temperatures are the "primitive" or "plesiomorphous" sequences, in the sense of Hennig (1966). A thermal stability curve is thus a phylogenetic description of the comparison between the two genomes that formed the DNA–DNA hybrid.

2.3.6. Viscosity

The rate of reassociation is inversely proportional to solvent viscosity. The effect of viscosity is eliminated by using the same solvent concentration for all experiments.

2.3.7. Genome Size

The haploid DNA content of 23 species of birds, representing seven orders and 17 families, ranged from 1.6–2.2 pg, with an average of 1.9 pg of DNA (Bachman et al. 1972), which is equivalent to ca. 1.7×10^9 NTP. Hinegardner (1976) reviewed the evolution of genome size in many organisms.

2.4. The Hydroxyapatite (HAP) Column Chromatography Procedure

Hydroxyapatite (HAP) is a form of calcium phosphate which, at 0.12 M sodium concentration, will bind double-stranded, but not single-stranded DNA. It may therefore be used to separate the rapidly reassociating repeated DNA sequences from the slowly reassociating single-copy DNA, and to separate double-stranded from single-stranded DNA during the thermal chromatography of DNA–DNA hybrids. Kohne and Britten (1971) and Britten et al. (1974) have described these procedures. Following is a synopsis of the procedure we have used for comparisons of avian DNAs.

DNAs were obtained from the nuclei of avian erythrocytes, purified according to the procedures of Marmur (1961) and Shields and Straus (1975), and "sheared" by sonication into fragments with an average length of ca. 500 nucleotides. Fragment size was determined by electrophoretic comparisons with DNA fragments of known size produced by the digestion of bacteriophage DNA with bacterial restriction endonucleases (Nathans and Smith, 1975).

The single-stranded DNA fragments of the species to be used as "tracers" were allowed to reassociate to a C_ot of 1000 at 50°C in 0.48 M sodium phosphate buffer. This permitted most of the rapidly reassociating repeated sequences to form double-stranded molecules while the slowly reassociating single-copy sequences remained single-stranded. The latter were recovered by chromatography on a hydroxyapatite column. This process produced a single-copy DNA preparation consisting of one copy per genome of each original single-copy sequence and *at least* one copy per genome of each different repeated sequence.

The single-copy DNA sequences of the "tracer" species were labeled with radioactive iodine (^{125}I) according to the procedures of Commorford (1971), Prensky (1976), and Chan et al. (1976). DNA–DNA hybrids were formed from a mixture composed of one part (= 200 ng) radioiodine-labeled single-copy DNA and 1000 parts (= 200 µg) sheared, whole DNA. The hybrid combinations were heated to 100°C for 10 min to dissociate the double-stranded molecules into single strands, then incubated for 120 hr (= $C_o t$ 16,000) at 60°C in 0.48 M sodium phosphate buffer to permit the single strands to form double-stranded hybrid molecules. After incubation, the buffer was diluted to 0.12 M and the DNA–DNA hybrids were bound to hydroxyapatite columns immersed in a temperature-controlled water bath at 55°C. The temperature was then raised in 2.5°C increments from 55°C to 95°C. At each of the 17 temperatures the single-stranded DNA produced by the melting of double-stranded molecules was eluted in 20 ml of 0.12 M sodium phosphate buffer.

The radioactivity in each eluted sample was counted in a Packard Model 5220 Auto-Gamma Scintillation Spectrometer, optimized for ^{125}I. A teletype unit connected to the gamma counter printed out the data and punched a paper tape, the entry to computer analysis.

The computer program uses a nonlinear regression least squares procedure to determine the best fit of the experimental data to one of four functions: (1) the Normal, (2) the dual-Normal, (3) the "skewed" Normal, or (4) a modified form of the Fermi–Dirac.

3. THE ANALYSIS OF DNA–DNA HYBRIDIZATION DATA

3.1. Definitions

1. *Homologous DNA hybrid.* A DNA–DNA hybrid formed from labeled and unlabeled DNA of the same species; a homoduplex molecule.

2. *Heterologous DNA hybrid.* A DNA–DNA duplex composed of one strand of radio-labeled DNA of one species and an unlabeled strand of DNA of a different species; a heteroduplex molecule. The labeled DNA is sometimes referred to as "tracer," the unlabeled as "driver," because its concentration promotes ("drives") the reassociation.

3. *Percentage of hybridization.* The percentage of the labeled DNA that forms hybrid molecules ("duplexes") with unlabeled DNA during incubation.

4. *Normalized percentage of hybridization (NPH).* The percentage of hybridization of a heterologous hybrid divided by that of its homolog times 100.

5. *Modal temperature.* The temperature in degrees Celsius at the mode of the frequency distribution of radioactive counts in a thermal dissociation curve. This value differs from the traditional "melting temperature" (T_m) of a DNA hybrid which is the *median* of a thermal dissociation curve.

6. $T_{50}H$ or $T_{50}R$. The temperature in an ideal, normalized, cumulative frequency distribution function at which 50% of all single-copy DNA sequences are in the hybrid form, and 50% have dissociated. Also designated T_mR (Benveniste and Todaro, 1976).

7. *Delta values*. The difference in degrees Celsius between the several parameters of the frequency distribution of radioactive counts of a homologous DNA hybrid and any heterologous hybrid formed from the same labeled DNA, for example, delta $T_{50}H$, delta mode, and delta T_m.

The mode is a true characteristic of any thermal elution curve, but the T_m is the true median only of the complete curve of a homologous hybrid. For heterologous hybrids, the T_m represents the median of the thermal elution curve only of those duplexes which are stable above 60°C. Hence, it indexes progressively less genetic information as the two species forming the heteroduplex are more divergent, i.e., less closely related. The mode has a slightly greater range than the T_m but both of these parameters are bound from below by the incubation temperature, and both are progressively "compressed" by the effects of sequence divergence. See Fig. 3 and Table I.

The normalized percentage of hybridization (NPH) has a range greater than that of the delta mode and is more nearly linear with respect to absolute time. The NPH values are not as reproducible as are the delta modes, but the branching pattern of taxa derived from a matrix of NPH values is the same as that based on delta modes (Sibley and Ahlquist, 1981a, p. 321).

The $T_{50}H$ (or $T_{50}R$) statistic was first suggested by Kohne (1970, p. 349) and has been used by Bonner et al. (1981), Sibley and Ahlquist (1981a–c, 1982a,c,f–h), and Sibley et al. (1982). In the calculation of $T_{50}H$ it is assumed that all of the single-copy sequences in the genomes of the two species being compared have homologs in the other species, that all single-copy sequences potentially can hybridize with their homologs, and that all degrees of divergence can be detected. For homologous hybrids the $T_{50}H$ and T_m values derived from normalized cumulative distributions are equal. For hybrids between more diverged taxa the percentage of hybridization declines and the thermal stability curve is progressively truncated by the effects of experimental conditions. It is possible to estimate the $T_{50}H$ by making a graphic extrapolation of the portion of the sigmoid curve that is most nearly linear. A better procedure is to calculate the best-fitting linear regression to the part of the curve having the most constant slope and to find the temperature corresponding to its intercept with the 50% hybridization level. The most objective procedure is to fit a cumulative distribution function

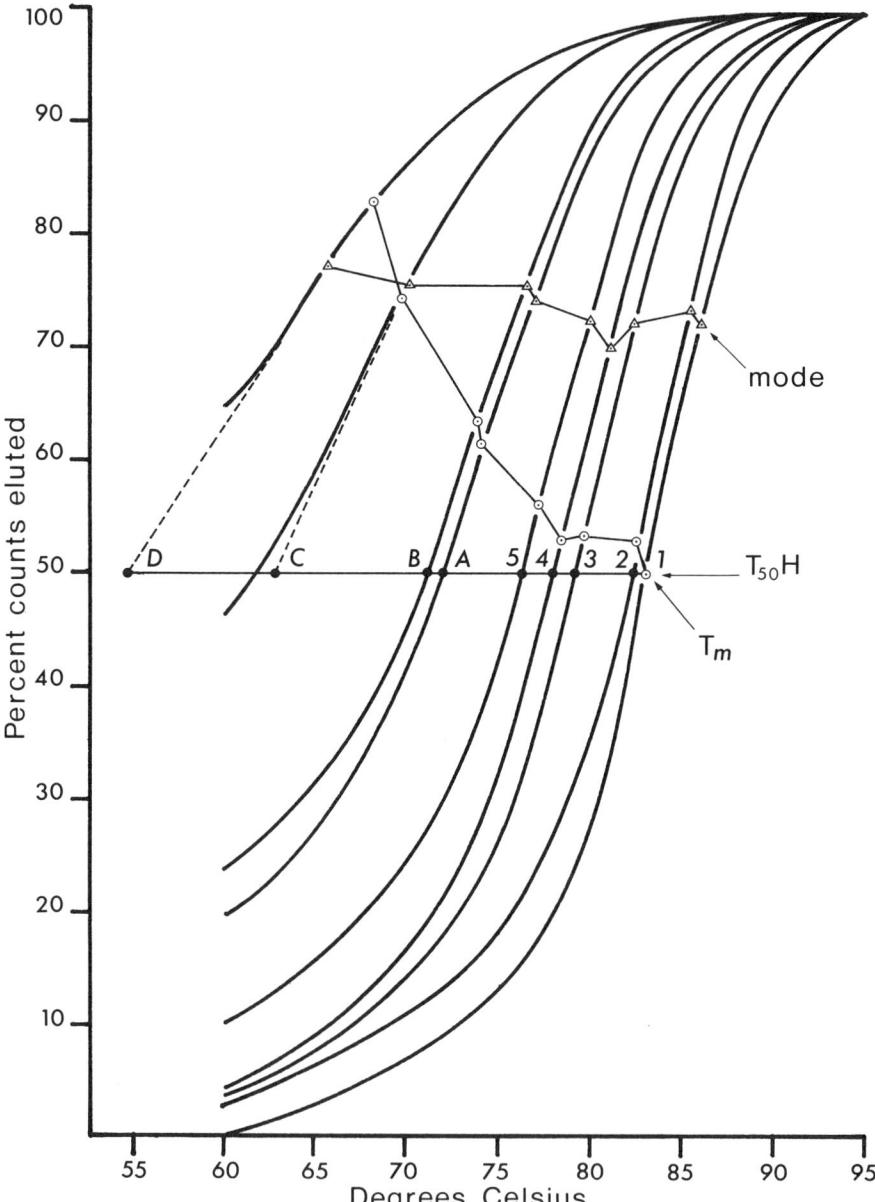

FIGURE 3. A set of thermal dissociation curves in which the tracer DNA of the Cape Weaver (*Ploceus capensis*) was hybridized with the driver DNAs of: (1) *Ploceus capensis* (=homologous hybrid), (2) Village Weaver (*Ploceus cucullatus*), (3) Red-billed Weaver (*Q. quelea*), (4) Red Bishop (*Euplectes orix*), and (5) Thick-billed Weaver (*Amblyospiza*

TABLE I
DNA–DNA Hybridization Data for Comparisons between the Cape Weaver and Other Species of Birds[a]

Name	$\Delta T_{50}H$	ΔMode	ΔT_m	NPH
A. Cape Weaver (*Ploceus capensis*)	0.0	0.0	0.0	100
Village Weaver (*Ploceus cucullatus*)	0.9	0.6	0.7	96.7
Red-billed Quelea (*Quelea quelea*)	4.3	3.7	3.6	95.9
Red Bishop (*Euplectes orix*)	5.1	5.1	4.8	91.3
Thick-billed Weaver (*Amblyospiza albifrons*)	6.8	6.2	6.0	89.6
B. Red-winged Starling (*Onychognathus morio*)	10.2	8.6	8.3	81.3
Golden-crested Myna (*Ampeliceps coronatus*)	10.4	8.9	8.6	82.0
Common Starling (*Sturnus vulgaris*)	10.7	8.7	8.5	79.8
Singing Starling (*Aplonis cantoroides*)	10.7	9.5	9.6	81.3
Pied Starling (*Spreo bicolor*)	11.1	9.4	10.1	80.0
C. Common Crow (*Corvus brachyrhynchos*)	12.0	8.9	8.7	78.0
Mexican Jay (*Aphelocoma ultramarina*)	12.0	9.1	9.0	75.1
Malaysian Treepie (*Dendrocitta occipitalis*)	12.2	9.6	9.2	74.8
Blue Jay (*Cyanocitta cristata*)	12.4	9.6	9.2	75.0
Black-billed Magpie (*Pica pica*)	12.5	9.6	9.8	75.1
D. Stout-billed Cinclodes (*Cinclodes excelsior*)	19.5	16.1	13.7	53.4
Black-tailed Leafscraper (*Sclerurus caudacutus*)	19.6	16.0	13.6	53.3
Striped Leafgleaner (*Hyloctistes subulatus*)	19.6	16.5	13.9	58.0
Pale-legged Hornero (*Furnarius leucopus*)	19.9	16.4	13.8	52.4
Chestnut-crowned Leafgleaner (*Automolus rufipileatus*)	20.2	16.3	13.7	51.7
E. Green Kingfisher (*Chloroceryle americana*)	27.9	21.3	15.3	36.5
Sacred Kingfisher (*Halcyon sancta*)	28.5	20.6	15.0	35.7
Hook-billed Kingfisher (*Melidora macrorhina*)	28.7	20.4	15.0	35.4
Common Paradise Kingfisher (*Tanysiptera galatea*)	28.8	19.7	15.2	34.6
Belted Kingfisher (*Ceryle alcyon*)	29.8	20.8	15.1	33.6

[a] A = Weaverbirds, Ploceidae; B = Starlings, Sturnidae; C = Crows, etc., Corvidae; D = Ovenbirds, Furnariidae; E = Kingfishers, Alcedinidae.

---

albifrons). (A) = average curve of five DNA hybrids between P. capensis and five species of starlings (Sturnidae). (B) = average curve of five hybrids between P. capensis and five species of crows, jays, and magpies (Corvini). (C) = average curve of five hybrids between P. capensis and five species of Neotropical ovenbirds (Furnariidae). (D) = average curve of five hybrids between P. capensis and five species of kingfishers (Alcedinidae). The $T_{50}H$ values (solid circles) lie on the 50% line, the Modes of the curves are indicated by triangles, the T_m values by open circles. Note that the T_m and $T_{50}H$ of the homologous hybrid (1) are identical. See Table I for the data and names of the species.

to the data and to find its intercept with the temperature axis at the 50% hybridization level. Because $T_{50}H$ incorporates the percent hybridization it is more nearly linear with true genetic distance, and with absolute time, than either delta mode or delta T_m values. These advantages recommend it over the mode and T_m for phylogenetic comparisons.

3.2. The Average Rate of Base Substitution

Experiments using synthetic polynucleotides of known composition have shown that a delta value of 1.0 indicates that ca. 1% of the base pairs are mismatched (Bonner et al. 1973). Since the single-copy tracer fraction in birds contains ca. 2×10^9 NTP a delta value of 1.0 indicates that ca. 20 million bases are mismatched. If we accept our rough calibration of delta $T_{50}H$ 1.0 = 4.5 MY the net average rate of nucleotide substitution is about 2 bases/yr per lineage for divergence times greater than ca. 50 MY. For divergences of less than 50 MY the net average rate is expected to be slightly greater because the regression between delta values and time is probably curvilinear due to the time-since-divergence probabilities of back mutations, multiple hits, and amplification events.

Because of the relationship between the melting temperature of a DNA–DNA hybrid and the percentage of base pairing, we can define "genetic distance" as a measure of that relationship. Thus a delta $T_{50}H$ value is an average genetic distance.

3.3. Mathematical Analysis of DNA–DNA Data

The raw data are plotted as a frequency distribution of radioactive counts vs. temperature and analyzed by computer, using a nonlinear regression least squares program developed by the University of Colorado Physics Department (under T. Bailey) and the Los Alamos Scientific Laboratory, and modified for use with DNA–DNA hybridization data by Dr. Temple F. Smith.

This program fits the following distribution functions to a DNA thermal elution curve. $Y(T)$ denotes the dependent variable as the amount of DNA duplex dissociated at temperature T. The P_is are the various distribution parameters over which the search for "best fit" is carried out.

1. The normal (a) and the dual-normal (b).

$$Y(T) = P_1 \exp\{-P_2 (T-P_3)^2\} \tag{a}$$

$$Y(T) = P_1 \exp\{-P_2 (T-P_3)^2\} + P_5 \exp\{-P_6 (T-P_4)^2\} \qquad (b)$$

in which P_1 is the maximum height of the distribution, P_2 is the reciprocal of twice the width, or spread (σ) squared, and P_3 is the mode. P_4, P_5, and P_6 have the same meaning as the corresponding terms in the single-normal function.

2. A modified form of the Fermi–Dirac equation:

$$Y(T) = P_1 \exp\{-P_2 [(T-P_3)^2 - P_4]\} / [1.0\text{-}\exp\{-P_2 ((T-P_3^2) - P_4)\}]^2$$

in which the denominator is squared. P_1 is the maximum, and P_3 the minimum, of the distribution (see Fig. 4). P_2 and P_4 are, respectively, the slopes of the leading (high temperature) and trailing (low temperature) sides of the distribution. The mode is given by $P_3 + (P_4)^{\frac{1}{2}}$.

Probably all thermal elution curves are normally distributed, but the presence of single-stranded fragments of tracer and poorly matched duplexes produce a skew toward the low temperature side. This skewness is most noticeable in the curves of homologous hybrids, and of heterologs genetically close to the homolog. The curves of hybrids at a middle distance from the homolog are more nearly symmetrical and can be fit equally well by the normal or the Fermi–Dirac function. The Fermi–Dirac is especially effective for fitting skewed distributions.

The curves of widely divergent hybrids present other problems because their $T_{50}H$ and modal values are near, or below, the 60°C incubation temperature. These curves may have a small peak near 85°C, which represents a small amount (1–3%) of self-hybridization by the tracer, and a broad peak at the low temperature end which represents the heteroduplexes. A bimodal distribution can be fitted to these data, the selfing peak subtracted, and the $T_{50}H$ computed from the remaining curve. By this method it is possible to obtain reliable delta values up to at least 30°C (see Fig. 5).

In our paper on the ratite birds (Sibley and Ahlquist, 1981a, pp. 213–313) we discussed the use of the dual-normal function to fit thermal elution curves which have a second peak due to the presence of a "low temperature component." We have found that this component can be eliminated from avian tracers by preparing single-copy DNA at 50°C in 0.48 M phosphate buffer and an equivalent C_0t of 1000.

3.4. A "Robust" Clustering Method for DNA Hybridization Data

The conversion of measurements of any kind into phylogenies requires a procedure to obtain a hierarchical clustering of taxa. Of the

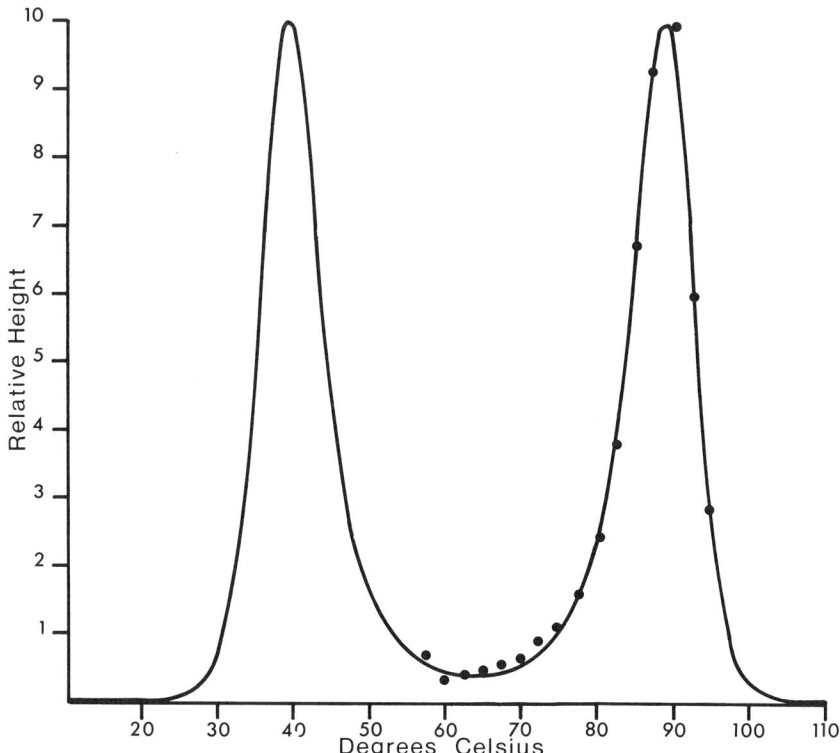

FIGURE 4. Modified Fermi–Dirac function of a homologous DNA–DNA hybrid with actual data points.

available procedures we believe that the simplest approach will produce the most conservative results. Dr. Temple F. Smith (personal communication) has developed a modified version of the well-known "average linkage" clustering method that uses a "robust" measure. The simplest of robust statistics is the "trimmed mean" (Huber, 1977) which is simply the mean value of a data set from which the highest and lowest values (the "outliers") have been excluded. Such a statistic is obviously less sensitive to single aberrant data points. This procedure allows for statistical noise, is robust for distant implied relationships, and produces a measure of uncertainty, as expressed by the standard deviation and/or standard error of the mean.

The average linkage procedure begins by clustering the closest pair or pairs of taxa. The next step links the taxa which have the smallest average distance to any existing cluster. This procedure continues until

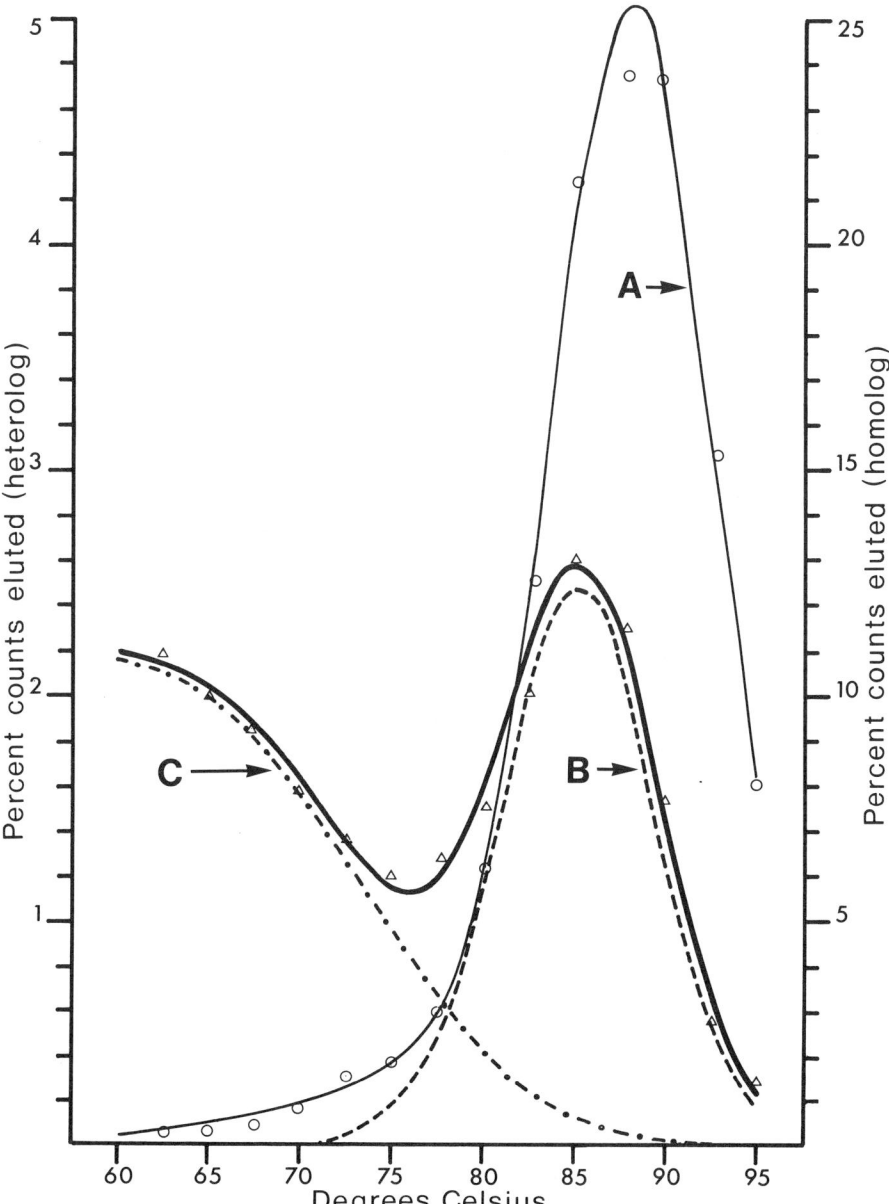

FIGURE 5. Thermal dissociation curves of bird–reptile DNA–DNA hybrids. Circles indicate the data points of a homologous hybrid of labeled Brown Kiwi (*Apteryx australis*), the thin solid line (A) is the curve of the modified Fermi–Dirac function fitted to the data. Triangles indicate the data points of a kiwi–crocodile hybrid, the heavy solid line is a dual-normal function fitted to the data. The dashed line (B) represents self-hybridization of the tracer. The dot-dash line (C) represents the true kiwi–crocodile hybrid.

all taxa are linked. At every step where five or more measurements are available the trimmed mean may be used. The underlying idea is that the only reason for "closeness" is true homology of characters, while "distance" may result from the failure to identify existing homologies. The DNA hybridization data are especially compatible with this view because there is no known reason, other than nucleotide sequence homology, to explain the thermal stabilities of DNA–DNA hybrids.

3.5. Reliability, Sensitivity, and Reciprocity

Fitch (1976, p. 162) noted that the accuracy with which the DNA hybridization melting point differences are measured is an important issue which "may be subdivided into reliability, sensitivity, and reciprocity." Reliability means "the extent to which one can rely upon the fundamental measurement." DNA hybridization is a "general measure" which provides an overall average value, thus a measure of reliability is the reproducibility of the parameters of the melting curves of DNA–DNA hybrids. Identical hybrids should have identical properties, thus deviations provide an index to reliability.

In Table II, example A presents the values for $T_{50}H$, mode, T_m, and NPH for ten individual DNA hybrids between the tracer DNA of an Australian warbler, the Yellow-rumped Thornbill (*Acanthiza chrysorrhoa*), and the driver DNA of an Australian honeyeater (*Anthochaera carunculata*). For each of the statistics, the range is moderately large, e.g., 0.6°C for delta $T_{50}H$, but the ten replicates make it possible to calculate the variances, e.g., the average $T_{50}H$ of 8.85 has an SE of ± 0.08 and an SD of ± 0.25. Because these ten hybrids are identical, the variation among them is clearly due to experimental error.

Example B in Table II is a comparison between the Yellow-rumped Thornbill and nine genera of meliphagid honeyeaters, the sister group of the thornbills. Because of the uniform average rate (UAR) of nucleotide substitution (see Section 5) we expect the delta values to be identical, except for experimental error. The range (1.7) and the variances are greater than for the ten identical replicates in example A but, again, an average $T_{50}H$ (9.06) with a relatively small SE (0.25) can be obtained.

Example C in Table II compares the Yellow-rumped Thornbill with 19 genera of other oscine passerines which are members of more distant taxa, all of which branched via a single divergence node from the thornbill-honeyeater lineage. Because of the UAR we expect them to be equidistant from the labeled species. The range for the 19 delta $T_{50}H$ values (2.3) is 0.6 larger than the 1.7 for the nine values in example B, but the SE (0.14) and the SD (0.62) are smaller, reflecting the larger sample size.

TABLE II
DNA–DNA Comparisons between the Yellow-rumped Thornbill and Other Passerine Taxa

Taxa	$\Delta T_{50}H$				ΔMode				ΔT_m				NPH			
	Range	$\bar{x}$	SE	SD	Range	$\bar{x}$	SE	SD	Range	$\bar{x}$	SE	SD	Range	$\bar{x}$	SE	SD
A. Ten identical DNA–DNA hybrids between the Yellow-rumped Thornbill (*Acanthiza chrysorrhoa*) and the Red Wattlebird (*Anthochaera carunculata*)	8.6–9.2 (0.6)	8.8	0.1	0.2	7.8–8.5 (0.7)	8.1	0.1	0.2	7.2–7.9 (0.7)	7.6	0.1	0.2	81.1–86.7 (5.6)	84.4	0.7	2.2
B. Nine DNA–DNA hybrids between the Yellow-rumped Thornbill and nine genera of honeyeaters (Meliphagidae)[a]	7.4–9.1 (1.7)	9.1	0.2	0.7	6.8–9.3 (2.9)	8.3	0.2	0.7	6.2–9.2 (3.0)	8.3	0.2	0.6	81.3–90.4 (9.1)	85.5	0.9	2.8
C. 19 DNA–DNA hybrids between the Yellow-rumped Thornbill and 19 genera of oscine passerines[b]	12.7–15.0 (2.3)	14.0	0.1	0.6	8.1–11.0 (2.9)	10.0	0.2	0.7	10.9–11.9 (1.0)	11.3	0.1	0.3	73.1–82.8 (9.7)	77.6	0.5	2.4

[a] *Prosthemadera, Acanthorhynchus, Conopophila, Plectorhyncha, Ramsayornis, Meliphaga, Certhionyx, Myzomela, Lichmera.*
[b] *Ficedula, Melaenornis, Parus, Stachyris, Turdoides, Pycnonotus, Thryothorus, Copsychus, Turdus, Sturnus, Toxostoma, Prunella, Motacilla, Anthreptes, Emberiza, Passer, Euplectes, Progne, Ammomanes.*

Thus, the range, or spread, of the delta $T_{50}H$ values is proportional to the diversity and the number of hybrids, but the accuracy of the averages is strongly influenced by sample size.

Sensitivity means "the extent to which one can pick up minor variants . . ." (Fitch, 1976, p. 164). DNA–DNA hybridization is relatively insensitive to "minor variants" but it can discriminate between closely related species. For example, by using several replicates we have obtained statistically significant reciprocal delta $T_{50}H$ values of 0.5 between two closely related grebes, the Western Grebe (*Aechmophorus occidentalis*) and Clark's Grebe (*A. clarkii*) (Ahlquist et al., in press). These morphologically similar species differ only slightly in several characters, but occasionally form mixed pairs and produce viable offspring (Ratti, 1979; Nuechterlein, 1981). Similarly, members of the Herring Gull complex (*Larus argentatus*, *L. hyperboreus*, *L. glaucoides*, *L. glaucescens*, *L. occidentalis*, *L. californicus*), some of which hybridize, differ by delta values between 0.5–1.0. We have many other examples of congeners which differ by less than delta $T_{50}H$ 1.0.

By using a solvent (2.4 M tetraethylammonium chloride) that suppresses the effect of base composition on melting temperature, Britten et al. (1978) found a 4% nucleotide pair mismatch between individual Purple Sea-urchins (*Strongylocentrotus purpuratus*) due to genetic polymorphisms. Using material supplied by us, Britten (personal communication) compared the DNAs of four individual Song Sparrows (*Melospiza melodia*) by the same method and found less than 0.5% polymorphism.

Another aspect of sensitivity is the ability to resolve the divergence nodes among closely related taxa. From the SE values in Table II, and our experience with many other data sets, it is clear that by using five or more replicates it is possible to separate nodes that differ by as little as delta $T_{50}H$ 0.5, or possibly less. This equates with ca. 1–2 million years by our preliminary calibration of delta values vs. absolute time. At the upper end of the divergence scale we have resolved $T_{50}H$ and NPH values between the most distantly related groups of birds which range up to ca. delta $T_{50}H$ 30, with NPH values down to 20%. Bird–reptile comparisons at ca. delta $T_{50}H$ 34 and NPH values below 20% have also been determined (see Fig. 5).

Reciprocity is the extent to which the evolutionary distance from taxon A to taxon B is the same as from B to A. It is usual to think of reciprocal taxa as single species but, because of the UAR, we can extend reciprocity to include measurements between monophyletic groups.

Table III contains the data for 14 reciprocal pairs ranging from closely related genera, such as *Paradisaea* and *Ptiloris*, to widely divergent families, such as the Maluridae (*Malurus*) and the Sylviidae (*Cincloramphus*). The discrepancies range from 0.0–0.3, and average 0.16, SE ± 0.02, SD ± 0.09. Most reciprocals differ by less than 0.5°C. Those that differ by more than 0.5 are due to excessive experimental error. As for all comparisons, average reciprocals based on several replicates, tend to be closest to equality.

The reciprocity between groups establishes the monophyly of clusters, determines branch nodes with accuracy, and provides relative rate tests in support of the uniform average rate. Because of the UAR, the reciprocity between two groups that diverged via a single node may be tested, using different taxa for the two sides of the reciprocal. For example, in our study of the New Zealand Wrens (see Section 5.2) we compared the suboscine (Oligomyodi) New Zealand Rifleman (*Acanthisitta chloris*) with 41 species of oscines (Passeres). The average delta $T_{50}H$ was 19.8 ± 0.13 SE. The reciprocal data are presented in Table I and Fig. 3 for comparisons between the tracer DNA of the Cape Weaver (*Ploceus capensis*), an oscine, and the DNAs of five species of the oligomyodian family Furnariidae. The average delta $T_{50}H$ and the SE, are identical to those of the *Acanthisitta* experiment, i.e., 19.8 ± 0.13, although completely different species were used! Although coincidence accounts for some of this precise reciprocity, this example demonstrates the ability of the technique to measure accurately large reciprocal distances, and it also is evidence in support of the uniform average rate of nucleotide substitution (see Section 5).

We submit that these examples, and others like them, indicate that the reliability, sensitivity, and reciprocity of DNA–DNA hybridization values are excellent for comparisons between taxa that diverged between ca. one million and 150 million years ago.

4. HOMOLOGY

"Homology is, without question, the most important principle in all comparative biology . . . homology is the only method of comparing attributes of different species . . . any comparative method in conflict with the concept of homology is . . . subject to question . . ." (Bock 1973, p. 386). Bock defined as homologous those features in two organisms which "can be traced phylogenetically to the same feature . . . in the immediate common ancestor of both organisms."

Fitch (1976, p. 161) defined homology "as similarity arising by virtue of common ancestry" and distinguished between two kinds of molecular homology. Orthologous genes are those "whose difference is a consequence of independence arising from speciation . . . because

TABLE III
Reciprocal DNA–DNA Hybridization Distances

Radio-labeled species	Unlabeled species	$\Delta T_{50}H$	Reference[a]
Paradisaea minor	P. paradiseus	1.5	5
Ptiloris paradiseus	P. minor	1.4	5
Chlamydera nuchalis	P. violaceus	2.7	5
Ptilonorhynchus violaceus	C. nuchalis	2.6	5
Daphoenositta chrysoptera	P. pectoralis	6.8	2
Pachycephala pectoralis	D. chrysoptera	6.9	4
Pachycephala pectoralis	P. paradiseus	7.3	4
Ptiloris paradiseus	P. pectoralis	7.2	5
Pachycephala pectoralis	D. brunneopygia	9.9	4
Drymodes brunneopygia	P. simplex	9.9	1
Malurus lamberti	S. frontalis	10.9	3
Sericornis frontalis	M. lamberti	10.7	5
Pachycephala pectoralis	A. inornata	11.7	4
Acanthiza chrysorrhoa	P. pectoralis	11.6	5
Daphoenositta chrysoptera	A. carunculata	11.9	2
Anthochaera carunculata	D. chrysoptera	11.7	5
Daphoenositta chrysoptera	E. albifrons	12.0	2
Ephthianura albifrons	D. chrysoptera	12.3	5
Malurus lamberti	G. cyanoleuca	12.8	3
Grallina cyanoleuca	M. lamberti	13.0	5
Malurus lamberti	D. brunneopygia	12.9	3
Drymodes brunneopygia	M. lamberti	13.0	1
Malurus lamberti	P. pectoralis	13.1	3
Pachycephala pectoralis	M. lamberti	12.8	4
Malurus lamberti	D. chrysoptera	13.5	3
Daphoenositta chrysoptera	M. lamberti	13.3	2
Malurus lamberti	C. cruralis	14.5	3
Cincloramphus cruralis	M. lamberti	14.7	5

[a]1 = Sibley and Ahlquist (1982e); 2 = Sibley and Ahlquist (1982f); 3 = Sibley and Ahlquist (1982h); 4 = Sibley and Ahlquist (1982g); 5 = unpublished data.

there is an exact phyletic correspondence between the history of the genes and the history of the taxa from which they derive." This definition applies to the data of DNA hybridization. Two genes derived from a common ancestral gene by gene duplication, and evolving side by side in the same organism, are defined by Fitch as *paralogous*. The distinction is useful with reference to gene products, e.g., proteins, but does not affect the interpretation of DNA hybridization data. In the broadest sense, all genes are homologous because it is probable that all genes, in all organisms, have originated by gene duplication and divergent evolution. This theory assumes that the genes in living species have been derived from pre-existing nucleotide sequences, back to the primordial gene that accompanied the origin of life on Earth.

As duplicate genes diverge, their nucleotide sequences eventually will acquire enough mismatched bases so that they will not form stable duplex molecules at the standard criterion of 60°C and 0.12 M sodium phosphate. At that point, such sequences may be said to have added to the sequence complexity of the genome and would be kinetically different under those conditions. They might still form stable duplexes under less stringent conditions, e.g., 50°C and 0.14 M salt, but further divergence eventually would prevent them from behaving as homologous pairs under any conditions. Thus, the degree of base pair complementarity, by which homologous genes are identified, gradually declines with time. However, homologies between genes that began their divergence hundreds of millions of years ago can still be recognized. For example, the immunoglobulin genes of the vertebrates are apparently the descendants of an ancestral gene of 330 base pairs that existed in some prevertebrate lineage more than 500 million years ago. By duplication and divergence this gene produced the immunoglobulin family which, in the mammals, consists of ten subunits which make up five (or six) immunoglobulin molecules. In birds, there are four immunoglobulins, in reptiles, three, and in amphibians, two. In the fishes there is only one, the "IgM" type, from which the others apparently evolved (Barker and Dayhoff, 1976, p. 165; Marchalonis, 1977). Indirect evidence of the presence of homologs of the insulin gene, and of the genes for other hormones, have been found in diverse organisms from vertebrates and insects to protozoa, fungi, and bacteria (Kolata, 1982). Doolittle (1979) cites many other examples.

Thus, totally new genes do not arise *de novo*, but are produced by the duplication and divergence of pre-existing genes. It is obvious that this is of paramount importance for the understanding and interpretation of the data of DNA–DNA hybridization, and for the concept of DNA sequence homology.

There is also evidence that genes may be "turned off" for long periods of time, but not eliminated from the genome. Kollar and Fisher (1980) showed that the embryonic epithelial tissue of the Domestic Fowl (*G. gallus*), when grafted to the embryonic molar mesenchyme of the House Mouse (*Mus musculus*), can produce a variety of dental structures, including perfectly formed mammalian teeth. Thus the loss of teeth in birds, ca. 100 million years ago, did not result from a loss of genetic coding for specific protein synthesis, but from an alteration in the tissue interactions required for odontogenesis. Phenotypic change need not involve loss of genetic information.

5. THE EVOLUTION OF DNA

5.1. The Uniform Average Rate of DNA Evolution

By "uniform average rate" (UAR) we mean that the *average* rate of nucleotide substitution, measured across the genome and over time, is the same in all lineages of birds. The UAR implies that the average genetic distance between any member of either of two sister groups, to and from any member of a third, more distant group, will be equal and that the average amount of genetic difference is proportional to the absolute length of time since the lineages diverged from one another.

DNA–DNA hybridization measures the average percent difference in base pair complementarity between taxa and we have found that the data fit the definition given above. To reveal the UAR, the design of experimental sets must meet certain criteria as follows: (1) each set must include one or more "relative rate" tests (Sarich and Wilson, 1967a,b), and (2) in each relative rate test, each monophyletic cluster must be represented by at least three species and/or several replicates so that average genetic distances can be calculated.

When these requirements are met it is found that all members of any monophyletic cluster are equidistant from any other taxon which diverged from the lineage leading to that cluster before the members of the cluster diverged from one another. This is a description of the relative rate test which, in its simplest form, consists of three taxa, A, B, and C, related as in Fig. 6. D is the most recent common ancestor of B and C, and E is the most recent common ancestor of A, B, and C. Assume that in three DNA hybridization experiments A, B, and C are the radio-labeled taxa which are compared as follows: AxB, AxC, BxA, CxA. If the same average rate of nucleotide substitution has occurred along D-B and D-C, the distances A-B and A-C should be equal and the reciprocal distances, B-A and C-A, should be equal to one another and to A-B and A-C. Our experimental sets contain 25 DNA hybrids and are more complex than Fig. 6, but the average distances corresponding to A-B, A-C, B-A, and C-A, *are always equal*, within the limits of experimental error. We have literally hundreds of experimental sets that follow these rules. However, since the autocorrelated portions, i.e., A-D, of the A-B and A-C pathways will always be greater than 50% of the total distances, any rate differences along D-B and D-C might be concealed by the variation caused by experimental error. Since all experimental sets have a spread, or range, of delta values it is important to obtain a measure of experimental error.

Table IV contains the results of several experiments designed to determine the magnitude of experimental error at three levels of diversity among driver DNAs, and four levels of evolutionary divergence among tracers. The same four radio-labeled species were used with three different driver preparations. In Table IV, group A, the four tracer DNAs were hybridized with the driver DNA from a single preparation from one bird, an American Robin (*Turdus migratorius*). Eleven homologous hybrids had an average $T_{50}H$ of 88.5, ± 0.2 SE, with a range of 1.7. The average delta $T_{50}H$ for eight heterologous hybrids between the *Myadestes townsendi* tracer and the *T. migratorius* driver was 7.9, ± 0.2, with a range of 1.1, and the other two sets in group A, using *Laniarius barbarus* and *Conopophaga castaneiceps* as tracers, also had narrow ranges of 0.7 and 0.8. (In Table II, group A, the ten identical *Acanthiza* x *Anthochaera* hybrids have a range of 0.6). Thus, it is clear that even sets of hybrids composed of identical tracer and driver DNAs exhibit some variation which must be ascribed to experimental error.

In group B of Table IV the same tracers were used, but the drivers were from up to 13 different individuals of *T. migratorius*. These had larger ranges (1.8–3.5), but this variation must also be due entirely to experimental error since all drivers were from the same species, thus there could be no detectable differences in the average rates of base substitution among them.

In group C the drivers were from different species of *Turdus* but the ranges of variation in the $T_{50}H$ values were about the same (2.1–3.3) as for group B. (See also the data in Table II).

It is apparent that the degree of genetic relationship between the tracer and driver species has no effect on the variation, and that the range does not increase as a function of sample size. This is indicated by the fact that the ranges for group B in Table IV, where N = 11 or 13, were virtually identical to group C where N = 20 or 22. Note that although the eight ranges for groups B and C average 2.6, the standard errors for the average $T_{50}H$ values are ± 0.2 or ± 0.3.

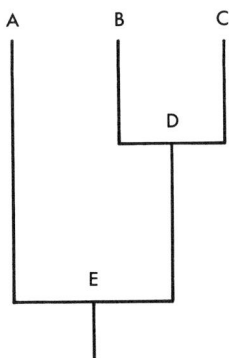

FIGURE 6. Diagram of the relative rate test.

The uniform average rate is demonstrated by the heterologous hybrids in Table IV. The three sets in which *Myadestes* is the tracer have average delta values of 7.9, 9.1, and 9.1. Those in which *Laniarius* is the tracer have averages of 13.9, 13.6, and 13.3, and for *Conopophaga*, 20.2, 21.0, and 20.6. The variation in each series must be due to experimental error because there are no changes in the averages, the standard deviations, or the ranges correlated with the increase in the diversity of the drivers.

Following are three more examples that demonstrate the uniform average rate of DNA evolution.

5.2. The New Zealand Wrens (Acanthisittidae)

The New Zealand Rifleman (*Acanthisitta chloris*) is a small passerine bird, endemic to New Zealand. Its relationships were uncertain until a DNA hybridization study (Sibley et al., 1982) showed that it is a member of the suboscine suborder Oligomyodi, the sister group of the oscine suborder, Passeres. The single-copy tracer DNA of *A. chloris* was compared with the driver DNAs of 57 other species of passerines representing the major groups of the order Passeriformes. The Passeres were represented by 41 species in 21 of the 54 living families recognized by Wetmore (1960). The delta $T_{50}H$ values between *A. chloris* and the 41 oscines ranged from 18.6–21.7 (range = 3.1) and averaged 19.8 with an SE of ± 0.13 and an SD of ± 0.8. The range of 3.1 is smaller than two of the ranges (3.5, 3.3) in Table IV, although the 41 driver species in the *A. chloris* comparisons were far more diverse than the species of *Turdus* in Table IV.

From a preliminary calibration of delta $T_{50}H$ values against absolute time (see Section 6) we estimated that the divergence between the

TABLE IV

DNA–DNA Comparisons to Determine the Magnitude of Experimental Error in $T_{50}H$ Values

	N	Average $T_{50}H$ or $\Delta T_{50}H$	SE	SD	Range	Low value	High value
A. DNA–DNA hybrids composed of the same sample of *Turdus migratorius* DNA as "driver", and the species below as "tracers"							
American Robin (*Turdus migratorius*)	11	88.5	0.2	0.6	1.7	87.6	89.3
Townsend's Solitaire (*Myadestes townsendi*)	8	7.9	0.2	0.5	1.1	7.3	8.4
Black-headed Bush-Shrike (*Laniarius barbarus*)	9	13.9	0.1	0.2	0.7	13.5	14.2
Chestnut-cr. Gnateater (*Conopophaga castaneiceps*)	9	20.2	0.1	0.3	0.8	19.9	20.7
B. DNA–DNA hybrids composed of "driver" DNAs from different individuals of *Turdus migratorius*, and the species below as "tracers"							
American Robin (*Turdus migratorius*)	13	87.1	0.2	0.6	2.1	86.3	88.4
Townsend's Solitaire (*Myadestes townsendi*)	11	9.1	0.2	0.7	2.4	8.0	10.4
Black-headed Bush-Shrike (*Laniarius barbarus*)	11	13.6	0.2	0.6	1.8	12.8	14.6
Chestnut-cr. Gnateater (*Conopophaga castaneiceps*)	11	21.0	0.3	0.9	3.5	19.1	22.6
C. DNA–DNA hybrids composed of "driver" DNAs from different species of the genus *Turdus* and "tracers" of the species listed below							
American Robin (*Turdus migratorius*)	22	1.1	0.2	0.8	2.8	0.0	2.8
Townsend's Solitaire (*Myadestes townsendi*)	20	9.1	0.2	0.7	2.1	8.1	10.2
Black-headed Bush-Shrike (*Laniarius barbarus*)	20	13.3	0.2	0.8	2.7	11.7	14.4
Chestnut-cr. Gnateater (*Conopophaga castaneiceps*)	20	20.6	0.2	0.9	3.3	18.9	22.2

Oligomyodi and the Passeres occurred ca. 95–100 million years ago (MYA). The 41 oscine taxa differ among themselves by delta $T_{50}H$ values ranging from 4–13. Therefore, we estimate that they diverged from one another during the Tertiary between 15–60 MYA. If the average rates of DNA evolution in the separate lineages of the 41 oscines had differed from one another it would be an astonishing coincidence that, after 15–60 million years of independent evolution, all 41 produced virtually the same delta value when compared with a member of a group from which the common ancestor of all 41 lineages had diverged nearly 100 MYA.

We conclude that the variation represented by the spread of 3.1 in the delta $T_{50}H$ values is due to experimental error, not to differences in the average rates of nucleotide substitution.

5.3. The Hawaiian Honeycreepers

The uniform average rate was also revealed in a study of the relationships of the Hawaiian honeycreepers (Fringillidae, Carduelinae, Drepanidini; Sibley and Ahlquist, 1982a). The single-copy DNA of the Apapane (*Himatione sanguinea*) was radio-labeled and hybridized with the driver DNAs of five species of carduelne finches (Carduelini). The average delta $T_{50}H$ for the hybrids was 4.3, the SE = ± 0.04, the SD = ± 0.1, and the range = 0.3. In this study, the tracer DNA of the Apapane was also hybridized with the drivers of 15 species of New World nine-primaried oscines of the tribes Cardinalini, Emberizini, Thraupini (including "coerebids"), Parulini, and Icterini. The average delta value for these comparisons was 7.3, SE = ± 0.07, SD = ± 0.3, range = 1.6. Finally, seven hybrids between the Apapane and species representing seven other "families" of oscines had an average delta $T_{50}H$ of 11.2, SE = ± 0.08, SD = ± 0.2, and the range = 0.7. These seven taxa differ among themselves by delta values from 7–12, indicating that they have been evolving independently from ca. 30–55 MY.

5.4. The Australo-Papuan Fairy-wrens (Maluridae)

In a study of the Fairy-wrens and their relatives (Sibley and Ahlquist, 1982h) we compared the tracer DNA of the Variegated Fairy-wren (*Malurus lamberti*) with the driver DNAs of 37 other species of oscine passerines. Four of these species, and the tracer species, were members of the same monophyletic cluster. As expected, each of the four species was a different delta distance from *M. lamberti*, viz., *M. splendens* 3.0, *M. alboscapulatus* 4.9, *Stipiturus malachurus* 5.9, and *Amytornis tex-*

tilis 8.1. These values indicate that each of these lineages branched from the *M. lamberti* lineage at different times, i.e., each diverged via a different node representing a different common ancestor shared with *M. lamberti*.

Eleven of the 37 species compared with *M. lamberti* were members of the Australian thornbill-honeyeater group, represented by the genera *Acanthiza, Sericornis, Phylidonyris, Aphelocephala, Ephthianura,* and *Lichmera*. These eleven species had an average delta $T_{50}H$ of 10.8, SE = ± 0.18, SD = ± 0.6, range = 1.7.

A second cluster of 15 species (and 15 genera) represented other Australian oscines. Compared with *M. lamberti* these 15 taxa had an average delta $T_{50}H$ of 13.0, SE = ± 0.10, SD = ± 0.4, and range = 1.5.

Figure 7 illustrates the preceding data. The diagram shows that all taxa on one side of a divergence node are equidistant from all taxa on the other branch of that node. The reciprocal values may be assumed to be equal, differing only by the amount of experimental error.

These examples illustrate what we have found in all of our DNA–DNA comparisons, representing many groups of birds, and discussed in the cited papers based on DNA hybridization. We conclude that the *average* rate of nucleotide substitution is the same in all avian lineages and, therefore, that DNA delta values are measures of relative time. We also conclude that the experimental error in our data has an average range of ca. 2.0, with a maximum of ca. 3.0. We believe that variation in fragment size accounts for most of the variation in delta values. By basing the average values for divergence nodes on several comparisons, the effect of experimental errors from all sources is minimized, as indicated by the small standard errors obtained by this method.

5.5. Genetic Rate vs. Morphological Rates

Some investigators using DNA–DNA hybridization have explained certain anomalies in their data as due to "slow-downs" or "speed-ups" in the average rate of DNA evolution (Kohne *et al.*, 1972; Bonner *et al.*, 1981). We have not encountered similar problems. For DNA–DNA hybridization data to reveal a true average rate "slow-down" in one lineage of an otherwise uniform average set of rates, it would require that a substantial percentage of the entire genome participate in the rate change over a considerable period of time. We fail to understand how this could be expected to occur since the average rate is determined by the rate of fixation of selectively neutral alleles, as proposed by Kimura (1968), and/or by natural selection as proposed by Van Valen (1973, 1974).

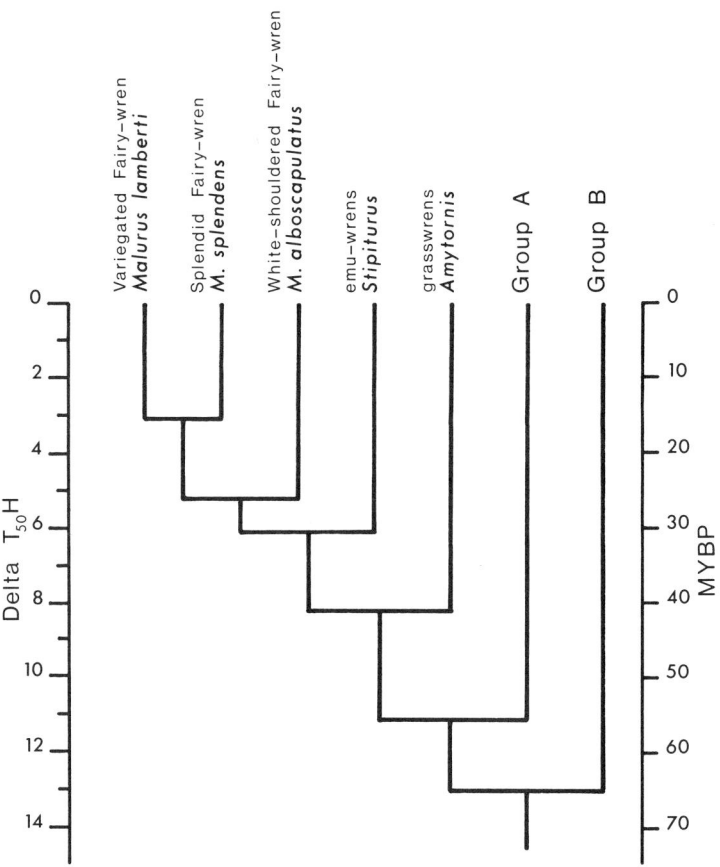

FIGURE 7. Dendrogram of the branching pattern of the Maluridae. Eight species of Australian warblers (Acanthizidae) = (A), and 15 genera of Corvoid passerines = (B).

Kimura and Ohta (1971, p 16) suggested that:

> We should expect that every species, including those that remain unchanged at the phenotypic level, are undergoing constant change at the molecular (DNA) level. Thus, genes in 'living fossils' such as coelacanths, horseshoe crabs, and *Lingula* may be expected to have undergone as many base substitutions as corresponding genes in more rapidly evolving species. In other words, underneath their unchanged morphology (and probably physiology as well) that have been kept remarkably constant by ... *natural selection* ... a steady stream of almost neutral genetic variations has flowed through *random gene frequency drift*, transforming their informational macromolecules tremendously. This will make comparative studies of amino acid sequences [and DNA] much more useful than if their evolutionary changes were mainly caused by natural selection. We ... look forward to

the days when . . . analyses of the phylogeny of . . . macromolecules supply much more useful and reliable information on the genealogy of existing organisms that can be attained by any other means based on phenotypes.

Van Valen's (1973, 1974) "Red Queen's hypothesis" arrives at essentially the same conclusion, although by a different path. It views the complex interactions (competition, predation, etc.) among organisms as the source of selection pressure which remains constant for all species through time and results in a "constant" rate of genetic evolution. This ecological treadmill is the driving force that maintains the uniform average rate of nucleotide substitution, although morphological evolution may appear to the human eye to proceed at various rates. Because no morphological "character," or suite of characters, is determined by more than a small percentage of the genome, the perceived rates of morphological ("organismal") evolution and the measured average rate of genetic evolution, are not closely coupled. In fact, there is ample evidence that morphological characters are not good indices to genetic evolution. For example, the DNAs of two morphologically identical species of the frog genus Xenopus differ by a delta T_m of 12 (Galau et al., 1976) which is greater than that (delta $T_{50}H$ 7.8) between humans and the Old World cercopithecid monkeys such as the Rhesus Macaque (Macaca mulatta; Sibley and Ahlquist, unpublished data). T_m and $T_{50}H$ are based on slightly different scales, but the above comparison is valid (see Table II).

In our DNA comparisons of avian taxa we have many similar results. For example, the Australo-Papuan nuthatch-like sittellas (Daphoenositta) have a delta $T_{50}H$ of 6.8 from the flycatcher-like whistlers (Pachycephala), but are 12.8 from the true nuthatches (Sitta; Sibley and Ahlquist, 1982f,g). Similarly, the vireos (Vireoninae), which are morphologically similar to the New World wood warblers (Fringillidae, Parulini) differ from them genetically by a delta $T_{50}H$ of 10.8, but are only an average delta of 8.4 from morphologically dissimilar members of the "corvine assemblage" (Sibley and Ahlquist, 1982c). Other examples of the discrepancy between morphological and genetic distances have been noted by Sibley and Ahlquist (1980, 1981a, 1982b,d,e,h).

Wallace et al. (1971) found that the serum albumins of morphologically similar frogs differ far more from one another than do the albumins of many morphologically diverse mammals. They concluded that "molecular evolution proceeds in a rather regular fashion with respect to time. By contrast, organismal evolution is . . . an irregular process, some species . . . changing rapidly, while others . . . change slowly." Thus, although the protein clocks may be "sloppy" and "erratic" (Fitch, 1976), when averaged over long periods of time they keep far better time than do morphological characters. Since single proteins are fairly good timekeepers it is hardly surprising that averages across the entire sequence complexity of the genome should produce the remarkable precision we have found in the DNA hybridization data.

5.6. Congruence between Morphological and DNA–DNA Hybridization Data

Although morphological and "molecular" data often yield conflicting results, there are many examples of congruence between the relationships indicated by the DNA comparisons and those based on morphology. In the more than 700 experimental sets we have examined, there are at least 500 examples in which DNA hybrids involving congeners of the radio-labeled taxon have been included. In every case the congeneric hybrids are the closest ones to the homolog.

Similarly, the DNAs of members of different genera within the same family are usually more alike than are those of different families. The exceptions have been shown to have been incorrectly classified and, in every case, we have found morphological characters that are congruent with the DNA data.

Congruence, as would be expected, becomes less consistent at the familial and ordinal levels where the taxonomic interpretation of morphological characters becomes more difficult.

Thus, we find that congruence is related to the degree of divergence, i.e., to the degree of morphological difference, and therefore to the ability of the morphologist to discern relationships. Morphological similarities are easily detected between closely related taxa, hence, the characters of congeneric species or closely related genera are seldom misinterpreted. However, whenever there have been major adaptive changes, even between closely related taxa, the congruence between the taxonomic opinions based on morphology and the DNA–DNA measurements declines. Examples are cited below in Section 7 (Adaptive Radiation).

It is also possible to examine congruence between distantly related taxa. Many of our experimental sets include one or more hybrids between the labeled taxon and taxa we can be confident are at certain relative distances from it. For example, in a set in which the tracer is an oscine passerine, and most of the comparisons are with other oscines, we may include one or more suboscines and one or two non-passerines. Without exception, the DNA distances are in accord with expectation, i.e., the oscines are closest to the homolog, the suboscines next, and the non-passerines are most distant.

Mickevich and Johnson (1976) demonstrated congruence between morphological and allozyme data in a study of five species of the fish genus *Menidia*. This agrees with our numerous examples of congruence between closely related taxa of birds. The question is, how congruent are similar comparisons among higher categories of fishes?

5.7. Generation Time

"It is a matter of controversy (Kohne et al., 1972; Sarich and Cronin, 1977) as to whether the extent of molecular evolution depends strictly on time or on the number of generations" (Bonner et al., 1981, pp. 296–297). Kohne (1970) and Kohne et al. (1972) found that a correction for different generation times made certain of their data compatible and they concluded (p. 627) that the rate of molecular evolution "appears to be faster in species with short generation times." Wilson and Sarich (1969), Sarich (1972), Sarich and Wilson (1973), and Sarich and Cronin (1977) have presented evidence against an effect of generation length on the rate of molecular evolution.

To test this hypothesis we compared the single-copy DNAs of long and short generation species of procellariiform birds using a relative rate test in which a Ringed Plover (*Charadrius hiaticula*) was the outside taxon. Tables V and VI contain the results. The albatrosses begin to breed when they are 6–10 years old and thereafter breed only every two years. Thus, they nest only six or seven times in 20 years. The smallest petrels, e.g., *Oceanodroma*, will nest about 18 times in 20 years, and the middle-sized species will fall between these extremes. Although the small petrels nest three times as frequently as the albatrosses, all of the procellariiforms in Table V are the same distance from

TABLE V

DNA–DNA Hybrid Values for Comparisons between the Ringed Plover and Six Species of the Order Procellariiformes

Species	$\Delta T_{50}H$
Ringed Plover (*Charadrius hiaticula*)	0.0
Galapagos Albatross (*Diomedea irrorata*)	11.7
Leach's Petrel (*Oceanodroma leucorhoa*)	11.7
Herald Petrel (*Pterodroma arminjoniana*)	11.8
Antarctic Petrel (*Thalassoica antarctica*)	11.8
Sooty Shearwater (*Puffinus griseus*)	11.9
Broad-billed Prion (*Pachyptila vittata*)	12.0

TABLE VI
DNA–DNA Hybrid Values for Comparisons between the Sooty Shearwater, Six Other Procellariiforms, and the Gray Plover

Species	$\Delta T_{50}H$
Sooty Shearwater (Puffinus griseus)	0.0
Cory's Shearwater (Calonectris diomedea)	1.8
Antarctic Petrel (Thalassoica antarctica)	3.6
Broad-billed Prion (Pachyptila vittata)	3.6
Herald Petrel (Pterodroma arminjoniana)	3.8
Galapagos Albatross (Diomedea irrorata)	5.9
Leach's Petrel (Oceanodroma leucorhoa)	6.6
Gray Plover (Pluvialis squatarola)	11.9

the Ringed Plover. For the six species the average $T_{50}H = 11.8$, $SE = \pm 0.05$, and $SD = \pm 0.12$.

The $T_{50}H$ values between the Sooty Shearwater and six other species of procellariiforms, plus the Gray Plover, are given in Table VI. As expected, the distances vary considerably (1.8–6.6) among the procellariiforms because each branched at a different time from the lineage leading to the Sooty Shearwater. Also as expected, the $T_{50}H$ for the Sooty Shearwater-Gray Plover DNA hybrid (11.9) is an excellent match for the average reciprocal of 11.8, noted above relative to the Ringed Plover.

A similar pattern of values was obtained when the DNA of a Great Blue Heron (Ardea herodias) was used as the outside taxon. We conclude that the number of generations does not affect the average rate of DNA evolution. The most likely explanation for the opposite conclusion by Kohne et al. (1972) is that the fossil datings they used for calibrating their data were inaccurate. The relative rate test, which is independent of divergence datings, is a more reliable way to demonstrate the uniform average rate in terms of relative time. Divergence dates are required only to calibrate the DNA distance values in absolute time.

Hartl and Dykhuizen (1979) used isogenic strains of the bacterium Escherichia coli to test the molecular clock hypothesis. They found that when averaged over a sufficiently long time, the rate of increase of fitness under selection is constant and independent of generation time. They concluded that "selectively driven gene substitutions ... do exhibit clocklike behaviour."

6. THE CALIBRATION PROBLEM

When two incipient species derived from a common ancestor cease to interbreed, their nucleotide sequences begin to diverge. The average rate of divergence is measurable by DNA–DNA hybridization and the resulting distance values can be used to reconstruct the phylogenies of living taxa in terms of relative time. The DNA–DNA data may be used directly to produce a dendrogram in which the relationships between monophyletic clusters and their divergence nodes will be correctly depicted. However, to convert the relative time scale to absolute time requires the calibration of the DNA–DNA values against an external dating source, viz., fossils, or geological events that have caused phyletic divergences.

We have discussed this problem elsewhere (Sibley and Ahlquist, 1981a, pp. 315–317) and will not review it here. The question is not what should be done, which is obvious, but to find trustworthy divergence dates in the fossil and/or geological records. For birds, there may be no dated divergences based on fossils that are accurate enough for this purpose, but the general constraints of the avian fossil record are useful. The largest delta values for comparisons among living groups of birds should equate with divergences in the early to mid-Cretaceous, ca. 100–130 MYA. The largest delta $T_{50}H$ values we have found are 25–28, which suggests that delta 1.0 = ca. 4–5 my for these oldest avian divergences. The same values were developed from our studies of the ratites (Sibley and Ahlquist, 1981a, p. 322) and the New Zealand wrens (Sibley et al., 1982). In the ratite study, we assumed that the "final separation between the ancestors of the living ostriches and rheas could have been about 80 MYA" when the Atlantic opening became too wide for the flightless ancestral birds to cross. We have recently recalculated the delta $T_{50}H$ for this comparison as 17.4, so delta 1.0 = 4.6 MY. We also have the data for 14 DNA hybrids between Old World suboscines (pittas, broadbills) and New World suboscines (antbirds, ovenbirds, tyrant flycatchers, cotingas, etc.). For these the average delta $T_{50}H = 16.9$ ($\pm$ 0.2 S.E., $\pm$ 0.5 S.D.) (Sibley and Ahlquist, in press b). Because the ancestors of the suboscines presumably could fly we may use 75 MY for the time when the Atlantic became impassable for them. This yields a value of 4.4 MY = 1.0 delta $T_{50}H$.

For the New Zealand wrens we obtained a value of delta $T_{50}H$ 1.0 = 4.5 MY, based on the opening of the Tasman Sea ca. 80 MYA and an average delta value of 17.7 for the divergence between *Acanthisitta* and the other suboscines (Sibley et al. 1982).

Although our assumptions are obviously uncertain, we suggest that a delta $T_{50}H$ of 1.0 = ca. 4.5 MY, at least for divergences older than ca. 50 MYA. The calibration for more recent divergences seems to be curvilinear, and additional datings are needed to obtain a better calibration curve.

We believe that a 1:4.5 ratio is reasonable for our data on birds, but Hall et al. (1980) estimated a ratio of 1:1 for sea urchins. We will not attempt to explain this discrepancy, but, if a 1:1 ratio is applied to our avian data, the oldest divergence between living groups would be ca. 28 MYA, in the late Oligocene. This cannot be correct because fossils of living groups are known from the late Cretaceous and Paleocene (Brodkorb, 1971, 1976). Clearly, the calibration problem is not yet solved.

7. ADAPTIVE RADIATION

Adaptive radiation is a primary principle of evolution but the recognition of the members of a monophyletic radiation is often difficult.

> The adaptive radiation so easily observed in the Galapagos finches ... and in the Hawaiian honeycreepers ... must also have occurred on the continents ... but the tidy picture becomes obscured because so many more species are involved, intermediate niches are present, barriers are less precise and a greater time span is involved. Being unable to sort out the genetically related groups using gross morphology we compromise by setting up groups based upon the feeding structures because they are the only characters which show significant variation ... we see what appear to be groups because the major feeding niches are fairly discrete ... where there are intermediate feeding niches we encounter intermediate species and the boundaries of taxa ... become blurred.
> Natural groups larger than genera must exist on the continents ... but we should not expect to be able to delineate them by ... any ... structure intimately concerned with feeding. A classification based upon such evidence cannot avoid being, in part, a classification of food niches (Sibley, 1970, pp. 107–108).

Our DNA–DNA comparisons have revealed several monophyletic radiations composed of closely related species previously considered to be members of different families. For example, in a study of the Australian oscine passerines (Passeres), the single-copy tracer DNAs of 31 genera were compared with one another and with many other genera of oscines. All groups that have been considered to be families or subfamilies were represented and ca. 2000 DNA hybrids were analyzed (Sibley and Ahlquist, in press, a).

From these data we conclude that the entire "old endemic" oscine fauna of Australia evolved from a single ancestral taxon. This radiation began ca. 65 MYA, i.e., near the Cretaceous-Tertiary boundary, and split into three major clusters during the early Tertiary, ca. 55–60 MYA. Each of these subdivided further as the birds adapted to most of the available ecological niches. More recently, representatives of a few "outside" groups have become established. An outline of the major groups of Australian passerines follows:

   I. The Lyrebird-Bowerbird Group
      A. Lyrebirds and Scrub-birds
      B. Bowerbirds
      C. Treecreepers
  II. The Fairy-wren-Honeyeater-Thornbill Group
      A. Fairy-wrens, Emu-wrens, Grass-wrens
      B. Honeyeaters, including Australian chats
      C. Thornbills, Scrub-wrens, Whitefaces, Bristlebirds, Pardalotes
 III. Australian Robin-Whistler-Monarch-Crow Group
      A. Australo-Papuan robins and flycatchers
      B. Log-runners
      C. Pseudo-babblers (*Pomatostomus*)
      D. Crows, Birds-of-Paradise, Monarchs, Whistlers, etc.
         a. Quail-thrushes and Whipbirds
         b. Australian Chough, Apostlebird
         c. Whistlers, Shrike-thrushes, Sittellas, Shrike-tits
         d. Monarchs, Drongos, Magpie-larks, Fantails
         e. Crows, B.-of-Paradise, Woodswallows, Currawongs, Orioles and Cuckoo-shrikes

The other Australian oscines are recent immigrants, viz., larks, swallows, thrushes, starlings, sylviine warblers (*Cincloramphus, Eremiornis, Megalurus*), white-eyes, wagtails, weavers, waxbills, sunbirds, and flowerpeckers (Sibley and Ahlquist, 1982e–h; in press, a).

Another major continental radiation, not previously recognized, includes the Neotropical tanagers, the Swallow-Tanager (*Tersina*), the Plush-capped Finch (*Catamblyrhynchus*), all of the Neotropical honeycreepers ("Coerebidae"), and the finch-billed genera *Diuca, Sicalis, Haplospiza, Volatinia,* and *Oryzoborus*. Other South American "finches" may also be tanagers, not emberizine finches, as they have always been assumed to be. Apparently, the tanagers captured most of the finch niches in South America during the Tertiary when South America was an island continent. Convergent evolution, mainly in bill shape, concealed their true relationships.

We also have DNA data showing that the ploceid radiation includes the accentors (Prunellinae), the wagtails and pipits (Motacillinae), and the sunbirds and flowerpeckers (Nectariniidae), as well as the waxbills (Estrildinae) and the sparrows (Passerinae).

Another oscine radiation consists of the waxwings, silky flycatchers and Palm Chat (Bombycillidae), the dippers (Cinclidae), the thrushes (Turdidae) which includes the true thrushes (Turdinae), and the closely related Old World flycatchers (Muscicapini) and chats (Erithacini). The family Sturnidae, composed of the starlings and the mockingbirds (Mimini) are also members of this radiation.

The sylvioid radiation includes the nuthatches (*Sitta*), the Northern creepers (*Certhia*), the wrens, gnatcatchers, Verdin, titmice, long-tailed tits and bushtits, swallows, kinglets, bulbuls, white-eyes, the sylviine warblers, and the babblers. The genus *Sylvia* is actually more closely related to the babblers (Timaliini) than to the leaf warblers (Phylloscopinae).

The wrens (Troglodytidae), long thought to be closely related to the mockingbirds and/or to the "Old World insect-eaters," are actually members of a radiation that includes the Verdin (*Auriparus*), the gnatcatchers (*Polioptila*), the creepers (*Certhia*), and two South American genera, *Microbates* and *Ramphocaenus*. On the basis of morphological characters, Rand and Traylor (1953) proposed that these two latter genera are related to *Macrosphenus* of Africa, rather than to *Polioptila*. We submit that convergent evolution, not recent common ancestry, has produced the similarities noted by Rand and Traylor.

Morphological convergence is a corollary of adaptive radiation and several other examples have been noted above, including *Sitta-Daphoenositta*, *Muscicapa-Monarcha*, and *Nectarinia-Meliphaga*.

An especially interesting example illustrating convergence and divergence in morphological characters was found in DNA–DNA comparisons among the kingfishers (Alcedinidae). The single-copy DNA of the Sacred Kingfisher (*Halcyon sancta*) of Australasia was radio-labeled and compared with the DNAs of five other kingfishers. The names, ranges, and delta values were as follows: Hook-billed Kingfisher (*Melidora macrorhina*) of New Guinea (4.9), a Paradise Kingfisher (*Tanysiptera galatea*) from New Guinea (5.6), the Woodland Kingfisher (*Halcyon senegalensis*) of Africa (7.6), and two New World genera, *Ceryle* (9.5), and *Chloroceryle* (11.2). Thus *Melidora* and *Tanysiptera* are more closely related to *H. sancta* than is *H. senegalensis*. We believe that the DNA data are correct and that they have revealed a taxonomic error, as follows.

Most kingfishers, and all species of *Halcyon*, have straight, pointed bills and short tail feathers. *Melidora* has a hooked bill, and *Tanysiptera* has long central rectrices with spatulate tips. These specialized characters are the bases for the generic separation of these taxa. *Melidora* and *Tanysiptera* are sympatric in New Guinea with several species of *Halcyon* and we suggest that the specialized bill of *Melidora*, and the long tail of *Tanysiptera*, evolved because of interspecific interactions among the closely related species in New Guinea. This example also illustrates the poor correlation between morphological characters and genetic relationships. *H. sancta* and *H. senegalensis* have similar bills and tails and have therefore been considered to be congeneric, but they have been diverging longer, and are genetically more different, than are the three morphologically diverse Australo-Papuan species.

8. CATEGORICAL EQUIVALENCE

Every systematist knows that taxonomic categories of the same rank in different groups do not represent the same degrees of evolutionary divergence. An Order of birds is not equivalent to an Order of reptiles, insects, or plants, and a Family of passerines is not equivalent to a Family of non-passerines. Although the boundaries of taxa can be defined, taxonomic categories have had to be arranged in a nested hierarchy of relative ranks within each group.

It has been impossible to define equivalent categories because a universal basis for defining evolutionary equivalence has not been available. Hennig (1966, p. 182) favored the age of origin as the basis for the absolute ranking of taxa, but concluded that "there is no single method with which the age of origin can be determined accurately." Mayr (1969, p. 72) dismissed Hennig's suggestion because "it ignores differences in the rates of evolution and confuses genealogical with genetic relationship."

We have discussed this problem elsewhere (Sibley and Ahlquist, 1982d, pp. 8–14). Because of the uniform average rate of genetic evolution, genealogical and genetic relationships *are identical* and the absolute ranking of taxonomic categories can, and should, be based upon the age of origin. DNA hybridization data can provide the absolute times for divergences back to ca. 150 MYA, and other nucleic acid techniques can extend the time scale to the earliest periods of life on Earth (see Woese, 1981). Fossil and geological evidence of divergence dates already provides the basic framework for the times of origin of major groups.

"We therefore propose that the major . . . dichotomies be dated by . . . fossil and/or nucleic acid sequence evidence and that DNA hybridization data be used to develop a genetic divergence-based, and hence time-based, system of taxonomic categories representing the dichotomies of the last 150 MY" (Sibley and Ahlquist, 1982d, p. 14).

This proposal is certainly controversial and will be difficult to implement. However, we have made a beginning by developing a modified absolute ranking of avian categories based upon DNA hybridization values, and therefore upon relative times of origin, as follows: Delta $T_{50}H$ values up to 4 = congeneric species, delta 5–7 = Tribe, 7–9 = Subfamily, 9–11 = Family, 11–13 = Superfamily, 13–15 = Parvorder, 15–17 = Infraorder, 17–20 = Suborder, and 20–25 = Order. We have used this scale to develop a classification of the Passeriformes (Sibley and Ahlquist, in press, a) and found it necessary to treat the boundaries as guidelines, not as hard barriers. The result proved satisfactory as a reflection of the DNA-based phylogeny.

9. CLASSIFICATION

"The central problem of evolutionary classification is that no satisfactory general measure of evolutionary differentiation has yet been found ... Since the structure of DNA ... molecules is comparable in all organisms ... such a breakthrough would make it possible [to use] descriptions of nucleotide structure as general indicators of the evolutionary differentiation of organisms" (Griffiths, 1973). DNA hybridization provides a "general measure of evolutionary differentiation" but how should we translate the data into a classification? We believe that the cladistic pattern of the phylogeny derived from DNA hybridization data is a suitable basis for classification. The uniform average rate of DNA evolution means that the phylogeny will be time-related, which should satisfy Hennig's (1966) criteria for a time-based ranking of taxa in which categories of equal age of origin are equivalent, and sister groups are of coordinate rank. Such a classification will express phylogenetic relationships and can embody the principles enunciated by Nelson (1973), including "the use of precise, nonarbitrary conventions for subordination and sequencing of units." We have used these principles in a new classification of the passerines.

The DNA data are measures of overall, or average similarity, and they do not provide the basis for the definition of "primitive-derived sequences" of characters, i.e., "synapomorphies" cannot be identified. However, the melting curve of a DNA–DNA hybrid is a plot of derived (= low melting temperature) to primitive (= high melting temperature) sequences. Delta $T_{50}H$ values are measurements of the overall *average* divergence between genomes and, should one wish to do so, the $T_{25}H$

values can be used to provide a measure of the average divergence between the "derived" sequences of different taxa. Griffiths (1973, p. 338), an avowed disciple of Hennig, seemed to find no conflict between cladistic principles and the use of DNA comparisons. We have found that the DNA data reflect cladistic relationships and it was this discovery that convinced us of the fundamental validity of Hennig's principles.

The "pheneticists" argue that organisms should be classified according to "overall similarity" based upon as many characters as possible with each character being given equal weight (Sneath and Sokal, 1973, p. 5). These authors (p. 73) equated "unit characters" with the "bits" of information theory and "tentatively identify ... taxonomic bits with the genetic code...." They further equated "taxonomic bits" with nucleotides by noting that "the number of bits in the genome ranges from around 10^4 for some viruses to around 10^{11} for many higher animals...." Colless (1967, p. 295) suggested that "the codon elements thus employed as attributes must, surely, be the ultimate approximation of our notion of 'unit attributes' ... and the rationale provides a valuable illustration of the ultimate identity of phylogenetic and phenetic procedures."

The "evolutionists" of the "Simpson–Mayr school" believe that biological classification should consider "the degree of genetic similarity between organisms; and the phylogenetic sequence of events in their history" (Bock, 1973, p. 375). We agree with these goals, but we do not accept the assumption that "greater phenotypical similarity implies greater genetical similarity and hence closer relationship" (Bock, 1973, p. 377). This idea was proposed by Simpson (1944, p. 3) as the practical solution to what he considered the ideal, namely, that the "Rate of evolution might most desirably be defined as amount of genetic change in a population per ... unit of absolute time." That is exactly what DNA hybridization measures.

There is, of course, a relationship between phenotype and genotype, but morphological characters appear to evolve at many different rates and this relationship is not simple and direct. However, the data of DNA hybridization should satisfy Simpson's and Bock's definitions because they directly measure genotypic divergence and can be calibrated against absolute time.

The only conflict we detect between our position and that of the proponents of "classical evolutionary classification" is that we do not consider it useful to try to express subjectively evaluated degrees of morphological differentiation ("grades") in a classification. Such attempts have led, and will continue to lead, to the recognition of arbitrary groupings which do not reflect genetic similarity. We have cited several examples in this paper. We suggest that a classification based on DNA hybridization data will reflect the degrees of genetic similarity among taxa and, therefore, will express the only general measure of evolutionary divergence, namely, nucleotide complementarity.

To whatever degree the DNA–DNA data reconcile the differences among the three schools the reasons are clear. Namely, all three are based upon the same principles of evolution and genetics and the information in the genotype is the common focal point. However, in the absence of data directly reflecting the genotype it has been necessary to use morphological characters and the different methods for doing so have produced disagreements.

The DNA hybridization technique is properly viewed as another method of data production, based on the same principles, but providing a direct measure of genotypic divergence. It thus avoids the problems that afflict all attempts to extract genetic information from gross morphology.

The use of genotypic data has long been approved in theory by the practitioners of the other methods, but it is unlikely that the DNA hybridization technique will be readily acknowledged as a source of such data. However, we are satisfied that the technique produces data of exceptional value for systematics and we propose to continue to act upon the advice of Colless (1967, p. 295) "by collecting new data while unstudied, or poorly-known, taxa are still with us." We hope that others will join us.

In this paper we have tried to provide a bridge between the biochemists who developed the DNA hybridization technique and the systematists who have untold numbers of problems awaiting solutions. We have been surprised to discover that the phylogeny and relationships of birds, a seemingly well-known group of organisms, have been misinterpreted and misunderstood to an astonishing degree. This is probably also true for other groups.

REFERENCES

Ahlquist, J. E., Bledsoe, A. H., Ratti, J. T., and Sibley, C. G., The relationship between the Western Grebe (*Aechmophorus occidentalis*) and Clark's Grebe (*A. clarkii*), as indicated by DNA–DNA hybridization. *Condor*, in press.

Bachmann, K., Harrington, B. A., and Craig, J. P., 1972, Genome size in birds, *Chromosoma* **37**:405–416.

Barker, W. C., and Dayhoff, M. O., 1976, Immunoglobulins and related proteins, in: *Atlas of Protein Sequence and Structure*, Volume 5, National Biomed. Res. Found, Silver Spring, Maryland, pp. 165–190.

Benveniste, R. A., and Todaro, G. J., 1976, Evolution of type E viral genes: Evidence for an Asian origin of man, Nature **261**:101–108.

Bock, W. J., 1973, Philosophical foundations of classical evolutionary classification, Syst. Zool. **22**:375–392.

Bonner, T. I., Brenner, D. J., Neufeld, B. R., and Britten, R. J., 1973, Reduction in the rate of DNA reassociation by sequence divergence, J. Mol. Biol. **81**:123–135.

Bonner, T. I., Heinemann, R., and Todaro, G. J., 1981, A geographic factor involved in the evolution of the single copy DNA sequences of primates, in: Evolution Today, Proc. 2nd Int. Cong. Syst. Evol. Biol. (G.G.E. Scudder and J.L. Reveal, eds.), Hunt Inst. Botan. Document. Pittsburgh, pp. 293–300.

Britten, R. J., 1971, Sequence complexity, kinetic complexity, and genetic complexity, Carnegie Inst. Wash., Yearbook **69**:503–506.

Britten, R. J., and Davidson, E. H., 1971, Repetitive and non-repetitive DNA sequences and a speculation on the origins of evolutionary novelty, Quart. Rev. Biol. **46**:111–138.

Britten, R. J., and Kohne, D. E., 1968, Repeated sequences in DNA, Science **161**:529–540.

Britten, R. J., Graham, D. E., and Neufeld, B. R., 1974, Analysis of repeating DNA sequences by reassociation, in: Methods in Enzymology, Volume 29 (L. Grossman and K. Moldave, eds.), Academic Press, London and New York, pp. 363–418.

Britten, R. J., Cetta, A., and Davidson, E. H., 1978, The single copy DNA sequence polymorphism of the sea urchin Strongylocentrotus purpuratus, Cell **15**:1175–1186.

Brodkorb, P., 1971, Origin and evolution of birds, in: Avian Biology, Volume 1 (D.S. Farner and J.R. King, eds.), Academic Press, New York, pp. 19–55.

Brodkorb, P., 1976, Discovery of a Cretaceous bird, apparently ancestral to the Orders Coraciiformes and Piciformes (Aves: Carinatae), in: Smithsonian Contrib. Paleo., 27 (S.L. Olson, ed.), Smithsonian Institution Press, Washington, D.C., pp. 67–73.

Chan, H.-C., Ruyechan, W. T., and Wetmur, J. G., 1976, In vitro iodination of low complexity nucleic acids without chain scission, Biochemistry **15**:5487–5490.

Colless, D. H., 1967, The phylogenetic fallacy, Syst. Zool **16**:289–295.

Commorford, S. L., 1971, Iodination of nucleic acids in vitro, Biochemistry **10**:1993–2000.

Doolittle, R. F., 1979, Protein evolution, in: The Proteins, Volume 4 (H. Neurath and R. L. Hill, eds.), Academic Press, New York, pp. 1–118.

Eden, F. C., Hendrick, J. P., and Gottlieb, S. S., 1978, Homology of single copy and repeated sequences in chicken, duck, Japanese quail and ostrich DNA, Biochemistry **17**:5113–5121.

Fitch, W. M., 1976, Molecular evolutionary clocks, in: Molecular Evolution (F. J. Ayala, ed.), Sinauer Associates, Sunderland, Massachusetts, pp. 160–178.

Galau, G. A., Chamberlain, M. E., Hough, B. R., Britten, R. J., and Davidson, E. H., 1976, Evolution of repetitive and nonrepetitive DNA, in: Molecular Evolution (F.J. Ayala, ed.), Sinauer Associates, Sunderland, Massachusetts, pp. 200–224.

Gillespie, D., 1977, Newly evolved repeated DNA sequences in primates, Science **196**:889–891.

Griffiths, G. C. D., 1973, Some fundamental problems in biological classification, Syst. Zool. **22**:338–343.

Grula, J. W., Hall, T. J., Hunt, J. A., Giugni, T. D., Graham, G. J., Davidson, E. H., and Britten, R. J., 1982, Sea urchin DNA sequence variation and reduced interspecies differences of the less variable sequences. Evolution **36**:665–676.

Hall, T. J., Grula, J. W., Davidson, E. H., and Britten, R. J., 1980, Evolution of sea urchin non-repetitive DNA, J. Molec. Evol. **16**:95–110.

Hartl, D., and Dykhuizen, D., 1979, A selectively driven molecular clock, *Nature* **281**:230–231.
Hennig, W., 1966, *Phylogenetic Systematics*, University of Illinois Press, Urbana.
Hinegardner, R., 1976, Evolution of genome size, in: *Molecular Evolution* (F.J. Ayala, ed.), Sinauer Associates, Sunderland, Massachusetts, pp. 179–199.
Huber, P. J., 1977, Robust statistical procedures, *Regional Conf. Ser. Appl. Math. Soc. Indust. Appl. Math.*, Philadelphia, Pa., pp. 56.
Kimura, M., 1968, Evolutionary rate at the molecular level, *Nature* **217**:624–626.
Kimura, M., and Ohta, T., 1971, On the rate of molecular evolution, *J. Mol. Evol.* **1**:1–17.
Kohne, D. E., 1970, Evolution of higher-organism DNA, *Quart. Rev. Biophys.* **33**:327–375.
Kohne, D. E., and Britten, R. J., 1971, Hydroxyapatite techniques for nucleic acid reassociation, in: *Procedures in Nucleic Acid Research*, Volume 2 (G.L. Cantoni and D. R. Davies, eds.), Harper & Row, New York, pp. 500–512.
Kohne, D. E., Chiscon, J. A., and Hoyer, B. H., 1971, Nucleotide sequence change in nonrepeated DNA during evolution, *Carnegie Inst. Yearbook* **69**:488–501.
Kohne, D. E., Chiscon, J. A., and Hoyer, B. H., 1972, Evolution of primate DNA sequences, *J. Hum. Evol.* **1**:627–644.
Kolata, G., 1982, New theory of hormones proposed, *Science* **215**:1383–1384.
Kollar, E. J., and Fisher, C., 1980, Tooth induction in chick epithelium: Expression of quiescent genes for enamel synthesis, *Science* **207**:993–995.
Marchalonis, J. J., 1977, *Immunity in Evolution*, Harvard University Press, Cambridge, Massachusetts.
Marmur, J., 1961, A procedure for the isolation of deoxyribonucleic acid from microorganisms, *J. Mol. Biol.* **3**:585–596.
Mayr, E., 1969, *Principles of Systematic Zoology*, McGraw Hill, New York.
McCarthy, B. J., and Farquhar, M. N., 1972, The rate of change of DNA in evolution, in: *Evolution of Genetic Systems* (H.H. Smith, ed.), *Brookhaven Symp. Biol.* **23**:1–43.
Mednikov, B. M., 1980, DNA × DNA hybridization in taxonomy, in: *Biology Review*, (*Physico-Chemical Aspects*), Volume 1 (V.P. Skulachev, ed.), Soviet Sci. Rev., Sect. D., pp. 447–476.
Mickevich, M. F., and Johnson, M. S., 1976, Congruence between morphological and allozyme data in evolutionary inference and character evolution, *Syst. Zool.* **25**:260–270.
Nathans, D., and Smith, H. O., 1975, Restriction endonucleases in the analysis and restructuring of DNA molecules, *Annu. Rev. Biochem.* **44**:273–293.
Nelson, G. J., 1973, Classification as an expression of phylogenetic relationships, *Syst. Zool.* **22**:344–359.
Nuechterlein, G. L., 1981, Courtship behavior and reproductive isolation between Western Grebe color morphs, *Auk* **98**:335–349.
Prensky, W., 1976, The radioiodination of RNA and DNA to high specific activities, in: *Methods in Cell Biology*, Volume 13 (D. M. Prescott, ed.), Academic Press, New York, pp. 121–152.
Rand, A. L., and Traylor, M. A., Jr., 1953, The systematic position of the genera *Ramphocaenus* and *Microbates*, *Auk* **70**:334–337.
Ratti, J. T., 1979, Reproductive separation and isolating mechanisms between sympatric dark- and light-phase Western Grebes, *Auk* **96**:573–586.
Rice, N. R., 1972, Change in repeated DNA in evolution, in: *Evolution of Genetic Systems* (H. H. Smith, ed.), Gordon and Breach, New York, *Brookhaven Symp. Biol.* **23**:44–79.

Sarich, V. M., 1972, Generation time and albumin evolution, *Biochem. Genet.* **7**:205–212.
Sarich, V. M., and Cronin, J. E., 1977, Generation length and rates of hominoid molecular evolution, *Nature* **269**:354–355.
Sarich, V. M., and Wilson, A. C., 1967a, Rates of albumin evolution in primates, *Proc. Natl. Acad. Sci. USA* **58**:142–148.
Sarich, V. M., and Wilson, A. C., 1967b, Immunological time scale for hominid evolution, *Science* **158**:1200–1204.
Sarich, V. M., and Wilson, A. C., 1973, Generation time and genomic evolution in primates, *Science* **179**:1144–1147.
Shapiro, H. S., 1976, Distribution of purines and pyrimidines in deoxyribonucleic acids, in: *Handbook of Biochemistry and Molecular Biology*, Volume II (G.D. Fasman, ed.), Chemical Rubber Co. Press, Cleveland, Ohio, pp. 241–275.
Shields, G. F., and Straus, N. A., 1975, DNA–DNA hybridization studies of birds, *Evolution* **29**:159–166.
Sibley, C. G., 1970, A comparative study of the egg-white proteins of passerine birds, *Bull. Peabody Mus. Nat. Hist.* **32**:1–131.
Sibley, C. G., and Ahlquist, J. E., 1980, The relationships of the "primitive insect eaters" (Aves: Passeriformes) as indicated by DNA–DNA hybridization, in: *Proc. 17th Int. Ornithol. Cong.* (R. Nöhring, ed.), Deutsche Orn. Gesellsch., Berlin, pp. 1215–1220.
Sibley, C. G., and Ahlquist, J. E., 1981a, The phylogeny and relationships of the ratite birds as indicated by DNA–DNA hybridization, in: *Evolution Today, Proc. Second Int. Cong. Syst. Evol. Biol.* (G.G.E. Scudder and J.R. Reveal, eds.), Hunt Inst. Botan. Document., Pittsburgh, pp. 301–335.
Sibley, C. G., and Ahlquist, J. E., 1981b, The relationships of the Accentors (*Prunella*) as indicated by DNA–DNA hybridization, *J. für Orn.* **122**:369–378.
Sibley, C. G., and Ahlquist, J. E., 1981c, The relationships of the wagtails and pipits (Motacillidae) as indicated by DNA–DNA hybridization, *L'Oiseau et R.F.O.* **51**:189–199.
Sibley, C. G., and Ahlquist, J. E., 1982a, The relationships of the Hawaiian honeycreepers (Drepaninini) as indicated by DNA–DNA hybridization, *Auk* **99**:130–140.
Sibley, C. G., and Ahlquist, J. E., 1982b, The relationships of the Wrentit (*Chamaea fasciata*) as indicated by DNA–DNA hybridization, *Condor* **84**:40–44.
Sibley, C. G., and Ahlquist, J. E., 1982c, The relationships of the vireos (Vireoninae) as indicated by DNA–DNA hybridization, *Wilson Bull.* **94**:114–128.
Sibley, C. G., and Ahlquist, J. E., 1982d, The relationships of the Yellow-breasted Chat (*Icteria virens*), and the alleged "slow-down" in the rate of macromolecular evolution in birds, *Postilla* **187**:1–19.
Sibley, C. G., and Ahlquist, J. E., 1982e. The relationships of the Australo-Papuan scrub-robins *Drymodes* as indicated by DNA–DNA hybridization, *Emu* **82**:101–105.
Sibley, C. G., and Ahlquist, J. E., 1982f, The relationships of the Australo-Papuan sittellas *Daphoenositta* as indicated by DNA–DNA hybridization, *Emu* **82**:173–176.
Sibley, C. G., and Ahlquist, J. E., 1982g, The relationships of the Australasian whistlers *Pachycephala* as indicated by DNA–DNA hybridization, *Emu* **82**:199–202.
Sibley, C. G., and Ahlquist, J. E., 1982h, The relationships of the Australo-Papuan fairy-wrens *Malurus* as indicated by DNA–DNA hybridization, *Emu* **82**:(in press).
Sibley, C. G., and Ahlquist, J. E., The phylogeny and classification of the passerine birds, based on comparisons of the genetic material, DNA, *Proc. 18th. Int. Ornithol. Cong.* (in press, a).
Sibley, C. G., and Ahlquist, J. E. The phylogeny and classification of the New World suboscine passerine birds (Passeriformes: Oligomyodi:Tyrannides), in: *Neotropical*

Ornithology (P.A. Buckley, E.S. Morton, R.S. Ridgely, and N.G. Smith, eds.), Ornithological Monographs, Amer. Ornithologists' Union, Washington, D.C. (in press, b).

Sibley, C. G., Williams, G. R., and Ahlquist, J. E., 1982, The relationships of the New Zealand Wrens (Acanthisittidae) as indicated by DNA–DNA hybridization, Notornis **29**(2):113–130.

Simpson, G. G., 1944, Tempo and Mode in Evolution, Columbia University Press, New York.

Sneath, P. H. A., and Sokal, R. R., 1973, Numerical Taxonomy, W. H. Freeman and Co., San Francisco, California.

Straus, N. A., 1976, Repeated DNA in eukaryotes, in: Handbook of Genetics, Volume 5 (R.C. King, ed.), Plenum Press, New York, pp. 3–39.

Van Valen, L., 1973, A new evolutionary law, Evol. Theory **1**:1–30.

Van Valen, L., 1974, Molecular evolution as predicted by natural selection, J. Mol. Evol. **3**:89–101.

Wallace, D. G., Maxson, L. R., and Wilson, A. C., 1971, Albumin evolution in frogs: A test of the evolutionary clock hypothesis, Proc. Natl. Acad. Sci. USA **68**:3127–3129.

Wetmore, A., 1960, A classification for the birds of the world, Smithson. Misc. Coll. **139**(11):1–37.

Wetmur, J. G., and Davidson, N., 1968, Kinetics of renaturation of DNA, J. Mol. Biol. **31**:349–370.

Wilson, A. C., and Sarich, V. M., 1969, A molecular time scale for human evolution, Proc. Natl. Acad. Sci. USA **63**:1088–1093.

Woese, C. R., 1981, Archaebacteria, Sci. Am. **244**:98–122.

CHAPTER 10

EXPERIMENTAL ANALYSIS OF AVIAN LIMB MORPHOGENESIS

J. R. HINCHLIFFE and M. GUMPEL-PINOT

1. INTRODUCTION

The bird limb bud gives rise to complex structures, the wing or the leg which represent extreme specializations of the pentadactyl limb, involving digital reductions, fusion processes, and differential growth of various parts of the skeleton (Hinchliffe, 1977a; Hinchliffe and Johnson, 1980). Interest attaches particularly to the question of how the limb structure, with its combination of repetitive pattern (the digits) and variation in pattern (none of the digits are morphologically identical) can arise from the limb bud with its apparent lack of organization (Fig. 2). Histologically, at 3 days, the chick limb bud is composed of a thin ectodermal coat, thickened distally into the apical ectodermal ridge (AER), and a core of mesenchyme of apparently uniform composition (Figs. 1 and 3). The mesoderm gives rise to muscle and to the wing skeleton, the latter by a complex process of condensation, chondrogenesis, and ossification (for reviews see Hinchliffe and Johnson, 1980; Ede, 1976; Thorogood, 1983). How does the diversity (at the level of both histogenesis and pattern) of the chick wing skeleton arise?

J. R. HINCHLIFFE • Zoology Department, University College of Wales, Penglais, Aberystwyth, Dyfed, SY23 3DA, Wales, United Kingdom   M. GUMPEL-PINOT • Institut d'Embryologie, Nogent-sur-Marne 94130, France.

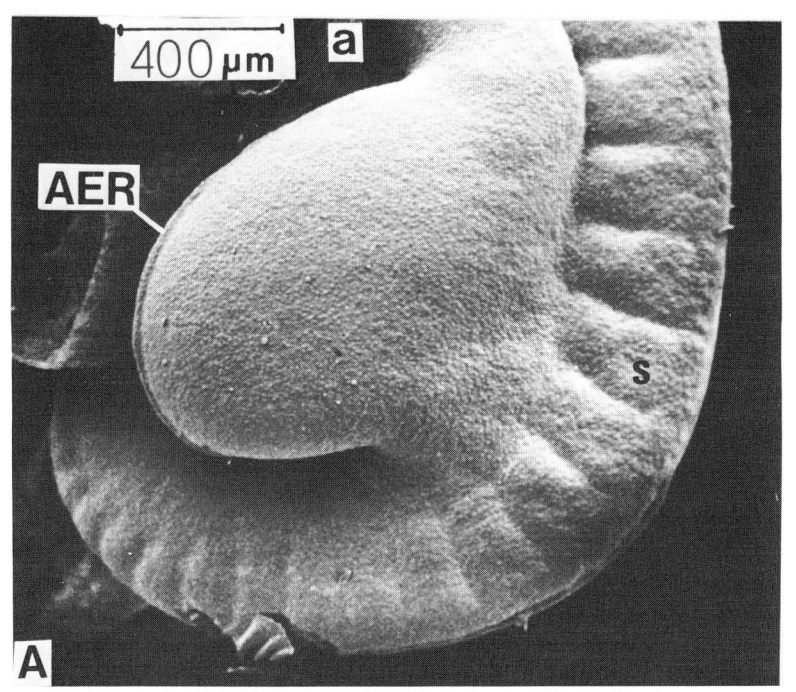

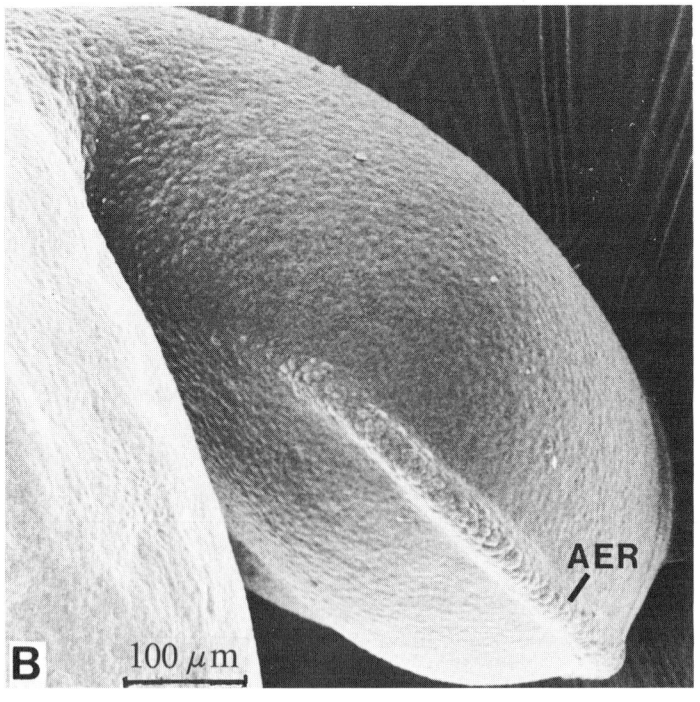

The chick wing bud is accessible to in ovo experimentation, such as removal or addition of limb bud parts. Limb buds can also be successfully grafted on the chorio-allantoic membrane, or into the flank region. The first wave of experimentation, aided by the discovery that by chemical treatment the mesodermal core and ectodermal coat could be separated, thus allowing many different recombinations, led to the so-called Saunders–Zwilling hypothesis in the 1960s (see review in Zwilling, 1961; Saunders, 1977). This hypothesis emphasized ectoderm–mesoderm interaction. The ridge induced mesodermal outgrowth, but was dependent in turn on the presence of a mesodermal factor, the apical ectodermal maintenance factor (AEMF). Limb type (wing or leg) was governed by the mesoderm.

Reversal of the limb bud tip on its base was discovered to lead to duplication of the skeleton of the wing tip (Fig. 2; Saunders and Gasseling, 1968; Amprino, 1965). This was first interpreted as due to asymmetrical distribution of the AEMF which was more concentrated postaxially. Eventually, this experimentation led to the discovery of the zone of polarizing activity (ZPA). The ZPA is a part of the posterior wing bud mesoderm, which, if implanted in the pre-axial wing bud border, provokes formation of a second wing tip whose antero-posterior axis it controls (Fig. 6B; see Section 4). Discovery of the ZPA led to a new wave of experimentation on the wing directed at the problem of pattern formation, which the original Saunders–Zwilling hypothesis did not attempt to explain.

Along with recent work on amphibian and insect limb regeneration, the analysis of the patterning role of the ZPA in the chick wing bud has produced a major shift of focus in the developmental world. While recent analysis of the molecular biology of gene action in development has provided valuable insight, it has left largely unresolved the problem of field phenomena, and it is to this issue that, after 50 years, many developmentalists have now returned. The field concept emphasizes the capacity of embryonic regions for self organization, and for organizing differentiation processes in definite spatial relationships. Of course, analytical methodology has progressed considerably. While classical grafting procedures remain essential for analysis of fields and patterning, other tools now include mathematical models and modern cell and organ culture techniques. Wolpert (1969, 1981) has contributed the concept of "positional information," popularized as "developing cells know where they are." Boundaries are important in the theoretical

FIGURE 1. Stereoscans of chick limb buds. (A) Stage 23 leg bud. (B) Stage 23 wing bud viewed anteriorly. (a, Anterior; AER, apical ectodermal ridge; S, somites.)

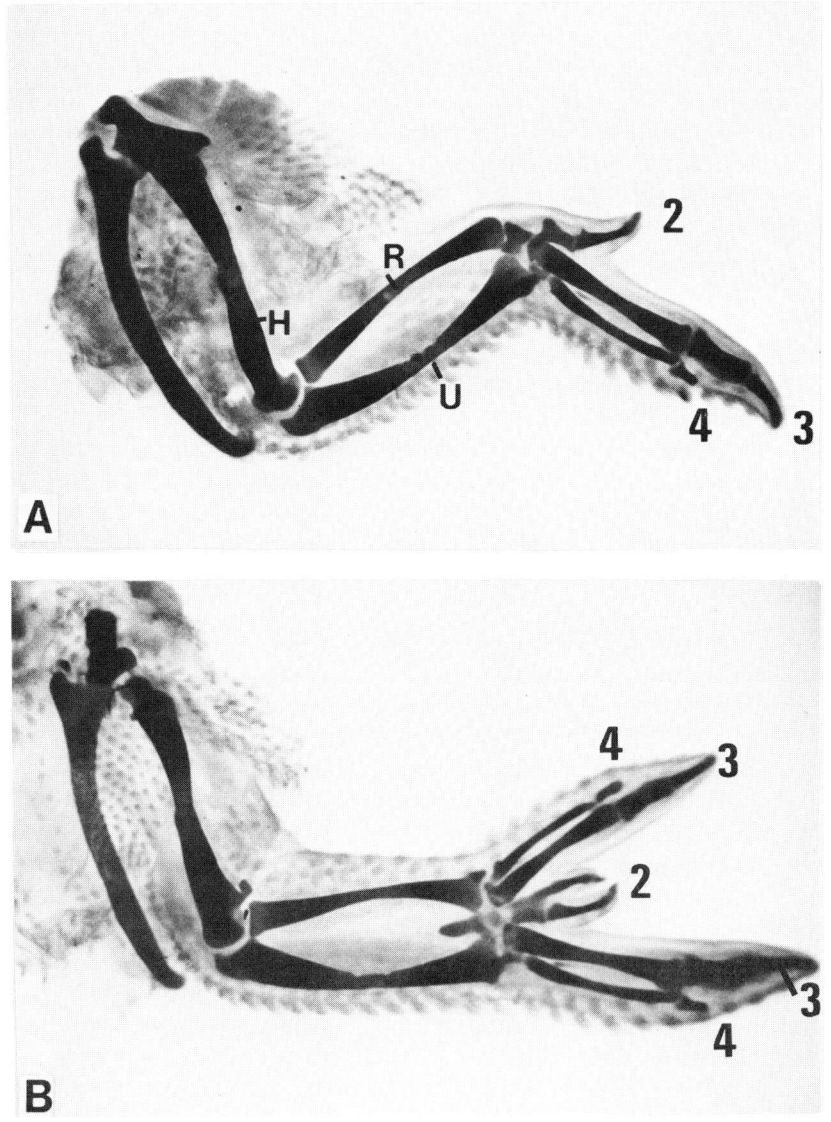

FIGURE 2. (A) Normal chick wing at 10 days of development stained for cartilage to show skeletal structure. (B) Duplicated wing skeleton following graft of zone of polarizing activity to anterior margin of 3-day wing bud (Fig. 6B). (H, humerus; R, radius; U, ulna; 2,3,4, digits; courtesy, Dr. C. Tickle.)

modeling of fields, and the chick limb ZPA appears to provide one of the two best examples of a boundary clearly demonstrable experimentally (the other is the amphibian dorsal lip originally analyzed by Spemann). Since one of the great hopes of the moment is that different fields may be controlled in similar ways, the ZPA and its mode of action is of consuming interest.

The current review is intended as a guide—far from comprehensive—to the considerable mass of recent experimentation on the chick limb. It looks in particular at work bearing on pattern formation: regulation phenomena, the ZPA and its mode of action (and alternative concepts) in patterning cartilage and muscle, and the patterning of the limb nerve supply. The recently advanced concept of a myogenic cell line of somitic origin is discussed. For further detailed accounts of recent work, the reader should consult the published accounts of two recent international conferences in Glasgow (Ede et al., 1977) and in Storrs, USA (Fallon and Caplan., 1983).

Inevitably, many aspects cannot be covered here. Recent interest in the developmental basis of evolutionary change (Gould, 1977) has given rise to much discussion of developmental constraints on evolutionary processes affecting the limb (Maderson, 1982). The recapitulationary view that development directly reflects, or simply repeats, earlier evolutionary stages has found expression in the identification of a pentadactyl limb archetype (Holmgren, 1933; Montagna, 1945; Jarvik, 1980) at the prechondrogenic condensation stage of the bird skeleton. The existence of such an archetype has been denied in recent studies (Hinchliffe, 1977b; Hinchliffe and Griffiths, 1983) which show that in the chick, even these early stages have highly specialized patterns. Another problem with implications for evolution is the differential growth of prechondrogenic condensations, which, though starting out more or less similar, become very different in size, such as the tibia and fibula in the chick leg. Such differential growth seems to be programmed very early (Hicks, 1982). Presumably, there is a growth program expressed in terms of differential rates of cell division, matrix synthesis, or recruitment from surrounding mesenchyme. The first attempts to analyze this are now being made (Hicks, 1982; Archer et al., 1982; also see review in Hinchliffe and Johnson, 1980, 1983).

Other aspects not covered include teratogenesis (Merker et al., 1980) and the analysis of the process of chondrogenesis, which though frequently involving the use of limb bud material does not necessarily illuminate limb morphogenesis itself. However, valuable studies have been made of the centripetal movement of cells (Ede, 1976) and of the role of fibronectin (Newman and Frisch, 1979) in condensation formation, and these processes may well be an essential part of pattern formation (see Conclusion). A final omission is cell death, which though such a striking feature of limb development, and known to shape the digital plate (Saunders and Fallon, 1967; Sanders et al., 1962), has been reviewed recently by the author (Hinchliffe, 1981, 1982).

2. MAPPING THE PROSPECTIVE SKELETAL AREAS

For interpretation of experiments on the chick wing bud it is essential to have a map of the prospective skeletal areas. This indicates which area of mesoderm gives rise to particular skeletal parts; it does not imply that the cells are already determined.

The first maps produced were those of Saunders (1948a) who used carbon particle marking to map the proximo-distal axis. Prospective wrist and digit mesenchyme is identifiable only from stage 20 onwards. The map does not indicate antero-posterior boundaries. Later, Stark and Searls (1973) drew up a map based on labeling small blocks of limb mesoderm with tritiated thymidine, and implanting these in a host limb bud. After development of the skeleton, the contribution of the implanted block to the skeleton was determined by autoradiography.

An attempt was made by the present authors to localize prospective areas accurately along the antero-posterior (A-P) axis by use of the labeling provided by the chick–quail system (Le Douarin, 1973). Anterior parts of the chick wing bud were replaced by corresponding parts of quail wing buds (Fig. 6A; Hinchliffe et al., 1981). Initially, halves were replaced, then anterior or posterior thirds. Such wing buds developed in most cases into normal wings. The quail contribution to the resulting skeleton was then determined. At stage 20, the line perpendicular to intersomite 17/18 (the midpoint in the wing bud) passes between digits 2 and 3, and between the radius and ulna (Fig. 3B). The prospective skeletal areas appear to occupy an area rather wider along the A-P axis than that estimated by Stark and Searls (1973) or by Summerbell (1979). The width is estimated as approximately $2\frac{1}{3}$ somite widths. All the maps agree that the anterior part (approximately $\frac{1}{3}$) of the limb bud makes no contribution to the digits.

3. REGULATION ALONG PROXIMO-DISTAL AND ANTERO-POSTERIOR AXES

Many developing systems have a capacity to regulate when parts are removed or added. Indeed, this property had lead to the concept of the embryonic field (see Waddington, 1956), which Wolpert (1981)

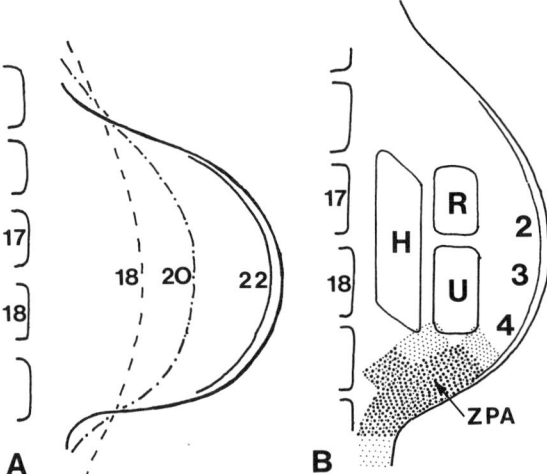

FIGURE 3. (A) Outline drawings of chick wing buds staged according to Hamburger and Hamilton (1951) and drawn in relation to somites. 18, 3 days; 20, 3½ days; 22, 3½–4 days. (B) The prospective areas at stage 20, from Saunders (1948a) and (for the A–P axis) from Hinchliffe et al. (1981). The ZPA (high activity: heavy stippling, intermediate activity: light stippling) is that mapped by MacCabe et al. (1973).

has recently defined as that set of cells which have their position specified with respect to the same boundary regions. Regulation may be interpreted as the specification of new positional values when part of the system is removed or added to.

There has been considerable discussion as to whether the limb bud will regulate along the proximo-distal (P-D) axis (reviewed in Hinchliffe and Johnson, 1980). Summerbell et al. (1973) have put forward the idea of a "progress zone" (reviewed in Wolpert, 1978) to explain the P-D differentiation of the limb, one striking feature of which is that more proximal parts are already differentiating as chondrogenic condensations while the distal mesoderm is histologically undifferentiated. The "progress zone" is defined as an area of mesoderm, approximately 300 μm deep and maintained in a labile state by the AER under which it lies. As the limb grows, and the cells of the "progress zone" divide, some cells start to leave the "progress zone" proximally and soon afterwards begin the formation of proximal structures. In the "progress zone," positional values of cells change depending on how long they have spent in the "progress zone." Initially they have proximal values, but these values become progressively more distal with the passage of time. Once cells leave the "progress zone," their positional value is regarded as fixed, i.e., they are now determined to form a particular structure, proximal or distal, according to their value.

This model fits the P-D pattern of differentiation quite well, as it does the finding that the AER does not play a role in specifying P-D levels of differentiation since its age (Rubin and Saunders, 1972) or even its species (Jorquera and Pugin, 1971) does not affect the ability of the underlying wing bud mesoderm to form a wing skeleton. The model predicts that differentiation along the P-D axis should be autonomous, and once cells have left the "progress zone," their development should be determined. Thus, if zeugopod* prospective areas are removed or duplicated, the stump should not influence the differentiation of the grafted wing bud tip.

The prediction has been tested in experiments in which prospective zeugopod material has been excised or duplicated (Fig. 4A and B; Summerbell et al., 1973; Summerbell and Lewis, 1975; Summerbell, 1977). The tip of a stage 19 wing bud pinned on a stage 24 wing bud base results in a composite wing bud with two prospective stylopods and zeugopods. When the tip of a stage 24 bud is added to a stage 19 base, the composite wing bud lacking prospective stylopod and zeugopod develops only wing digits. In both these experiments there is a considerable difference in developmental age as well as P-D level between stump and graft; by stage 24 the zeugopod condensations are already formed. But in other experiments, doubling the prospective zeugopod areas by pinning stage 19–20 wing buds to a stage 20 or 22 stump was reported as producing two sets of stylopods and/or zeugopods (Summerbell and Lewis, 1975). There appeared to be no regulation along the P-D axis.

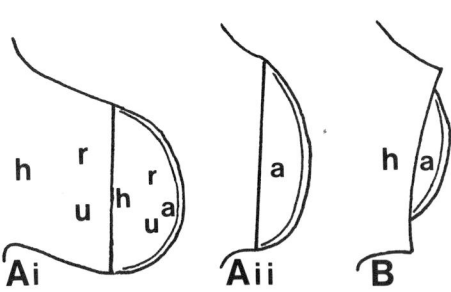

FIGURE 4. Regulation experiments along the proximo-distal axis, (A) by Summerbell et al. (1973), Summerbell and Lewis (1975), and (B) by Kieny (1964). (Ai) is an excess: a stage 19/20 wing bud grafted onto a stage 24 base. (Aii) is a deficiency: a stage 24 tip grafted on to a stage 19 base. (B) is a deficiency experiment by Kieny at stage 20 (host) and stage 22 (donor from leg bud). Prospective areas are indicated (a, autopod; h, humerus; r, radius; u, ulna).

---

*The zeugopod is the radius-ulna or the tibia-fibula section of the limb in wing or leg respectively. The stylopod is the humerus (wing) or femur (leg).

By contrast, French workers found considerable evidence of regulation along the P-D axis. Hampé (1959), working on stage 18–22 leg buds found that excesses or deficiencies of zeugopod material were regulated. The work was carried further by Kieny (1964, 1977) who found that regulation was better in leg than wing. But even in wing buds she found good regulation of both excesses and deficiencies at stages 18–22. The use of chick–quail combinations (quail wing buds grafted on to chick stumps) allows the relative contribution of graft and host to be established. In excesses it is clear that the stump tissues have their fate shifted to produce structures more proximal than normal. Thus, up to a certain developmental stage, cells which have left the progress zone can still change their normal development pathway. According to this work, limb buds possess regulative capacity.

The different results obtained by the two groups may partly be reconciled. Wing buds regulate less well than leg buds, which were used mainly by Kieny. Maximum graft–stump contact is necessary for regulation. Many of the experiments by Summerbell and his co-workers involve late stages, e.g., 24, at which cytodifferentiation of the zeugopod has begun. It is not surprising if such tissue has lost its regulative capacity. Regulation only takes place when graft and host have more or less the same early developmental stage. The situation has been clarified by recent work by Hornbruch (1980) on zeugopod deficient wing buds. Ability to regulate to a normal limb length was gradually lost from stage 19 (when it was nearly perfect) to stage 25 (when it was nil).

Regulation along the A-P axis has been neglected until recently. Early work of Warren (1934) seemed to show mosaic development on either side of an impermeable barrier. Experimental amputations by Hinchliffe and Gumpel-Pinot (1981) showed that if anterior halves were removed, posterior halves developed mosaically and were unable to regenerate the missing anterior part. But, under different experimental conditions, such as pre-axial ZPA grafting (Fig. 6B), the mesoderm seems to be labile along the A-P axis.

The regulation of excesses and deficiencies along the A-P axis has now been examined (Yallup and Hinchliffe, 1983). Limb buds were created in which a central $1\frac{1}{2}$ somite width section was either removed or duplicated (Fig. 5). According to our fate maps, the section affected contains most of the prospective skeletal areas. Good regulation of both pattern and size was found in stages 19–22, but by stage 24 size regulation was poor and pattern regulation had virtually ceased. The relative contribution to regulation of anterior and posterior parts was studied in both excesses and deficiencies by substituting for the anterior

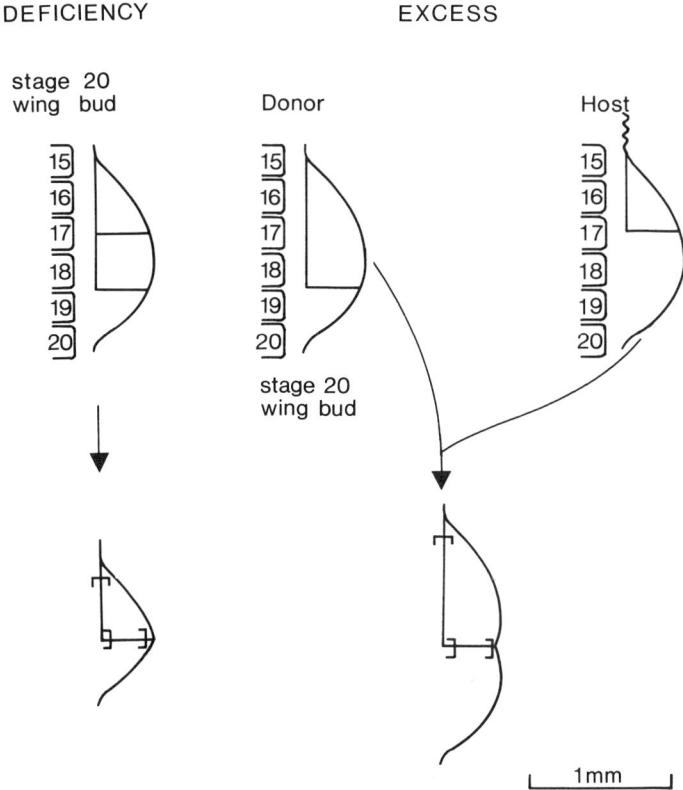

FIGURE 5. Regulation experiments along the antero-posterior axis by Yallup and Hinchliffe (1983).

graft the corresponding area from a quail wing bud. A contribution to regulation came from both graft and host tissues. In excesses, there is considerable loss of prospective areas on both sides of the interface, and with deficiencies a rather greater contribution to making up the lost prospective area is made from the anterior part.

These results have implications for both the progress zone hypothesis and for the interpretation of ZPA action. In our results, proximal parts of the limb bud such as the prospective zeugopod and stylopod still have regulative properties at stage 23, in spite of having left the "progress zone" well before. As discussed in the following section, fundamentally different views of limb development are represented by the ZPA theory suggesting long range ZPA control of the A-P axis of the limb field, and the polar coordinate theory which considers that in regulation, existing tissues do not change their present positional values and hence determination. Our results are consistent with the theory of ZPA control, which allows for respecification (or regulation) of limb bud mesoderm, rather than the polar coordinate theory, which does not.

Regulation thus appears to be a property of wing buds along both P-D and A-P axes until about stage 23, a result which argues against the "progress zone" but for the ZPA long-range control theory. The "progress zone" theory in its original form, which insists on absolute determination of the mesenchyme once it has left the "progress zone," seems to require modification to accommodate Kieny's and Yallup and Hinchliffe's results. Nonetheless, the theory incorporates an important conclusion that under certain experimental circumstances, e.g., when stump and graft are of different developmental stages, differentiation is autonomous along the P-D axis and is without reference to more proximal parts. Fundamental work still remains to be done on the cell cycle aspects of regulation. But the molecular basis of this cellular dialogue which must go on within the limb field remains enigmatic.

4. THE ZPA AND EXPERIMENTAL ANALYSIS OF CONTROL OF ANTERO-POSTERIOR DIFFERENTIATION OF THE WING

The discovery of the zone of polarizing activity (ZPA) and its effect on limb bud development has aroused great interest in the developmental world and initiated a wave of experimentation designed to evaluate the significance of the ZPA. Its discovery springs from earlier experiments by Saunders et al. (1958) and by Amprino and Camosso (1959) in which rotation of the wing bud tip through 180° on its base was followed by duplication of the wing tip skeleton. Saunders and Gasseling (1968) then found that transplanting a small piece of mesoderm out from the posterior margin of the wingbud into a preaxial site (Fig. 6B) produced the same effect as wing bud tip rotation. The host AER adjacent to the graft thickened and elongated, and an excess of mesoderm underlying this transformed anterior AER soon appeared. From this additional pre-axial host mesoderm a second wing tip developed, which was always orientated with its posterior digit (no. 4) adjacent to the graft. Control experiments in which mesoderm from anterior or central regions of the wing bud was transplanted to the same site failed to initiate any host response. Though experimental studies have usually concentrated on skeletal duplication (which is conveniently revealed by staining of whole mounts), it is clear that the pattern of muscles and tendons is also duplicated (Shellswell and Wolpert,

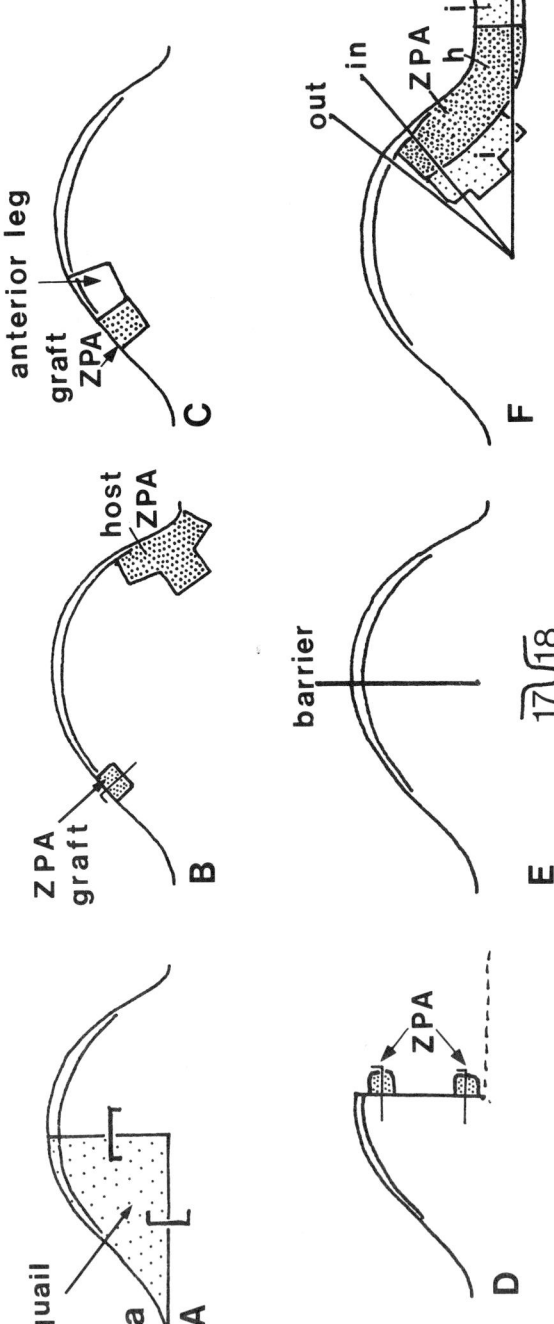

FIGURE 6. Experiments on the chick wing bud. (A) To map prospective areas, a quail anterior half has been substituted for the normal chick anterior (a) half (Hinchliffe et al., 1981). (B) The classic wing duplication experiment in which a ZPA graft is made to the pre-axial position. (C) Experiment by Honig (1981) in which anterior leg bud tissue is placed posterior to the ZPA graft and forms toes III and II. (D) ZPA grafted to an anterior half initiates limb development only if it is in contact with the AER (upper position). In the lower position it has no effect. The two positions are alternatives (Wilson, unpublished). (E) Barrier insertion in the midline (Summerbell, 1979). (F) A posterior wedge of tissue is removed, either extirpating all the ZPA ("out"), or allowing a small part to remain ("in"). Only "in" permits normal wing development (Hinchliffe et al., 1981; h, high and i, intermediate ZPA activity).

1977). From the grafting experiments, it appears that the ZPA graft controls the A-P axis of the additional wing tip which it has initiated. Other experiments have established that the effect is due to the transplanted mesoderm, the effect of transplanted ectoderm can be discounted.

Using transplantation experiments, the area of the wingbud which had polarizing activity was painstakingly mapped out by MacCabe et al., 1973. ZPA activity was found in posterior mesoderm between stages 19–28 (3–6 days). Initially, at stage 17 there is a lower level of activity possessed by much of the posterior mesoderm, but by stage 21 it is restricted to the posterior border, although there is an area just anterior to this which has intermediate activity (Fig. 3B). Eventually, by stage 27 activity is restricted to a small fraction of mesoderm at the posterior end of the AER. Recently, Summerbell and Honig (1982) have remapped the ZPA, finding maximum activity at stages 21–23 after which activity declines.

The ZPA is a feature found not only in the chick wing bud. The leg bud possesses a similar ZPA which produces duplication of the wing bud when grafted pre-axially, and the skeletal duplication is always of the host type. Indeed, the ZPA appears to be a general feature of the amniote limb bud. Polarizing zones have been found in the limb buds of other species of bird, e.g., quail, in rodents, humans, and even reptiles (turtles; Fallon and Crosby, 1977). ZPAs from all these will produce wing-type duplication when transplanted pre-axially to a chick wing bud.

Attempts have been made to identify the difference between the ZPA and the other limb mesoderm in molecular terms with a view to identifying a ZPA factor active morphogenetically. MacCabe has developed an in vitro assay for ZPA activity which involves testing factors on stage 21–22 anterior wing bud tissue cultured in vitro. In a successful test (or when combined with a ZPA graft) the AER remains thick, and the underlying mesoderm healthy, while in the absence of ZPA factor the AER flattens and the mesoderm dies (Calandra and MacCabe, 1978). The culture method does not allow skeletal development, and thus an active factor is one shown to maintain limb mesoderm rather than one to control pattern formation. MacCabe et al. (1977) identified two active factors. One was a low molecular weight (LMW), soluble component found in culture medium conditioned by exposure to posterior limb bud tissue. Cells from the same region when homogenized in high-salt-phosphate buffered saline yielded a soluble high molecular weight (HMW; 300,000 daltons) component which was morphogenetically active. Further analysis suggests that the HMW form, which is trypsin sensitive and binds concanavalin A, is probably a glycoprotein (MacCabe and Richardson, 1982). It is absent from anterior mesoderm. The LMW form can be converted to the HMW by incubation with a high molecular weight fraction from anterior mesoderm. The HMW factor appears to owe its activity to the binding of a small active molecule with a large inactive one (MacCabe et al., 1983).

Much experimentation on the ZPA has been designed to analyze its mode of action: how quickly and for how long it acts, and over what distance and in conjunction with what other limb bud parts. Initial experimentation made it clear that the duplicated wing skeletal structures which appeared were a host response and not due to self differentiation of the ZPA graft. This is clear by implication from the pre-axial grafts of mouse or chick leg ZPA, since the duplicated structures were always of the wing type, and therefore from the host. In fact, Smith (1979) has shown that when a pre-axial graft of quail ZPA is irradiated it can still provoke limb duplication without itself making any cellular contribution to the limb.

Since pre-axial grafting of the ZPA initiates a pre-axial mesenchymal excess in the host, it is reasonable to suppose that cell division rates are increased in the host mesoderm adjacent and posterior to the graft. Cooke and Summerbell (1980) demonstrated that the host cells show significant increases in cell labeling index 9 hr after a ZPA graft, and that by 17 hr the mitotic index in several areas has doubled. The increased cell division is found across the whole A-P axis of the host wing bud, suggesting that the ZPA is not simply affecting the host mesoderm immediately adjacent to it. This is borne out by an elegant experiment by Honig (1981). He grafted the ZPA pre-axially in a wing bud in the classic way, but inserted a second graft of anterior leg tissue of known width just posterior to the ZPA (Fig. 6C). The combination produced a duplicated limb skeleton, whose parts closest to the ZPA graft were recognizable as leg digits, e.g., III and II, while those further away were wing digits, e.g., 2). It was thus possible to show that the ZPA was acting at a distance on host mesoderm through the intervening leg donor tissue. Honig estimates the area of mesoderm affected by the ZPA as being about 300 μm, which represents approximately the width of the hand field in the normal wing bud at stage 18. Since the width of the limb bud is approximately 1 mm, the part forming hand skeleton, variously estimated as 325 μm (Stark and Searls, 1973) or 400 μm (Summerbell, 1979) wide, is a fairly small fraction of the whole width.

The action of the ZPA is thus not simply to impose a pattern on adjacent limb mesoderm independently of any effect on cell division. If the system were morphalactic (positional values changed without growth), then the grafted ZPA would create a second antero-posterior axis which would lead to both duplicated and normal digits being small. The mesoderm normally used for forming three digits would now have to be used for forming six. But as has already been mentioned, a preaxial ZPA graft rapidly provokes an excess of mesoderm from which the supernumerary digits arise. Both sets of digits (normal and duplicated) are of more or less normal size. It is not clear whether the ZPA has separate effects on mesoderm cell division and on pattern formation. But the two must be intimately linked since, as Smith and Wolpert (1981) clearly show, the greater the widening of the host antero-posterior width following a ZPA graft, the greater the degree of polydactyly.

It is not necessary to graft the whole of the ZPA to obtain full reduplication. Small blocks of ZPA tissue of a volume of approximately 0.01 mm^3 are usually taken. Tickle (1980) has investigated by cell dilution experiments (ZPA cells disaggregated and mixed at a controlled concentration with non-ZPA anterior cells to produce a pellet which can then be implanted) how few ZPA cells are necessary to produce a duplication. Alternatively, she implanted small pieces of plastic on which a small number of ZPA cells had been plated out. As few as 100 ZPA cells may provoke a digit 4 duplication. Tickle has also established that such cells must be adjacent to the AER in order to be effective. Wilson (unpublished results) confirmed that ZPA–AER contact is necessary for ZPA grafts to be effective. A ZPA graft will only initiate digit formation in an anterior half wing bud if it is implanted adjacent to the AER. ZPA implantation in the base of the half wing bud fails to save the anterior half from subsequent regression (Fig. 6D).

The controlling role of the ZPA in limb development is also demonstrated by a number of experiments resulting in poor development of anterior parts of the wing bud if isolated from the ZPA. Summerbell (1979) inserted a barrier into the mid position of a wing bud isolating the anterior half (Fig. 5E), and found that in general, no digit 2 developed anterior to the barrier, but digits 3 and 4 developed normally posteriorly. At Aberystwyth we find that by 24 hr after operation, there is greatly increased mesenchymal cell death accompanied by AER regression anterior to the barrier (Hinchliffe, 1981; Griffiths, unpublished observation). Such cell death would explain the loss of digit 2 since it removes its prospective area. While such barrier experiments may be criticized on the grounds of lack of specificity [Rowe and Fallon (1981, 1982), believe the effect is due to the barrier separating the AER into two], they are consistent with the view that the anterior part of the limb bud requires a factor from the ZPA for its survival and/or differentiation. This point is emphasized in amputation experiments carried out by Hinchliffe and Gumpel-Pinot (1981) which showed that anterior parts of the wing bud regressed when posterior parts including the ZPA were removed. The critical experiments showed dramatically different results depending on whether all the ZPA (according to the maps of MacCabe et al., 1973) was amputated, when the limb bud regressed and no digits developed, or whether a fraction of the ZPA was allowed to remain, after which a normal wing skeleton developed (Fig. 6F; Hinchliffe et al., 1981) The wing bud appears to depend on the ZPA for development during stages 18–22, but its excision at stages 23 and 24 fails to prevent a normal skeleton (including digits) forming (D. Wilson, unpublished observations). Even though digital condensations have still to form, by this stage the distal mesoderm no longer requires developmental instruction from the ZPA.

All the work just reviewed points to a controlling role for the ZPA in limb morphogenesis. There has been much speculation as to how the ZPA acts. Impressed by the action of the ZPA at a distance, several workers have assumed the ZPA is the source of a signal interpreted by the remaining limb bud mesenchyme. An attractive and plausible model is that put forward by Tickle et al. (1975) according to which the ZPA is the source of a hypothetical morphogen whose concentration declines anteriorly, due to the presence of an anterior "sink." Thus, a morphogen gradient is formed whose different levels specify the different digits (Fig. 7; high level, digit 4; low level, digit 2). Such a model emphasizes two separate processes: provision of a signal, and reading and interpretation of the level of the signal as a developmental instruction. Thus, the signal may be from any ZPA, but interpretation ("form wing digit 4") depends on the particular type, e.g., wing or leg, or species of limb bud mesenchyme. Indeed, Wolpert (1981) believes that signal mechanisms may be universal in different embryonic fields, only the interpretation varying according to genome and developmental history of the responding tissue. An experimental result which lends some support to this view is Hornbruch's unpublished finding that Henson's node will substitute effectively for a transplanted ZPA. Wolpert (1981) regards the limb mesoderm cells as acquiring "positional information" (knowledge of their relative position in the limb field) from the ZPA,

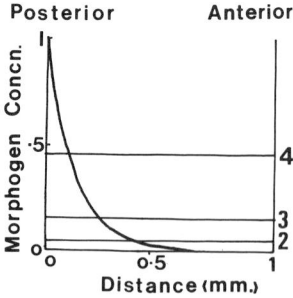

FIGURE 7. Hypothetical morphogen profile which could specify the digit pattern (Wolpert, 1981). The ZPA is assumed to be the morphogen source, and a gradient is set up since the anterior of the wing bud is a sink. The A–P digital pattern is 2,3,4, the different digits being specified by particular thresholds (2, low; 4, high).

possibly via the supposed concentration profile of morphogen it sets up. It should be emphasized that there are models of generating positional values by the ZPA alternative to the widely discussed hypothesis of a free diffusion gradient set up by a source–sink arrangement.

4.1. The ZPA and Normal Development.

Despite the undoubted properties of the ZPA in causing wing duplication following grafting pre-axially, its role in normal development has been questioned by several workers (Saunders, 1977, 1982; Saunders and Gasseling, 1982; Rowe and Fallon, 1981, 1982). On the one hand, normal development does not always appear to depend on continued ZPA presence, and on the other, ZPA properties are found in non-limb mesoderm.

Removal of the ZPA from the early wing bud can be followed by the development of a normal wing (MacCabe et al, 1973; Fallon and Crosby, 1975). Fallon was able to rule out the possibility that the ZPA had been regenerated following its removal. But, as described in the previous section, Hinchliffe and Gumpel-Pinot (1981) made amputations showing that if the ZPA region were removed, the remaining wing bud mesoderm died and failed to develop distal structures such as the digits. The difference between our results and those of Fallon and Crosby (1975) is that according to the MacCabe et al. (1973) maps, we excised all the regions of ZPA activity, whether high or intermediate, while in the latter, only the region of high ZPA activity was excised. In Fallon's experiments, sufficient ZPA activity may have been left in the limb to permit normal development. We have already seen that the total ZPA is far in excess of the minimum fraction of ZPA required for polarization of the limbs (Tickle, 1980).

ZPA activity has been found in non-limb mesoderm, such as flank (Saunders, 1977) and tail bud tip mesoderm, which Saunders and Gasseling (1982) claims is as morphogenetically active as the ZPA itself. This evidence seems to present less of a problem. In experimental situations, tissues frequently have morphogenetic effects on cells with which they do not normally make contact. The point surely is that within the limb bud mesoderm the ZPA has unique morphogenetic properties.

Recently, Fallon has argued that leg development is in some ways quite different from the wing. Insertion of a barrier in the mid point of the leg bud does not result in the loss of structures anterior to the barrier (Rowe and Fallon, 1982), although it does in the wing (Summerbell, 1979). Fallon considers the wing bud barrier results to be due to the effect of dividing the ridge, since he also found that removal of posterior wing AER was followed by regression of the remaining anterior part

(Rowe and Fallon, 1981). In the leg, by contrast, ridge survives if the posterior part is excised. Fallon considers there is still no clear evidence of a ZPA requirement for normal limb development. But this issue must be regarded as still open since preliminary experiments at Aberystwyth show that wing anterior AER survives and anterior skeletal structures are formed even if the posterior AER is removed, while posterior half amputation in the leg bud is followed by substantial mesodermal cell death and loss of structures anteriorly (Griffiths, unpublished observations).

A more fundamental challenge to the ZPA has been made by Iten and her co-workers who do not accept the idea of the uniqueness of the ZPA. Instead, they claim that chick limb development is governed by local differences in positional value within the mesoderm, in which all the mesoderm has a role to play. This approach is discussed next.

4.2. The Polar Coordinate Model and the Chick Wing Bud

Recent work on amphibian and insect limb regeneration has resulted in the polar coordinate model for embryonic fields (French et al., 1976; Bryant et al., 1981), and Iten has proposed that this model is applicable to the chick limb (Iten and Murphy, 1980a,b; Iten, 1982). This model, like the ZPA morphogen profile model, is expressed in terms of positional information. According to this hypothesis, cells around the circumference of the limb bud have unique positional values, and whenever normally non-adjacent tissue surfaces are brought into contact in an experiment, tissue is intercalated at the interface with the positional values that would normally be found between the two original surfaces. The intercalated tissue will then generate new structures appropriate to its positional value. This model "explains" rather well regeneration in urodele limbs or regulation of deficiencies in limb buds, since the original structure is restored through the intercalation of tissue with the missing values. It differs from the ZPA morphogen profile model in emphasizing that interactions are short range, that positional values of graft or host tissue do not change, and it attributes no overriding importance to the ZPA which represents just one of a number of positional values.

As Iten points out, the classic pre-axial ZPA graft and the consequent duplication of the wing skeleton can be explained on both models. Using the polar coordinate model, a growth zone would form between ZPA graft and anterior host tissue which would intercalate the missing values between anterior and posterior mesoderm. The intercalated tissue would then have the values necessary for the formation of supernumerary digits.

Iten and Murphy (1980a) have supported their theory with grafts of anterior wedges of tissue into slits made in more posterior positions in the host wing bud (Fig. 8A). The experimental limb produces supernumerary digits, and the greater the degree of graft–host positional disparity, the greater the degree of polydactyly. In further experiments, ZPA tissue was removed from the posterior wing bud and the cut surface pinned back to the base; the wing bud proceeded to form a normal wing, even though apparently lacking a ZPA (Fig. 8B; Iten and Murphy, 1980b). The result is attributed to tissue intercalated at the wound interface.

In fact, most of Iten and Murphy's results with anterior wedges fit the ZPA model quite well, since they involve the splitting of the host ZPA into two, and the insertion of the donor graft between the two ZPAs. The supernumerary limbs can be ascribed to the action on the graft of the adjacent host ZPAs in the same way that with the classic ZPA graft, anterior tissue is stimulated into differentiating into posterior digits that it does not normally form. When anterior wedges are inserted into central positions (away from direct contact with host ZPA) the incidence of polydactyly is low. In the experiments in which ZPA tissue is removed from the wing bud and the cut surface pinned back (Fig. 8B), normal development may be due to the influence of ZPA activity in the posterior body wall, since the MacCabe et al. (1973) maps show high activity in this region.

It is clearly important to know the origin, whether graft or host, of the supernumerary limbs in the wedge experiments. This has now been

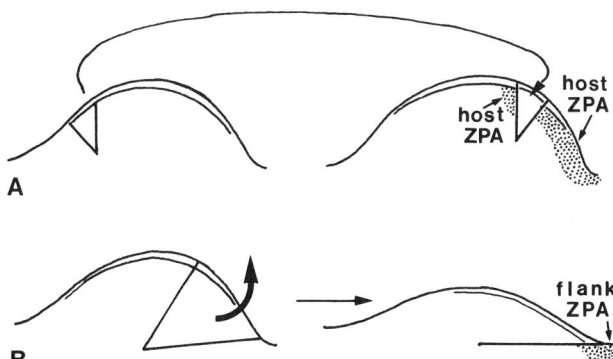

FIGURE 8. Experiments by Iten and Murphy (1980a,b). (A) An anterior wedge is inserted in a posterior slit in the host wing bud. Polydactyly results. (B) A posterior wedge is removed and the cut edges then pinned together. A normal wing is formed. According to maps of the ZPA (MacCabe et al., 1973), activity is found in the remaining flank region.

examined by Honig (1983) who has repeated Iten and Murphy's experiments using naturally labeled wedges of quail tissue. The supernumerary limbs are formed entirely from the graft, rather than from a mixed host–graft zone, and this is consistent with the interpretation that they have been formed by the response of the graft to the adjacent ZPA.

But there are a number of other experiments difficult to reconcile with the polar coordinate theory. This theory insists that where there are experimental confrontations between cells of different positional values, the existing positional values are not changed (Iten et al., 1981). In the regulation of excesses along the antero-posterior axis described earlier and which involve positional disparity at the host–graft interface, positional value (or presumptive fate) is respecified, and a normal 3-digit limb results (Yallup and Hinchliffe, 1983). Application of the polar coordinate theory to these excesses would predict polydactyly, probably of the form 2, 3, 3, 2,③,4 along the A-P axis, with existing positional values being maintained and 3 being formed from intercalated tissue formed at the interface of graft and host (Fig. 5).

Our ZPA amputation experiments (Hinchliffe and Gumpel-Pinot, 1981) also point to a special role for the ZPA. ZPA amputation is followed by the death of much of the remaining wing bud mesoderm, but removal of anterior mesenchyme has no effect on the normal development of the remaining posterior tissue. If positional values remain unchanged (polar coordinate theory) ZPA amputation should not interfere with normal limb development. The barrier experiments of Summerbell (1979), in which wing bud tissue anterior to the barrier regresses and fails to develop digits in accordance with its prospective fate, also suggest changes in positional value can take place. Metabolic attenuation of ZPA activity and the measurements of the area of mesoderm which responds to a ZPA graft discussed in the previous section (Cooke and Summerbell, 1980; Honig, 1981) all provide a considerable weight of evidence for a long-range ZPA-initiated signal.

It must be admitted that some of the experimental evidence is conflicting. Iten (1982) has been able to obtain duplicated limb skeletal structures following tip reversals in wing buds from which the ZPA has been removed. In the wedge insertions the degree of duplication of both skeleton and musculature has now been shown to be increased by reversal of the graft dorso-ventral surface with respect to the host (Javois and Iten, 1982), thus providing evidence for the importance of positional disparity. The testing of the two models is being actively carried out at present. In the mass of experimental data now available on the limb bud there are bound to be inconsistencies. The novel interpretation of Iten has prompted a closer examination of the nature of the action of the ZPA and its supposed signaling. But at present, most data are consistent with the hypothesis that the ZPA exerts overall long-range control of limb bud morphogenesis.

5. SOMITE AND LIMB BUD DEVELOPMENTAL RELATIONS

While there have never been doubts that the somatopleure contributed to limb bud mesenchyme, the question of a somitic contribution in birds has been more controversial. Clear histological descriptions have been given of a somitic contribution of tongues of tissue into limb buds in amphibians, reptiles, and mammals, but this contribution seems less well defined in birds and was even ruled out by some workers (Saunders, 1948b).

Two problems are involved. Does the limb bud depend on the presence of somites for its development? Do the somites make a cellular contribution to the limb bud and to particular developing tissues, e.g., the muscles?

5.1. Developmental Dependence

Early work was contradictory. Hunt (1932) obtained good development on the chorio-allantoic membrane of 2-day limb prospective areas taken with adjacent somites. But Hamburger (1938) found limbs developed from transplanted areas excised at 25–34 somites. He claimed that the explants lacked endoderm and practically all somites, so that limbs developed independently from these structures.

Murillo-Ferrol (1963,1965), by grafting a large barrier (heart, optic vesicle, mica) between somites and limb bud at stage 13–18, was able to block limb development completely. But, if the barrier were a Millipore filter, the limb developed normally. The author concluded that the somitic mesenchyme influenced limb development via a diffusible chemical "somitic factor." The chemical nature of this factor remains unknown.

Pinot (1970) grafted in the coelomic cavity the limb prospective area (wing or leg) taken either with or without the adjacent somites from the stage of 15 somites stage onwards. Her conclusions were that the limb area is able to differentiate autonomously and completely normally in these conditions from the stage of 15 somites onwards. Before this stage the presence of somites is necessary. Kieny (1971) reached similar conclusions about the developmental dependence of the limb on the presence of somites. However, Kieny believes the somites have a directive role in differentiation, wing somites inducing *wing* development (Dhouailly and Kieny, 1972). From associations of wing or leg level somites with presumptive limb areas, Pinot assumes that the presumptive limb areas are already determined as wing or leg before being able to self-differentiate, the role of somites being to provide a permissive stimulation.

5.2. The Somitic Contribution of the Musculature

As early as 1895, Fischel had described the migration of a few cells from the somites to the chick limb bud. But Saunders (1948b), using carbon particle marking, claimed that there was no somitic participation in wing formation in the chick. Electron microscopy studies and optical observation of semi-thin sections persuaded Grim (1970) that a few cells are released from the ventral edge of the dermomyotome and are integrated in the adjacent limb mesenchyme. Gumpel-Pinot (1974), after associating limb territory with somites either labeled with tritiated thymidine or using the quail–chick system (Le Douarin, 1973), demonstrated that somitic cells did migrate into the developing wing bud. This work did not, however, establish the developmental fate of the somitic cells within the wing.

This problem of somitic contribution to limb formation was systematically studied by two groups in France and West Germany, using heterospecific chick–quail recombinations and very similar experimental procedure, exchange of somites or lateral plates at different levels in ovo (Fig. 9; Christ et al., 1974, 1977; Christ and Jacob, 1980; Jacob et al., 1978; Jacob and Christ, 1980; Chevallier et al., 1976, 1977a,b; Kieny, 1980). The main conclusions of these authors are that in the developing limb that skeletal muscle cells are of somitic origin, whereas cartilage, smooth muscles, tendons, and connective tissue are of somatopleural origin. Moreover, whatever their level of origin, somites can form the limb muscles. No antero-posterior regionalization exists for myotomal derivatives (Kieny and Chevallier, 1980). Myocytes already in the limb, and even differentiated muscles grafted in replacement of somites, are able to migrate and develop into limb muscle (Mauger and Kieny, 1980).

Thus, the limb mesenchymal cells appear to derive from different lineages, each of which has a different fate in limb development. The somitic and somatopleural lineage view for the differentiation of muscle and cartilage is still disputed. For some authors, the early mesenchymal cells are not already determined by their different origin. Differentiation as muscle or cartilage, depending on intrinsic conditions such as position in the limb bud, is proposed by Zwilling (1966), Searls and

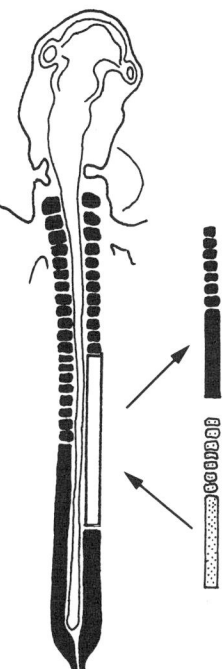

FIGURE 9. Substitution of corresponding quail somites (white) in place of brachial somites excised (black) from a developing chick embryo. In the wing that forms, the musculature is derived from the implanted quail somites. (After Christ et al., 1977.)

Janners (1969), Nathason et al. (1978), and Caplan and Koutroupas (1973).

In abnormal conditions, for instance if a limb primordium not yet invaded by the somitic cells is grafted in the coelomic cavity (McLachlan and Hornbruch, 1979), somatopleural cells appear able to form striated muscles. But Christ and his co-workers claim that the somitic–myogenic lineage is strict and that both subpopulations (somitic and somatopleural) are committed to either myogenic or non-myogenic lineage (Christ and Jacob, 1980).

Recent work from Kieny (1980), however seems to suggest that in certain experimental conditions a developmental plasticity among different mesenchymal cells is not precluded. However, the double lineage view during *normal* development of the limb is now very widely accepted.

6. INNERVATION AND LIMB DEVELOPMENT

Limb innervation in tetrapods involves several spinal segments in the brachial plexus (forelimb) and the lumbar-sacral plexus (hindlimb).

For instance, in the adult White Leghorn the major contribution to innervation of the forelimb is from spinal segments 14, 15, and 16 (Pettigrew et al., 1979; Stirling and Summerbell, 1977). The 14th segmental nerve innervates the anterior proximal part of the wing, the 16th the posterior distal area. The territory of the 15th lies in between but overlaps considerably those of 14 and 16.

The main trunk nerves penetrating into the limb bud after about 4 days of incubation are mixed, formed by motor axons of the spinal ventral root and sensory axons of the dorsal root (Fig. 10). The sensory neurons originate from the neural crest, as do the Schwann cells which form myelin in the peripheral nervous system. In the chick limb bud the nerve pattern, as described by Roncali (1970), is remarkably invariant in a particular strain. This pattern becomes established very early when muscle and cartilage are just beginning to differentiate. Using electro-physiological methods, Landmesser and Morris (1975) have described the normal motor innervation in the chick hindlimb bud at $5\frac{1}{2}$ days of incubation. At these early stages of development, when the boundaries of the various tissues are still not sharply defined, it is difficult to understand how the patterning of the nerve can be so precisely laid out. Lewis (1980) explains this by the tendency of nerves to fasciculate. The individual pioneer axons take erratic and divergent routes, but more proximally the axons are bound in a single fascicule. The direction of growth thus represents the resultant of the divergent tendencies of the individual growth cones.

6.1. Role of Limb Pattern in Establishing Innervation.

Before the establishment of innervation it is relatively easy to modify experimentally the skeleton–muscle pattern of a developing limb bud by operations resulting in truncation, intersegmental deletion, or reduplication of a segment on the P-D axis (Section 3), or mirror image reduplication across the A-P axis (Section 4). These experiments change the environment into which the invading axons have to migrate and thus provide good systems to study the factors which guide peripheral nerves to their destination. Using these types of operations, Stirling (1976) and Stirling and Summerbell (1977) showed that the route fol-

---

FIGURE 10. (A) Embryonic spinal cord and dorsal ganglion of 8-day chick embryos. d, Dorsal sensory root; m, ventral motor column; sg, spinal ganglion. (B) Dorsal nerve pattern in normal (i) and stylopod- and zeugopod-deleted (iii) wing. Generalized "wiring" diagrams (dorsal and ventral nerves) are given for (ii) normal and (iv) deleted wings. (B, after Stirling and Summerbell, 1977.)

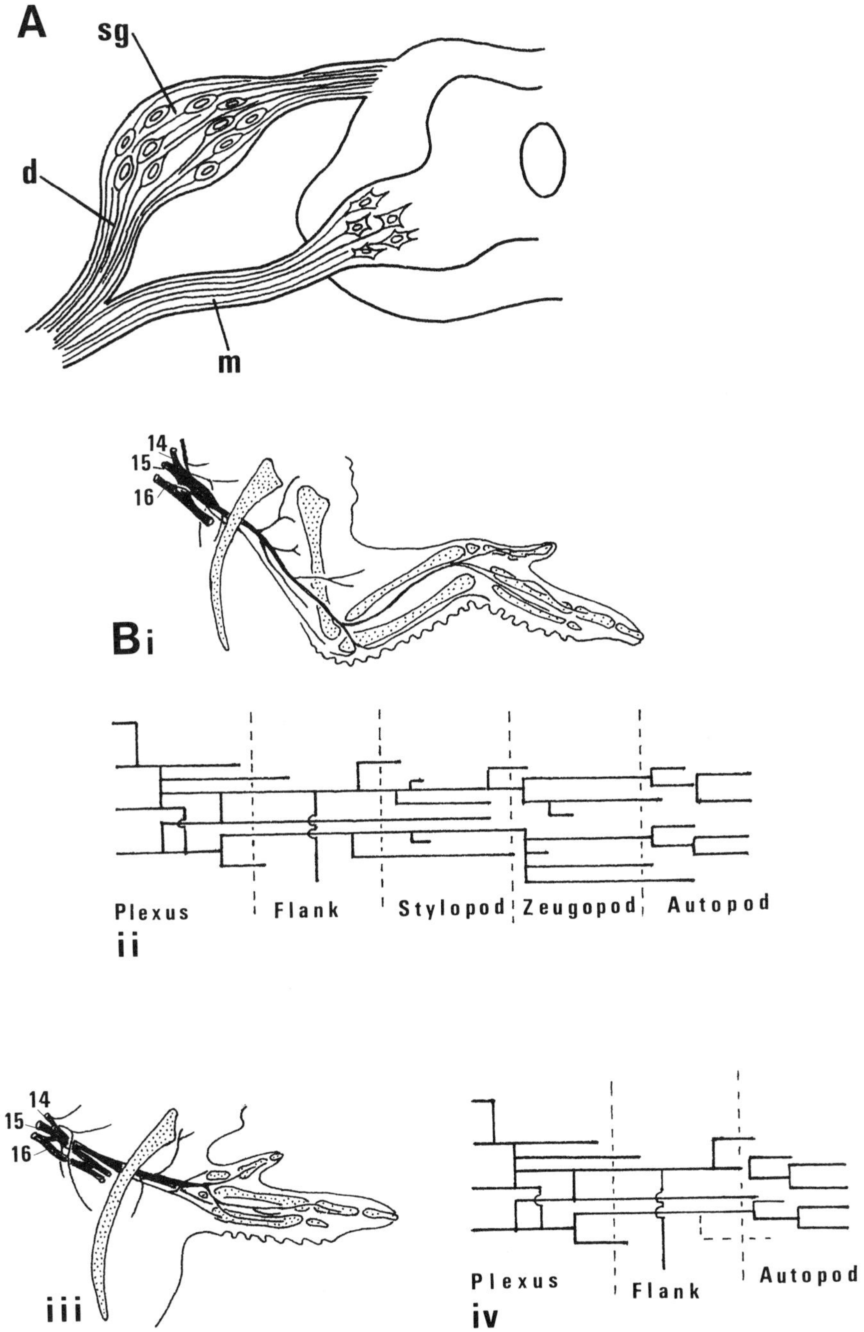

AVIAN LIMB MORPHOGENESIS

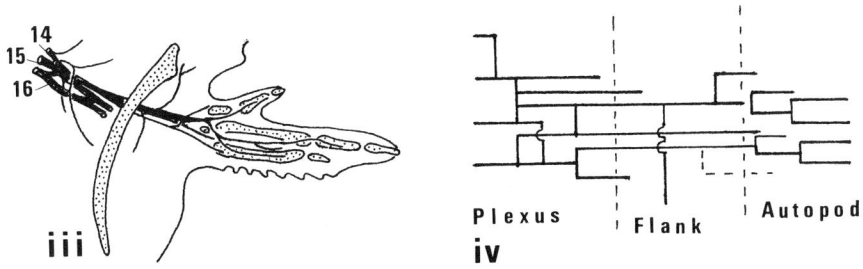

lowed by the nerves is characteristic of the skeletal level they traverse (Fig. 10). If a segment is missing, the corresponding branch of the nerve is missing too. The nerve pattern develops as a result of interactions between axons and limb tissues. Similar results are obtained when an uninnervated limb bud is grafted in a heterotopic site such as the head or neck where it becomes innervated by nerves normally foreign to it. The nerve pattern in these limbs is almost the same as the normal one (Hamburger, 1939; Lewis, 1980). The paths of nerve outgrowth are thus evidently defined by the intrinsic structure of the limb (Horder, 1978).

6.2. Relationship between Neurons and Peripheral Conditions

In normal conditions it appears that neurotization of a system is assured by the production of a large excess of motor axons and degeneration of those that fail to make peripheral connections (reviewed in Prestige, 1970; Jacobson, 1978; Hollyday, 1980). The system appears to have great plasticity during development.

After removal of the limb in the chick embryo, a reduction in number of motor neurons in developing spinal cord at the level of the limb bud has been observed (Hamburger, 1934, 1958; Hamburger and Keefe, 1944; Jacobson, 1978). An increase in number of motor neurons occurs after grafting an additional limb bud (Hamburger, 1939; Hollyday and Hamburger, 1976). At the level of the spinal ganglia, a decrease in number of sensory neurons has been observed following limb amputation. A similar decrease has been found by Fallon et al. (1983) in the *limbless* mutant of the chick, in which limb buds regress at an early stage of development. By contrast, an increase in number of sensory neurons follows graft of a supernumerary limb bud.

6.3. Relationship between Axons and Their Targets

The muscles are the target of the motor neurons. In experimental, completely muscleless limbs (Lewis et al., 1981) the main nerve trunk (mixed motor and sensory) develops normally, but no muscle branches are visible.

Thus, the motor nerve pattern seems to be controlled by the muscle pattern (Lewis, 1980). The first axon to reach an uninnervated muscle will neurotize it and at the same time will exclude other axons and constrain them to terminate elsewhere. The muscles are thus innervated in an orderly sequence (Roncali, 1970). The inability of an axon terminal to form a stable connection in a muscle results in the death of the motor neuron (Pettigrew et al., 1979).

Experimentally obtained nerveless wings show an essentially normal pattern of muscle, but these muscles gradually atrophy (Straznicky, 1967; Shellswell, 1977; Lewis, 1980). Similar observations have been reported by Eastlick (1943). In coelomic grafts, uninnervated limb primordia develop a normal muscle pattern, but these muscles degenerate rapidly. The motor innervation is necessary for the maintenance of muscle structure.

After partial elimination of the sensory component of the limb innervation by destroying the neural crest, Lewis (1980) has obtained limbs with a normal muscle pattern, all the muscle nerve branches present and in normal location, but with almost all the cutaneous nerve branches missing. Thus, motor axons do not depend on sensory axons or on Schwann cells to guide them to their target.

6.4. Innervation and the Pattern of Skeletal Differentiation

The conventional view is that the skeletal pattern of the limb is independent of innervation (Hamburger, 1939). However, Kieny and Fouvet (1974) found that excision of a portion of the spinal cord in 2-day chick embryos produced a high frequency of proximal hemimelia and a lower proportion of post-axial hemimelia. But the authors themselves suggest that this result could be a consequence of the surgical operation.

The view that innervation does not play any role in the establishment of the normal skeletal pattern has been recently disputed by Čihak et al. (1978), who produced deformities of the wing by precocious excision of part of neural tube, and also by McBride (1979) and McBride et al. (1980) after the same type of experiment at the level of the leg. Moreover, McCredie et al. (1978) suggested that a relationship might exist between neural crest (sensory innervation) and limb malformations.

Very precise experiments from Lewis (1980) seem to provide a solution since he avoids surgical operations that could be responsible for secondary factors affecting morphogenesis. By focusing UV irradiation the author attempted to destroy sensory or motor innervation. The result was often completely nerveless wings, whose skeletons were complete and with normal proportions.

In conclusion, innervation does not seem to play a role in the establishment of the skeletal and muscular pattern of the limb. But it is important in the maintenance of its targets, muscle and sensory structures. It is the intrinsic structure of the limb that imposes the pathways of axon migration. Survival of motor neurons in the spinal cord and the spinal ganglia sensory neurons is dependent on their axons terminating in muscles or sensory receptor cells. Initially, an excess of neurons is produced whose final number is controlled by peripheral conditions ("peripheral loading").

7. CONCLUSIONS

From recent experiments on chick limb buds, the following picture emerges. The limb bud will regulate until quite late stages, in fact until cytodifferentiation actually begins along both A-P and P-D axes, a result that does not support the "progress zone" hypothesis in its original form. The ZPA plays an important, though disputed, role in the control of pattern along the A-P axis, and in stimulating cell division in adjacent distal mesoderm. The first simple model of ZPA action, involving a freely diffusible morphogen establishing a gradient, is still consistent with much of the experimental data, but is far from being proved. The posterior part of the limb bud is the source of a chemically defined factor that maintains anterior mesoderm and AER. The pattern of innervation is imposed by the tissue invaded and is not inherent in any particular part of the nervous system. Somites contribute cells that normally constitute a myogenic line.

Many questions remain unanswered. The molecular basis of limb morphogenesis is almost totally unknown. There are problems concerning the relationship of the Saunders–Zwilling hypothesis of AER–mesoderm interaction, and the hypothesis of ZPA control of the A-P axis. Is the apical ectodermal maintenance factor of the first hypothesis the same as the ZPA factor of the second. This is a possibility discussed by MacCabe and Richardson (1982). The AER and ZPA seem to require mutual contact to be effective: perhaps the ZPA operates via the AER, and not, as is generally assumed, via the mesoderm. Fallon et al. (1983) have rightly insisted that the AER requires closer attention, and have shown that active AER cells are physiologically coupled (Fallon, Sheridan, and Clark, in preparation). Many individual factors or components have been shown to be important in limb morphogenesis, but the difficulty is to understand the temporal and spatial interrelationships of the various factors.

The ZPA–morphogen profile hypothesis has been influential in stimulating experimental work, partly because, unlike other models that can generate spaced cartilages (Ede, 1976; Newman and Frisch, 1979), it seems to explain the origin of the individual differences in skeletal elements. However, this hypothesis loads the whole responsibility for generating complex patterns onto the mechanisms of interpretation of the profile. This mechanism is not understood at all, and (unlike the signaling region) is difficult to approach experimentally.

While limb bud experimentation has made a crucial contribution to theories of pattern formation, our understanding of even the first principles of limb morphogenesis remains disturbingly inadequate.

REFERENCES

Amprino, R., 1965, Aspects of limb morphogenesis in the chicken, in: *Organogenesis* (R. L. De Haan & H. Ursprung, eds.), Holt Rinehart and Winston, New York, pp. 255–281.

Amprino, R. and Camosso, M., 1959, Observations sur les duplications expérimentales de la partie distale de l'ébauche de l'aile chez l'embryon de poulet, *Archives d'Anatomie microscopique et de Morphologie expérimentale* **48**:261–305.

Archer, C. W., Rooney, P., and Wolpert, L., 1982, The early growth and morphogenesis of limb cartilage, *Limb Development and Regeneration*, A (J. F. Fallon and A. I. Caplan, eds.), A. Liss, New York, pp. 267–278.

Bryant, S. V., French, V., and Bryant, P. J., 1981, Distal regeneration and symmetry, *Science* **212**:993–1002.

Calandra, A. J., and MacCabe, J. A., 1978, The in vitro maintenance of the limb bud apical ridge by cell free preparations, *Dev. Biol.* **62**:258–269.

Caplan, A. I., and Koutroupas, S., 1973, The control of muscle and cartilage development in the chick limb: The role of differential vascularization, *J. Embryol. Exp. Morphol.* **29**:571–583.

Chevallier, A., Kieny, M., and Mauger, A., 1976, Sur l'origine de la musculature de l'aile chez les Oiseaux, *C.R. Acad. Sci.* **282**:309–311.

Chevallier, A., Kieny, M., and Mauger, A., 1977a, Limb-somite relationship: Origin of the limb musculature, *J. Embryol. Exp. Morphol.* **41**:245–258.

Chevallier, A., Kieny, M., Mauger, A., and Sengel, P., 1977b, Developmental fate of the somitic mesoderm in the chick embryo, in: *Vertebrate Limb and Somite Morphogenesis* (D.A. Ede, J.R. Hinchliffe, and M. Balls, eds.), Cambridge University Press, Cambridge, pp. 421–432.

Christ, B., and Jacob, H. T., 1980, Origin, distribution and determination of chick mesenchymal cells in: *Teratology of the Limbs* (H-J. Merker, H. Nau, and D. Neubert, eds.), de Gruyter, Berlin, pp. 67–77.

Christ, B., Jacob, H. J., and Jacob, M., 1974, Uber den Ursprung der Flügel-muskulatur. Experimentelle Untersuchungen mit Wachtel-und Hühnerembryonen, *Experientia* **30**:1446–1448.

Christ, B., Jacob, H. J., and Jacob, M., 1977, Experimental analysis of the origin of the wing musculature in avian embryos, *Anat. Embryol.* **150**:171–186.

Čihak, R., Doskocil, M., and Seichert, V., 1978, Some morphogenetic aspects of the aneurogenic and explanted limb primorida, in: *XIXth Morphological Congress Symposia*, Charles University, Prague, pp. 119–125.

Cooke, J., and Summerbell, D., 1980, Growth control early in embryonic development: The cell cycle during experimental pattern duplication in the chick wing, *Nature (London)* **287**:697–701.

Dhouailly, D., and Kieny, M., 1972, The capacity of the flank somatic mesoderm of early bird embryos to participate in limb development, *Dev. Biol.* **28**:162–175.

Eastlick, H. L., 1943, Studies on transplanted embryonic limbs of the chick. I, The development of muscles in nerveless and uninnervated grafts, *J. Exp. Zool* **93**:29–49.

Ede, D. A., 1976, Cell interaction in vertebrate limb development, in: *The Cell Surface* in Animal Embryogenesis and Development (G. Poste, and G.L. Nocolson, eds.), Elsevier, Amsterdam, pp. 495–543.

Ede, D. A., Hinchliffe, J. R., and Balls, M., 1977, Vertebrate Limb and Somite Morphogenesis, Cambridge University Press, Cambridge.

Fallon, J. F., and Crosby, G. M., 1975, Normal development of the chick wing following removal of the polarizing zone, J. Exp. Zool. **193**:449–455.

Fallon, J. F., and Crosby, G. M., 1977, Polarizing zone activity in limb buds of amniotes, in: Vertebrate Limb and Somite Morphogenesis (D. A. Ede, J.R. Hinchliffe, and M. Balls, eds.), Cambridge University Press, Cambridge, pp. 55–69.

Fallon, J. F., and Caplan, A. I., 1983, Limb Development and Regeneration, A and B, A. Liss, New York.

Fallon, J. F., Rowe, D. A., Frederick, J. M., and Simandl, B. K., 1983, Epithelial-Mesenchymal Interactions in Development (R.H. Sawyer and J.F. Fallon, eds.), Praeger Scientific, New York, (in press).

Fallon, J. F., Frederick, J. M., Carrington, J. L., Lanser, M. E., and Simandl, B. K., 1983, Studies on a limbless mutant chick embryo, in: Limb Development and Regeneration, A (J.F. Fallon, et al., eds.), A. Liss, New York, pp. 33–43.

Fischel, A., 1895, Zur Entwicklung der Verbraten Rumpf und der Extremitaten muskulatus der Vögel und Sängetiere, Morph. Jb. **23**:544–561.

French, V., Bryant, P. J., and Bryant, S. V., 1976, Pattern regulation in epimorphic fields, Science **193**:969–981.

Gould, S. J., 1977, Ontogeny and Phylogeny, Belknap Press, Cambridge.

Grim, M., 1970, Differentiation of myoblasts and the relationship between somites and the wing bud of the chick embryo, Z. Anat. EntwGesch **132**:260–271.

Gumpel-Pinot, M., 1974, Contribution de mésoderme somitique a la genèse du membre chez l'embryon d'Oiseaux, C.R. Acad. Sci **279**:1305–1308.

Hamburger, V., 1934, The effects of wing bud extirpation on the development of the central nervous system in chick embryos, J. Exp. Zool. **68**:449–494.

Hamburger, V., 1938, Morphogenetic and axial self-differentiation of transplanted limb primordia of 2-day chick embryos, J. Exp. Zool. **77**:379–400.

Hamburger, V., 1939, The development and innervation of transplanted limb primordia of chick embryos, J. Exp. Zool. **80**:347–389.

Hamburger, V., 1958, Regression versus peripheral control of differentiation in motor hypoplasia, Am. J. Anat. **102**:365–410.

Hamburger, V., and Hamilton, H. L., 1951, A series of normal stages in development of the chick embryo, J. Morphol. **88**:49–92.

Hamburger, V., and Keefe, E. L., 1944, The effects of peripheral factors on the proliferation and differentiation in the spinal cord of the chick embryo, J. Exp. Zool. **96**:223–242.

Hampé, A., 1959, Contribution à l'étude du développement et de la régulation des déficiences et des excédents dans la patte de l'embryon de poulet, Archives d'Anatomie microscopique et de Morphologie expérimentale **48**:345–478.

Hicks, M. J., 1982, Analysis of the differential growth of chondrogenic elements in the embryo chick limb, Ph.D. thesis, University College of Wales, Aberystwyth.

Hinchliffe, J. R., 1977a, "Rudimentation," reduction and specialisation in the development and evolution of the bird wing, in: Mécanismes de la Rudimentation des Organes chez les Embryons de Vértebres, Éditions du C.N.R.S. 266, Paris, pp. 411–414.

Hinchliffe, J. R., 1977b, The chondrogenic pattern in chick limb morphogenesis: A problem of development and evolution, in: Vertebrate Limb and Somite Morphogenesis (D. A. Ede, J. R. Hinchliffe, and M. Balls, eds.), Cambridge University Press, Cambridge, pp. 293–309.

Hinchliffe, J. R., 1981, Cell death in embryogenesis, in: *Cell Death* (I. D. Bowen and R. A. Lockshin, eds.), Chapman and Hall, London, pp. 35–78.

Hinchliffe, J. R., 1982, Cell death in vertebrate limb morphogenesis, *Prog Anat.* **2**:1–17.

Hinchliffe, J. R., and Griffiths, P. J., 1983, The prechondrogenic patterns in tetrapod limb development and their phylogenetic significance, in: *Development and Evolution* (B. Goodwin, N. Holder, and C. Wylie, eds.), Cambridge University Press, Cambridge, pp. 99–121.

Hinchliffe, J. R., and Gumpel-Pinot, M., 1981, Control of maintenance and anteroposterior skeletal differentiation of the anterior mesenchyme of the chick wing bud by its posterior margin (the ZPA), *J. Embryol. Exp. Morphol.* **62**:63–82.

Hinchliffe, J. R., and Johnson, D. R., 1980, *The Development of the Vertebrate Limb*, Oxford University Press, Oxford.

Hinchliffe, J. R., and Johnson, D. R., 1983, The growth of cartilage, in: *Cartilage*, Volume 2 (B.K. Hall, ed.), Academic Press, New York, pp. 255–295.

Hinchliffe, J. R., Garcia-Porrero, J. A., and Gumpel-Pinot, M., 1981, The role of the zone of polarising activity in controlling the maintenance and anteroposterior differentiation of the apical mesoderm of the chick wing bud: Histochemical techniques in the analysis of a developmental problem, *J. Histochem* **13**:643–658.

Hollyday, M., 1980, Motoneuron histogenesis and the development of limb innervation, *Curr. Top. Dev. Biol.* **15**:181–215.

Hollyday, M., and Hamburger V., 1976, Reduction of the naturally occurring motor neuron loss by enlargement of the periphery, *J. Comp. Neurol.* **170**:311–320.

Holmgren, N., 1933, On the origin of the tetrapod limb, *Acta Zool. (Stockholm)*, **14**:185–295.

Honig, L. S., 1981, Positional signal transmission in the developing chick limb, *Nature (London)* **291**:72–83.

Honig, L. S., 1983, Does Anterior (Non-Polarizing Region) Tissue Signal in the Developing Chick Limb, *Dev. Biol.* (in press).

Horder, T. J., 1978, Functional adaptability and morphogenetic opportunism, the only rules for limb development, *Zoon.* **6**:181–192.

Hornbruch, A., 1980, Abnormalities along the proximo-distal axis of the chick wing bud: The effect of surgical intervention, in: *Teratology of the Limbs* (H-J. Merker, H. Nau, and D. Neubert, eds.), De Gruyter, Berlin, pp. 191–197.

Hunt, E. A., 1932, The differentiation of chick limb buds in chorioallantoic grafts, with special reference to muscles, *J. Exp. Zool.* **62**:57–91.

Iten, L. E., 1982, Pattern specification and pattern regulation in the embryonic chick limb bud, *Am. Zool.* **22**:117–129.

Iten, L. E., and Murphy, D. J., 1980a, Pattern regulation in the embryonic chick limb: Supernumerary limb formation with anterior (non-ZPA) limb bud tissue, *Dev. Biol.* **75**:373–385.

Iten, L. E., and Murphy, D. J., 1980b, Supernumerary limb structures with regenerated posterior wing bud tissue, *J. Exp. Zool.* **213**:327–335.

Iten, L. E., Murphy, D. J., and Javois, L. C., 1981, Wing bud with three ZPAs, *J. Exp. Zool.* **215**:103–106.

Jacob, H. J., and Christ, B., 1980, On the formation of the muscular pattern in the chick limb, in: *Teratology of the Limbs* (H-J. Merker, H. Nau, and D. Neubert, eds.), de Gruyter, Berlin, pp. 89–97.

Jacob, M., Christ, B., and Jacob, H. T., 1978, On the migration of myogenic stem cells into the prospective wing region of chick embryos. A scanning and transmission electronmicroscopy study, *Anat. Embryol.* **153**:179–193.

Jacobson, M., 1978, *Developmental Neurobiology*, Plenum Press, New York, pp. 309–344.
Jarvik, E., 1980, *Basic Structure and Evolution of Vertebrates*, Volume 2, Academic Press, London.
Javois, L. C., and Iten, L. E., 1982, Supernumerary limb structures after juxtaposing dorsal and ventral chick wing bud cells, *Dev. Biol.* **90:**127–143.
Jorquera, B., and Pugin, E., 1971, Sur le comportement du mésoderme et de l'ectoderme du bourgeon de membre dans les éschanges entre le poulet et le rat, *C.R. Acad. Sci.* **272:**522–525.
Kieny, M., 1964, Étude du mécanisme de la régulation dans le développement du bourgeon de membre de l'embryon de Poulet, *J. Embryol. Exp. Morphol.* **12:**357–371.
Kieny, M., 1971, Les phases d'activité morphogènes du mésoderme somatopleural pendant le développement précoce du membre chez l'embryon de poulet, *Annales d'Embryologie et de Morphogenèse* **4:**281–298.
Kieny, M., 1977, Proximo-distal pattern formation in avian limb development, in: *Vertebrate Limb and Somite Morphogenesis* (D. A. Ede, J.R. Hinchliffe, and M. Balls, eds.), Cambridge University Press, Cambridge, pp. 87–103.
Kieny, M., 1980, The concept of the myogenic cell line in developing avian limb buds, in: *Teratology of the Limbs* (H-J. Merker, H. Nau, and D. Neubert, eds.), de Gruyter, Berlin, pp. 79–88.
Kieny, M., and Chevallier, A., 1980, Existe-t-il une relation spatiale entre le niveau d'origine des cellules somitiques myogenes et leur colonisation terminale dans l'aile? *Archs. Anat. Microsc. Morph. Exp.* **69:**35–46.
Kieny, M., and Fouvet, B., 1974, Innervation et morphogènese de la patte chez l'embryon de Poulet. II. Anomalies consécutives a l'excision d'un tronçon de tube neural a deux jours d'incubation, *Arch. Anat. Microsc. Morph. Exp.* **63:**281–288.
Landmesser, L., and Morris, D. G., 1975, The development of functional innervation in the hindlimb of the chick embryo, *J. Physiol. (Lond.)* **249:**301–325.
Le Douarin, N., 1973, A biological cell labelling technique and its use in experimental embryology, *Dev. Biol.* **30:**217–222.
Lewis, J., 1980, Defective innervation and defective limbs: Causes and effects in the developing chick wing, in: *Teratology of the Limbs* (H-J. Merker, H. Nau, and D. Neubert, eds.), de Gruyter, Berlin, pp. 235–242.
Lewis, J., Chevallier, A., Kieny, M., and Wolpert, L., 1981, Muscle nerve branches do not develop in chick wings devoid of muscle, *J. Embryol. Exp. Morphol.* **64:**211–232.
MacCabe, A. B., Gasseling, N. T., and Saunders, J. W., Jr., 1973, Spatiotemporal distribution of mechanisms that control outgrowth and antero-posterior polarization of the limb bud in the chick embryo, *Mech. Aging Dev.* **2:**1–12.
MacCabe, J. A., and Richardson, K. E. Y., 1982, Partial characterization of a morphogenetic factor in the developing chick limb, *J. Embryol. Exp. Morphol.* **67:**1–12.
MacCabe, J. A., Calandra, A. J., and Parker, B. W., 1977, In vitro analysis of the distribution and nature of a morphogenetic factor in the developing chick wing, in: *Vertebrate Limb and Somite Morphogenesis* (D.A. Ede, J.R. Hinchliffe, and M. Balls, eds.), Cambridge University Press, Cambridge, pp. 25–39.
MacCabe, J. A., Leal, K. W., and Leal, C. W., 1982, The control of axial polarity. I. A low molecular weight morphogen affecting the ectodermal ridge. II. Ectodermal control of the dorso-ventral axis, in: *Limb Development and Regeneration*, A (J. F. Fallon and A. I. Caplan, eds.), A. Liss, New York, pp. 237–244.
Maderson, P. F. A., 1982, The role of development in macroevolutionary change, in: *Evolution and Development* (J.T. Bonner, ed.), Springer-Verlag, Berlin, pp. 279–312.

Mauger, A., and Kieny, M., 1980, Migratory and organogenetic capacities of muscle cells in bird embryos, *W. Roux's Arch. Dev. Biol.* **189**:123–234.
McBride, W. G., 1979, The inductive influence of neural tissue in limb development in the chick, *Aust. J. Zool.* **27**:673–679.
McBride, W. G., Stokes, P. A., and Vady, P. H., 1980, The influence of nerve supply in limb development, in: *Teratology of the Limbs* (H-J. Merker, H. Nau, and D. Neubert, eds.), de Gruyter, Berlin, pp. 223–234.
McCredie, J., Cameron, J., and Shoobridge, R., 1978, Congenital malformations and the neural crest, *Lancet* **2**:761–763.
McLachlan, J. C., and Hornbruch, A., 1979, Muscle forming potential of the non-somitic cells of the early avian limb bud, *J. Embryol. Exp. Morphol.* **54**:209–217.
Merker, H-J., Nau, H., and Neubert, D., 1980, *Teratology of the Limbs*, de Gruyter, Berlin.
Montagna, W., 1945, A re-investigation of the development of the wing of the fowl, *J. Morphol.* **76**:87–113.
Murillo-Ferrol, N. L., 1963, Analisis experimental de la paticipacion del mesoblasto paraxial sobre la morfogenesis de los miembros en el embrion de las Aves, *An. Descarrollo* **11**:63–76.
Murillo-Ferrol, N. L., 1965, Étude causale de la differenciation la plus precoce de l'ebauche morphologique des membres. Analyse expérimentale chez les embryons d'Oiseaux, *Acta Anat.* **62**:80–103.
Nathanson, M. A., Hilfer, S. R., and Searls, R. L., 1978, Formation of cartilage by nonchondrogenic cell types, *Dev. Biol.* **64**:99–117.
Newman, S. A., and Frisch, H. L., 1979, Dynamics of skeletal pattern formation in developing chick limb, *Science* **205**:662–668.
Pettigrew, A. G., Lindeman, R., and Bennet, M. R., 1979, Development of the segmental innervation of the chick fore limb, *J. Embryol. Exp. Morphol.* **49**:115–137.
Pinot, M., 1970, Relations entre le mésenchyme somitique et la plaque des membres, chez le Poulet, *Année Biologique* **9**:277–284.
Prestige, M. C., 1970, Differentiation, degeneration and the role of the periphery: Quantitative considerations, in: *The Neurosciences* (F.O. Schmidt ed.), Rockefeller University Press, New York, pp. 73–82.
Roncali, L., 1970, The brachial plexus and the wing nerve pattern during early developmental phases in chick embryos, *Monit. Zool. Ital. N.S.* **4**:81–98.
Rowe, D. A., and Fallon, J. F., 1981, The effect of removing posterior apical ectodermal ridge of the chick wing and leg on pattern formation, *J. Embryol. Exp. Morphol.* (Suppl.) **65**:309–325.
Rowe, D. A., and Fallon, J. F., 1982, Normal anterior pattern formation after barrier placement in the chick leg: Further evidence on the action of polarizing zone, *J. Embryol. Exp. Morphol.* **69**:1–6.
Rubin, L., and Saunders, J. W., Jr., 1972, Ectodermal-mesodermal interactions in the growth of limb buds in the chick embryo: Constancy and temporal limits of the ectodermal induction, *Dev. Biol.* **28**:94–112.
Saunders, J. W., Jr., 1948a, The proximo-distal sequence of origin of the parts of the chick wing and the rol of the ectoderm, *J. Exp. Zool.* **108**:363–404.
Saunders, J. W., Jr., 1948b, Do the somites contribute to the formation of the chick wing? *Anat. Rec.* **100**:746.
Saunders, J. W., Jr., 1977, The experimental analysis of chick limb development, in: *Vertebrate Limb and Somite Morphogenesis* (D.A. Ede, J.R. Hinchliffe, and M. Balls, eds.), Cambridge University Press, Cambridge, pp. 1–24.
Saunders, J. W., Jr., 1982, *Developmental Biology*, Macmillan, New York.
Saunders, J. W., Jr., and Fallon, J. F., 1967, Cell death in morphogenesis, in: *Major*

Problems in Developmental Biology (M. Locke, ed.), Academic Press, London, pp. 289–314.

Saunders, J. W., Jr., and Gasseling, M. T., 1968, Ectodermal-mesenchymal interactions in the origin of limb symmetry, in: *Epithelial-Mesenchymal Interactions* (R. Fleischmajor and R.F. Billingham, eds.), Williams & Wilkins, Baltimore, pp. 78–97.

Saunders, J. W., Jr., Gasseling, M. T., and Gfeller, M. D., Sr., 1958, Interactions of ectoderm and mesoderm in the origin of axial relationships in the wing of the fowl, *J. Exp. Zool.* **137**:39–74.

Saunders, J. W., Jr., Gasseling, M. T., and Saunders, L. C., 1962, Cellular death in morphogenesis of the avian wing, *Dev. Biol.* **5**:147–178.

Saunders, J. W., Jr. and Gasseling, M. T., 1982, New insights into the problem of pattern regulation in the limb bud of the chick embryo, in: *Limb Development and Regeneration*, A (J. F. Fallon and A. I. Caplan, eds.), A. Liss, New York, pp. 67–76.

Searls, R. L., and Janners, M. Y., 1969, The stabilisation of cartilage properties in the cartilage-forming mesenchyme of the embryonic chick limb bud, *J. Exp. Zool.* **170**:365-376.

Shellswell, G. B., 1977, The formation of discrete muscles from the chick wing dorsal and ventral muscle masses in the absence of nerves, *J. Embryol. Exp. Morphol.* **41**:269–277.

Shellswell, G. B., and Wolpert, L., 1977, The pattern of muscle and tendon development in the chick wing, in: *Vertebrate Limb and Somite Morphogenesis* (D.A. Ede, J.R. Hinchliffe, and M. Balls, eds.), Cambridge University Press, Cambridge, pp. 71–86.

Smith, J. C., 1979, Evidence for a positional memory in the development of the chick wing bud, *J. Embryol. Exp. Morphol.* **52**:105–113.

Smith, J. C., and Wolpert, L., 1981, Pattern formation along the antero-posterior axis of the chick wing: The increase in width following a polarizing region graft and the effect of X-irradiation, *J. Embryol. Exp. Morphol.* **63**:127–144.

Stark, R. J., and Searls, R. L., 1973, A description of chick wing bud development and a model of limb morphogenesis, *Dev. Biol.* **33**:138–153.

Stirling, R. V., 1976, Functional innervation of abnormally constructed limbs in chicken embryos, *Neurosci. Lett.* **3**:110.

Stirling, R. V., and Summerbell, D., 1977, The development of functional innervation in the chick wing bud following truncations and deletions of the proximo-distal axis, *J. Embryol. Exp. Morphol.* **41**:189–207.

Straznicky, K., 1967, The development of the innervation and the musculature of wings innervated by thoracic nerves, *Acta Biol. Hung.* **18**:437–448.

Summerbell, D., 1977, Regulation of deficiencies along the proximo distal axis of the chick wing-bud: A quantitative analysis, *J. Embryol. Exp. Morphol.* **41**:137–159.

Summerbell, D., 1979, The zone of polarizing activity: Evidence for a role in normal chick limb morphogenesis, *J. Embryol. Exp. Morphol.* **50**:217–233.

Summerbell, D., and Honig, L. S., 1982, The control of pattern across the antero-posterior axis of the chick limb bud by a unique signalling region, *Am. Zool.* **22**:105–116.

Summerbell, D., and Lewis, J. H., 1975, Time, place and positional value in the chick limb bud, *J. Embryol. Exp. Morphol.* **33**:621–643.

Summerbell, D., Lewis, J. H., and Wolpert, L., 1973, Positional information in chick limb morphogenesis, *Nature (London)* **244**:492–496.

Thorogood, P. V., 1983, The morphogenesis of cartilage, in: *Cartilage*, Volume 2 (B.K. Hall, ed.), Academic Press, New York, pp. 223–254.

Tickle, C., 1980, The polarising region and limb development, in: *Development in Mammals 4* (M.H. Johnson, ed.), Elsevier, North Holland, pp. 101–136.

Tickle, C., Summerbell, D., and Wolpert, L., 1975, Positional signalling and specification of digits in chick limb morphogenesis, *Nature (London)* **254**:199–202.

Waddington, C. H., 1956, *Principles of Embryology*, Allen and Unwin, London, pp. 23–28.

Warren, A. E., 1934, Experimental studies on the development of the wing in the embryo of *Gallus domesticus*, *Am. J. Anat.* **54**:449–485.

Wolpert, L., 1969, Positional information and the spatial pattern of cellular differentiation, *J. Theoret. Biol.* **25**:1–47.

Wolpert, L., 1978, Pattern formation and the development of the chick limb, *Birth Def.* **14**:547–559.

Wolpert, L., 1981, Positional information and pattern formation, *Phil. Trans. R. Soc. Lond. B* **295**:441–450.

Yallup, B. L., and Hinchliffe, J. R., 1983, Regulation along the antero-posterior axis of the chick wing bud, in: *Limb Morphogenesis and Regeneration* (J.F. Fallon, et al., eds.), A. Liss, New York, pp. 131–140.

Zwilling, E., 1961, Limb morphogenesis, *Adv. Morphol.* **1**:301–330.

Zwilling, E., 1966, Cartilage formation from so called myogenic tissue of chick embryo limb buds, *Ann. Med. Exp. Biol. Fenniae* **44**:134–139.

CHAPTER 11

VARIATION IN MATE FIDELITY IN MONOGAMOUS BIRDS

NORMAN L. FORD

1. INTRODUCTION

Ornithologists recently have devoted considerable attention to avian mating systems. However, the primary focus of both theoretical papers and field studies has been on the exceptional systems and species, namely the various forms of polygamy and cooperative breeding. Monogamy, the predominant avian mating system (Lack, 1968), and "typical" monogamous species have received comparatively little attention, especially in the theoretical papers. Exceptions to the above are the many studies of monogamous colonial birds, the recent paper by Wittenberger and Tilson (1980) on the evolution of monogamy, the review of avian mating systems by Oring (1982), and two reviews of extrapair copulations in monogamous species (Gladstone, 1979; McKinney et al., in press). The purpose of this review is to extend our knowledge of monogamous birds by examining the extent to which individuals of monogamous species practice various strategies of mate infidelity as a means to increase their reproductive success.

NORMAN L. FORD • Department of Biology, St. John's University, Collegeville, Minnesota 56321.

2. WHICH BIRDS ARE MONOGAMOUS?

In preparing this review, I encountered two difficulties in distinguishing between monogamy and polygamy (used here to include polygyny and polyandry). The first concerns the mating relationships of a single individual in the course of one breeding season. Typically, an individual is considered to be monogamous if it forms a pair bond with only one member of the opposite sex, and polygamous if it forms pair bonds either simultaneously or successively with more than one individual of the opposite sex (Lack, 1968; Wittenberger, 1979). Polygamy is thus defined on the basis of one individual having overlapping pair bonds with two or more individuals of the other sex. However, there are no uniform standards for degree of required overlap, or for deciding when a pair bond has been dissolved. Some authors label mate-switching between breeding attempts as polygamy (Radabaugh, 1972; Fraga, 1972; Burns, 1982), while others label such events as successive monogamy (Carey and Nolan, 1975). Actually, successive monogamy has the same genetic consequence as polygamy (one male fertilizes the clutches of two females), but since polygamy and successive monogamy are decidedly different behavioral phenomena, I believe it is appropriate to distinguish between them.

To circumvent the problem of deciding when a pair bond has been dissolved, I use overlap in breeding attempts to distinguish between successive monogamy and polygamy. As used here, a breeding attempt begins with the onset of nesting activities and terminates when the young leave the nest. The justifications for considering fledging to be the end point of a breeding attempt are (1) that it is usually difficult to determine when young have reached a stage of independence, and (2) fledging frequently appears to be the point at which one member of a pair that has been supplying biparental care can cease to care for the young, e.g., in double-brooded species the female typically ceases to care for fledglings when she begins renesting. When a breeding attempt is interrupted by loss of eggs or young, renesting is considered to be a distinct, non-continuous event, and mate-switching at this point would be considered successive monogamy.

The second problem is to determine which *species* of birds are monogamous. The problem arises from the fact that the populations of many species of birds consist of spatially and temporally varying mixtures of monogamous and polygamous mating units. In fact, as Oring (1982) and others have pointed out, there appears to be a continuum among birds from species in which only monogamy is known to species in which polygamy is the predominant mating pattern. There are even species in which some populations are purely monogamous while others are highly polygynous, e.g., the Winter Wren (*Troglodytes troglodytes*; Armstrong, 1955). Recognition of this continuum and analysis of the factors contributing to the differences among populations and species is, of course, what is important to the development of an understanding of the sources of variation in mating patterns. However, the continuum does have the unfortunate consequence of making it impossible to assign many species to one of the traditional mating system categories without misrepresenting to some degree their actual mating habits. As a case in point, Verner and Willson (1969) use a 5% or greater frequency of polygyny to categorize a species as "normally polygynous." Thus, only one of 20 males in an otherwise monogamous sample need have two mates for the species' mating system to be labeled as "polygynous." Such labeling emphasizes the unusual at the expense of the usual and misrepresents the mating habits of the group. A low rate of polygynous matings may simply indicate that some polygyny regularly occurs in an otherwise monogamous species. I am not aware of any criteria having been proposed or employed to attach the label "polyandrous" to a species' mating system. In general, authors reviewing the occurrence of polyandry in birds (Jenni, 1974; Oring, 1982) have not attached a mating system label to species in which polyandry occurs in low frequency.

In this review, I consider a species to be monogamous if (1) reported rates of polygamy do not regularly exceed approximately 20%, and (2) there is considerable temporal and spatial variation in the rate below that percentage value. Thus, for the mating system of a species to be labeled polygynous or polyandrous, consistently more than 20% of the known mating units must be polygamous. Although these criteria are no less arbitrary than Verner and Willson's (1969) 5% rule, I believe they are preferable because it is more appropriate to view low rates of polygamy as deviations from monogamy than to view high rates of monogamy as deviations from polygamy. Species that regularly evidence the higher rates of polygamy clearly have well-developed behavior for practicing polygamy and occupy niches in which there is consistently favorable Environmental Potential for Polygamy (EPP; Emlen and Oring, 1977). Species that evidence low rates of polygamy are either behaviorally less well adapted to be polygamous or occupy niches that present low EPP. Either way, these latter species are limited to relatively small deviations from a basically monogamous breeding system.

3. TYPES OF VARIATIONS IN MATE FIDELITY

There are only two types of mate infidelity. Either member of a mated pair can engage in extrapair copulations (EPC), or either member can form polygamous bonds with additional individuals. However, the potential for increasing reproductive success (RS) via infidelity is different for the two sexes. A male engaging in EPCs is limited only by the number of females whose eggs he can fertilize, provided those females can either raise the young alone or can acquire help to raise the young. A female's RS, on the other hand, is limited to the number of fertilized eggs she can produce and for which she can either provide or acquire care. Thus, a mated male may increase his RS by engaging in EPCs with the mates of other individuals, but a mated female cannot do so (although she may gain in fitness if there is an advantage to producing a brood sired by more than one male).

Males also have an advantage in exploiting polygamous strategies, not only because some degree of male emancipation from parental care is more common than female emancipation, but also because a female's RS has, at the most, a production limit of one fertilized egg per day while a male has the capability to fertilize many eggs per day. However, females may gain in RS by forming polyandrous bonds. In cooperative polyandry, a female can gain as a result of increased fledging success, e.g., Harris' Hawk (*Parabuteo unicinctus*; Mader, 1975). In simultaneous or sequential polyandry, a female can produce more than one clutch, provided the care of at least one clutch is assumed by one of her mates, e.g., Spotted Sandpiper (*Tringa macularia*; Oring and Knudson, 1972).

Thus, the subject of this review is a consideration of the occurrence of EPCs and polygamy among species considered to be normally monogamous. I have not attempted to review all the literature on the breeding biology of all species considered to be monogamous. I have instead relied extensively on some major reviews (Lack, 1968; Verner and Willson, 1969; von Haartman, 1969; Wittenberger and Tilson, 1980; Hunt, 1980; Oring, 1982; McKinney *et al.*, in press) and have focused my own review on the English language literature on passerines. However, the non-passerines and foreign language literature have not been completely ignored.

3.1. Extrapair Copulations

McKinney *et al.* (in press), in their review of sperm competition in monogamous birds, distinguish between two types of EPC, those that are forced on the female (FEPC) and those in which the female appears to either solicit or not offer resistance (UEPC). Distinguishing between the two is clearly important, since the former seems to indicate a strategy that may be advantageous only to males, while the latter indicates a strategy that also may be advantageous to females. Unfortunately, it is often difficult to distinguish between the two, either in direct observation or from the literature. Since McKinney et al. (in press) and Buitron (1983) have discussed this topic to the extent of current knowledge, it will not be further considered here, and I will use the neutral term EPC to refer to any copulation between a mated bird and another to whom it is not mated.

Perhaps the most difficult problem associated with EPCs is evaluating how successful they are in introducing sperm into the female reproductive tract, and in fertilizing eggs. It is virtually impossible to determine by observing copulation whether insemination occurs, let alone whether fertilization occurs. However, there is some evidence that EPCs can be successful. Burns et al. (1980), using genetic plumage markers, demonstrated that forced copulations can be effective in achieving fertilization in captive Mallards (Anas platyrhynchos). Copulations were not observed in any of the following examples but they are either indicative or suggestive that successful EPCs occurred. Bray et al. (1975) and Roberts and Kennelly (1980) demonstrated that female Red-winged Blackbirds (Agelaius phoeniceus) mated to vasectomized males produce some fertilized eggs. Gowaty and Karlin (in McKinney et al., in press) using electrophoretic techniques showed evidence of mixed paternity in broods of Eastern Bluebirds (Sialia sialis). The presence of fertile eggs in the nests of female–female pairs of Western Gulls (Larus occidentalis; Hunt and Hunt, 1977), California Gulls (L. californicus; Conover et al., 1979), and Ring-billed Gulls (L. delawarensis; Ryder and Somppi, 1979) imply EPCs. Finally, the presence of a nestling hybrid Barn-Cave Swallow (Hirundo rustica X H. fulva; Martin, 1980) in a nest tended by a pair of Barn Swallows also indicates a successful EPC. In regard to the latter, one wonders how many of the unusual interspecific hybrids reported for birds are a result of EPCs rather than the formation of pair bonds. Montagna (1942) observed a copulation between a male Sharp-tailed Sparrow (Ammodramus caudacutus) and a female Seaside Sparrow (A. maritimus) that was probably an EPC. It would seem reasonable to expect that interspecific barriers might be crossed more readily by EPC (especially forced) than by the formation of a pair bond.

McKinney et al. (in press) list 104 species of monogamous birds in 26 families, and six species of "partially monogamous" birds in four additional families, for which they found acceptable published documentation of EPCs or attempted EPCs. EPCs have also been observed in the Common Greenshank (*Tringa nebularia*; Nethersole-Thompson, 1979), Chipping Sparrow (*Spizella passerina*; Keller, 1979), and Indigo Bunting (*Passerina cyanea*; R.B. Payne, personal communication). Gladstone (1979) lists two species of colonial birds not included in McKinney et al. (in press) in which EPCs have been reported: Great Cormorant (*Phalacrocorax carbo*) and House Martin (*Delichon urbica*). Thus, there may be as many as 115 species of "monogamous" birds in 30 families for which there is reasonably good evidence that EPCs occur. This number is small compared to the number of species of birds presumed to be monogamous, but the diversity of kinds, both taxonomically and behaviorally, for which EPCs have been reported may indicate more widespread occurrence among monogamous birds. The species include colonial nesters and dispersed nesters; among the latter are included cooperative breeders, species that hold small territories, species that hold large territories, and species that practice female-defense territoriality. The number of colonial nesters among the 115 species is disproportionately large, but it should be borne in mind that an investigator working with these species can observe the behavior of a much larger sample of breeding birds than can one working on dispersed nesters. However, it also seems probable that EPCs will ultimately prove to be more common among colonial birds owing to the greater opportunity afforded males by the proximity of large numbers of breeding females.

3.2. Circumstantial Evidence for an EPC Strategy

Besides direct observation of EPCs or EPC attempts and the other evidence presented in Section 3.1, there is additional circumstantial evidence supporting the conclusion that EPC may be part of the male reproductive strategy in other species. Two lines of evidence are discussed below.

3.2.1. Mate-Guarding

The term mate-guarding (Birkhead, 1979) has been used to refer to the male behavior of remaining close to a mate and actively preventing approach by other males during the time when copulations might result in fertilization of eggs. The presumption is that the male is guarding his mate from possible insemination by other males. Since mate-guarding behavior has been described in a number of species in which EPCs or EPC attempts have been observed (McKinney et al., in press; Nolan, 1978; Keller, 1979), similar behavior by males of species in which EPCs have not been observed suggests that EPCs are a threat. In fact, behavior considered to be mate-guarding (or other anti-cuckoldry tactics) has been described for Ring Doves (*Streptopelia risoria*; Zenone et al., 1979), European Starlings (*Sturnus vulgaris*; Power et al., 1981), Mountain Bluebird (*Sialia currucoides*; Power and Donner, 1980), and Malachite Sunbirds (*Nectarinia famosa*; Wolf and Wolf, 1976), and there are numerous reports of males of many species remaining close to their mates during the period when the females may presumed to be fertilizable. For example, such behavior has been noted in Yellow Wagtails (*Motacilla flava*; Smith, 1950), Twites (*Acanthis flavirostris*; Marler and Mundinger, 1975), Cassin's Finches (*Carpodacus cassinii*; Samson, 1976), Snow Buntings (*Plectrophenax nivalis*; Tinbergen, 1939), Lapland Longspurs (*Calcarius lapponicus*; Seastedt and MacLean, 1979), Yellow Warblers (*Dendroica petechia*; Ford, unpublished observation), and Gray Catbirds (*Dumetella carolinensis*; Ford, unpublished observation; see also Table I, Nest Building Period, in Verner and Willson, 1969).

If close attendance of the female during her presumed fertile period is correctly interpreted as an anti-EPC strategy, this behavior is further indication of the widespread occurrence among birds of the practice of EPC. However, the validity of this conclusion is dependent upon the correct interpretation of the function of behavior labeled as mate-guarding. If EPCs were a continual threat throughout the time a female is receptive, then it seems reasonable to assume that males would have been selected to guard their mates continuously during this period. In fact, males leave their mates unguarded some of the time. In some species, failure to guard continuously may result from one member of a pair always having to remain at or near the nest to protect it against usurpation or predation while the other member leaves to feed or gather nest material. However, in species in which continual nest-guarding does not occur, males sometimes appear to voluntarily leave their mates unguarded for varying periods. Buitron (1983) observed that male Black-billed Magpies (*Pica pica*) were often out of sight of their mates when an intruding male entered the territory and courted the female, and occasionally a male was observed on the territory of another pair during the period his own female was presumably fertilizable. Similarly, Nolan (1978) reported that 26 of 146 extraterritorial explorations by male Prairie Warblers (*Dendroica discolor*) were made during a mate's preincubation period. Fifty-eight of 146 departures from territories by male Yellow Warblers observed by me in 1977 and 1978 (Ford, unpublished observation) were made during a mate's building or laying stage (35%

of departures when adjusted to number of male-days with mates in these two stages). Keller (1979) also observed male Chipping Sparrows leaving their territories during the periods when their mates were presumably receptive.

Why all males do not guard continuously during the period their mates may be vulnerable remains unanswered. It may be that EPCs are not a serious threat and that "mate-guarding" has other functions, e.g., predator detection. However, since the evidence indicates that a male strategy of seeking EPCs is widespread among birds, there must be some other explanation for males failing to guard continuously. It may be, for example, that in some species the intervals when copulation can result in fertilization are intermittent and brief. A male may be aware of the high risk periods for his mate and guard closely only at those times.

3.2.2. Territorial Intrusions

In recent years, evidence has been accumulating that mated males of a number of species generally considered to hold "Type A" (Nice, 1941) territories regularly engage in extraterritorial excursions. Some of these excursions are onto unoccupied areas but, at least in some species, they are mostly intrusions onto the territories of other males. Apparently successful EPCs between an intruding male and a territory holder's mate have been observed in the American Redstart (*Setophaga ruticilla*; Ficken, 1962), Ovenbird (*Seiurus aurocapillus*; Hann, 1937), Pied Flycatcher (*Ficedula hypoleuca*; von Haartman, 1951), House Wren (*Troglodytes aedon*; Kendeigh, 1941), Chipping Sparrow (Keller, 1979), Indigo Bunting (R. B. Payne, personal communication), Savannah Sparrow (*Passerculus sandwichensis*; Welsh, 1975), and Chaffinch (*Fringilla coelebs*; Marler, 1956).

Attempted EPCs by intruding males have been reported in the Black-billed Magpie (Birkhead, 1979), Yellow-billed Magpie (*Pica nuttalli*; Verbeek, 1973), Australian Magpie (*Gymnorhina tibicen*; Robinson, 1956), Yellow Bunting (*Emberiza citrinella*; Howard, 1929) see additional references in McKinney et al. (in press). Unfortunately, it is usually not possible to determine from the literature whether the intruding males observed engaging in successful or attempted EPCs are mated. However, since mated males greatly outnumber unmated males in monogamous populations and mated males are known to regularly make territorial intrusions (see below) it is probable that mated males are responsible for some, if not most, of the records of EPCs and EPC attempts in territorial species.

The timing of territorial intrusions and the behavior of intruding males in species for which there are no records of EPCs suggest that

EPCs are the motivation behind intrusions. In 1977, while studying a color-banded population of ten mated male Yellow Warblers on a 14 ha upland field in central Minnesota, I discovered that males regularly left their own territories to intrude on the territories of other males. During the 1977 breeding season I recorded 59 intrusions by males, most of which were identified and known to be mated (there were no unmated males on the study area and there were only two nearby unbanded males whose mating status was unknown). Forty-four (74.6%) of the 59 intrusions occurred on territories where the resident female was building a nest and presumably fertilizable (four of five copulations I have seen between mated birds occurred during nest-building). On the average, a female was nest-building on only 26% of the days of the 1977 season, suggesting a positive correlation between intrusions and the potential receptivity of the female being visited.

During the 1978 season I spent 125.9 hours on timed observations of eight Yellow Warbler territories during all phases of the breeding cycle and recorded 136 territorial intrusions by males. The intruder was identified in 103 cases and was a mated male in 92 of them. The intrusions were distributed according to nesting stage as shown in Table I. Assuming that the resident female is potentially fertilizable during nest-building and at least early egg-laying stages, the data show that intrusions are preferentially directed to territories occupied by a receptive female.

Intrusions were also recorded incidental to other studies during the seasons of 1979, 1980, and 1981, and show the same pattern. One hundred and sixty-one (91%) of 177 intrusions occurred during the nest-building and egg-laying stages.

The behavior of intruding male Yellow Warblers also indicates that the resident female is the focus of their interest. Most records of intrusion consist of the resident male chasing another male from the vicinity of the female or her nest. However, I had some opportunities to observe

TABLE I
Territorial Intrusions Relative to Nesting Stage in the Yellow Warbler[a]

	Building	Laying	Incubation	Nestlings	Fledglings
Hours observed	59.5	16.0	22.7	13.6	14.1
Intrusions	105	18	3	2	8
No./hour	1.76	1.13	0.13	0.15	0.57

[a]Rearrangement of the data on the basis of occurrence or non-occurrence of an intrusion during 5 min intervals of sexually receptive stages (building, laying) vs. non-receptive stages (incubation, nestlings, fledglings) permit analysis by chi-square test of independence; $\chi^2 = 54.5$, $df = 1$, $p < .001$.

the behavior of intruders during periods when they were undetected or when the resident male was absent. Intruders seldom crossed a territorial boundary by direct flight in the open. They usually entered behind a screen of trees in which they perched briefly before making a direct flight toward the female or nest site, or dropping into lower vegetation through which they made their approach. Males sometimes made intrusions via a circling maneuver, flying around the margin of an adjacent territory and approaching the nest area from the side opposite their own territory. Except when making an overt move toward the female, intruders behaved furtively. They were silent, moved short distances at a time in thick cover, and frequently remained motionless and tensely alert, especially when watching the resident male and/or female. If the resident male was absent or not in view, the intruder invariably moved directly toward the female, or, were she not in evidence, toward the nest site. I have seen intruders follow females closely as they moved through low vegetation, chase the female without catching her, and fly from an elevated perch into a bush where the female was working on a nest. The latter incidents were followed by "chipping" vocalizations and fluttering motion that I could not see clearly. Twice I saw intruders fly from elevated perches and capture and grapple with a flying female. Both times the male and female tumbled together into ground cover and out of view. Chipping and fluttering followed briefly before the birds reappeared and moved away through bushes, the intruder following the female.

The latter episodes suggest forced copulation attempts but are difficult to distinguish from the sexual chases and "pouncing" behavior (Nolan, 1978) characteristic of mated pairs. I have never observed an unequivocal copulation between an intruder and another male's mate, nor have I seen a female behave receptively toward an intruder. Female Yellow Warblers usually appear to ignore intruders that follow them unless the male approaches closely. They then usually turn and appear to threaten the male with an open-bill display. The males back off but continue to follow the female until out of view or until the intruder is detected and chased by the female's mate. I have seen females give this "threat" display to their own mates when they approached closely.

Most intrusions by male Yellow Warblers (78% in 1977–1978) are on adjacent territories, but males are occasionally discovered on territories that are over 700 m and two or more territories distant from their own.

The only studies reporting extensive extraterritorial activities by males are those of Nolan (1978) on the Prairie Warbler and Buitron (1983) on the Black-billed Magpie. Nolan (1978) did not directly link territorial intrusions with possible EPC attempts but of 155 cases of exploration recorded by him, 140 of them were on the territories of other males. At least 58 of these occurred during periods when the resident female was probably receptive; another 24 occurred when fledglings were present and the female may have been receptive since this species is frequently double-brooded. Only 27 intrusions occurred during incubation and nestling stages, and I cannot categorize the remainder (31) from the information given.

Buitron (1983) has reported that territorial intrusions by mated male Black-billed Magpies occur regularly, and that the intruding males direct courtship behavior toward the resident female when her mate is away. As with Yellow Warblers and Prairie Warblers, the timing of intrusions is strongly skewed toward the periods the resident female is building or laying. Birkhead (1979) has observed EPC attempts in this species in Europe.

Several times in 1977 through 1980, while attempting to capture Yellow Warblers in mist nets, I caught three adult Gray Catbirds at the same time, and then discovered that the net had been set near a Gray Catbird nest. Suspecting that the trios represented a mated pair and an intruding male, we color-banded Gray Catbirds and observed them in the 1981 and 1982 seasons. As a result of direct observation or capture in mist nets, we obtained 23 records of territorial intrusions near nests by males whose identity was known; all were mated. In 17 (74%) of the cases the resident female was either building or laying. Since the average territory during these two seasons had a female in these stages on only 27% of the days, the presence of a potentially fertilizable female appears to correlate positively with the occurrence of intrusions. Other intrusions were noted but the identity of the intruder was not determined.

A correlation between territorial intrusions and the presence of a fertile female has been reported for the European Blackbird (*Turdus merula*; Snow, 1956), Red-backed Shrike (*Lanius collurio*; Durango, 1956), Chipping Sparrow (Keller, 1979), Lapland Longspur (Seastedt and MacLean, 1979), and Twite (Marler and Mundinger, 1975).

Territorial intrusions or departures from territories have been noted as usual behavior of males in the Sedge Wren (*Cistothorus platensis*; J. Burns, personal communication), Blackpoll Warbler (*Dendroica striata*; B. Eliason, personal communication), Palm Warbler (*Dendroica palmarum*; B. Falls, personal communication), Wilson's Warbler (*Wilsonia pusilla*; Stewart, 1973), Bluethroat (*Luscinia svecica*), and Brambling (*Fringilla montifringilla*; Cederholm et al., 1974). In fact, published "life history" studies of many additional species either imply or specifically mention that territorial intrusions occur, but such behavior is typically reported in an incidental way and few details are provided.

Overall, territorial intrusion by mated males appears to be a fairly widespread phenomenon, at least among passerines. The probable motivation of these intrusions is the seeking of copulations with the mates of other males. Future workers studying individually marked populations of monogamous birds can test this hypothesis by looking closely at the extraterritorial activities of mated birds.

4. OPPORTUNISTIC AND FACULTATIVE POLYGYNY

As was discussed earlier (Section 2), in this review I consider a species to be monogamous if no more than approximately 20% of the individuals regularly have more than one mate, and there appears to be considerable spatial and temporal variation in the rate below that value. At present, it appears that the species practicing polygyny at low rates generally fall into one of two groups. In one group polygyny is rare, often with only one or two cases reported for a species, even though it may have been fairly well-studied. Furthermore, polygyny in this group frequently seems to arise from unusual circumstances, e.g., a mated male acquires the territory and mate of a vanished neighbor, or from unusual behavior, e.g., two females lay in the same nest. In the other group, low rates of polygyny appear to occur regularly, at least in some populations, and the circumstances and behavior associated with the polygynous matings are similar to those of species practicing polygyny at much higher rates. In many of the species in this group emancipation of the male from parental care appears to be as great as in many more highly polygynous species.

It is useful to distinguish these two groups of "monogamous" species practicing limited polygyny from species that are either entirely monogamous or highly polygynous. I refer to the group in which polygyny is rare as "opportunistically polygynous" and the group in which polygyny occurs regularly as "facultatively polygynous." The terms *opportunistic* and *facultative* do not have entirely distinctive meanings, but the former implies capitalizing on an unusual opportunity while the latter suggests the exercising of an option (as in "facultative anaerobe" or "facultative parasite" referring to organisms that, depending on circumstance, can exercise an option in mode of existence). Unfortunately, for most species that appear to practice polygyny at low rates, the sample size of known mating units is either unknown or limited through both time and space. Often only one or two cases of polygyny have been reported for a species, and the number of monogamous units is either not known or not given. And, for most of the species that are known to practice polygyny regularly, only one population has been studied, typically for only 1–3 years. Neither the mating habits of other populations nor the long term variation in the studied population are known. Consequently, designating a species as either opportunistically polygynous or facultatively polygynous (not to mention as "monogamous" or "polygynous") can be done only on a tentative basis.

4.1. North American Passerines

Verner and Willson (1969) reviewed the literature on the breeding biology of 278 species (based on 34th Supplement to the AOU Check-List of North American Birds, 1982) of native North American passerines and reported records of polygyny for 36 species. They assumed polygyny occurred in three others, and listed three species as having promiscuous mating systems. I was unable to substantiate records of polygyny for two species: Cliff Swallow (*Hirundo pyrrhonota*; Emlen, 1954) and American Redstart (Barney, 1929).

Table II is an updated list of species breeding in North America for which there are acceptable records of polygyny. Included are six species for which the only records are from European populations, and the two *Quiscalus* species Verner and Willson (1969) listed as being "promiscuous." Based on Selander's (1972) description of the mating system of these species, it seems more appropriate to categorize them as polygynous. Of the 61 species listed in Table II, the reported rates of polygyny for 14 are sufficiently high that they can be considered to have polygynous mating systems. The remaining 47 species appear to be basically monogamous (as defined here). Omitting European records of polygyny, low rates of polygyny are known to occur in 42 (16%) of 264 species that have neither polygynous nor promiscuous mating systems. This is an impressively large number when one considers that of the remaining 222 species, six are known to practice polygyny in Europe, and little is known about the mating habits of most of the rest. In fact, Verner and Willson (1969) considered the breeding biology of only 21 of these species well enough known to warrant labeling them monogamous. Currently, this number is somewhat higher, since for a number of species studied since 1969, only monogamous mating units have been found, e.g., Blue Jay (*Cyanocitta cristata*; W. Hilton, personal communication), Gray Catbird (Darley et al., 1971; Ford, unpublished observations), Clay-colored Sparrow (*Spizella pallida*; Knapton, 1978), and Field Sparrow (*S. pusilla*; Best, 1977). However, the fact remains that the mating habits of well over half of the species of North American passerines are poorly known.

It appears likely that careful study of these species will produce

TABLE II
Species of North American Passerines for Which Polygyny Has Been Reported

Species	Extent of polygyny[a]	Source[b]
Eastern Phoebe (Sayornis phoebe)	Opp	Sherman (1952)
Western Wood Pewee[d] (Contopus sordidulus)	Opp	Eckhardt (1976)
Acadian Flycatcher[c] (Empidonax virescens)	Opp	Mumford (1964)
Violet-green Swallow (Tachycineta thalassina)	Opp	Bent (1942)
Tree Swallow (Tachycineta bicolor)	Opp	Shelley (1935)
Bank Swallow[c] (Riparia riparia)	Opp	Stoner (1942)
Barn Swallow[e] (Hirundo rustica)	Opp	Richardson (1956)
Purple Martin (Progne subis)	Fac	Southern (1959), Brown (1975, 1979)
Scrub Jay[d] (Aphelocoma coerulescens)	Opp	Woolfenden (1976)
Common Crow (Corvus brachyrhynchos)	Opp	Patch (1918)
Black-capped Chickadee (Parus atricapillus)	Opp	Smith (1967)
Dipper[d] (Cinclus mexicanus)	Fac	Price and Bock (1973)
House Wren (Troglodytes aedon)	Fac	Kendeigh (1941)
Winter Wren[e] (Troglodytes troglodytes)	Py	Armstrong (1955)
Marsh Wren (Cistothorus palustris)	Py	Welter (1935), Verner (1965), Verner and Engelsen (1970)
Sedge Wren[d] (Cistothorus platensis)	Fac	Crawford (1977), Burns (1982)
Mockingbird (Mimus polyglottos)	Opp	Lasky (1946), Logan and Rulli (1981)
Robin (Turdus migratorius)	Opp	Howell (1942)
Eastern Bluebird (Sialia sialis)	Opp	Bryens (1925), Wetherbee (1933)
Wheatear[e] (Oenanthe oenanthe)		von Haartman (1969)
Bluethroat[e] (Luscinia svecica)		von Haartman (1969)
White Wagtail[e] (Motacilla alba)		von Haartman (1969)
Yellow Wagtail[e] (Motacilla flava)		von Haartman (1969)
Loggerhead Shrike (Lanius ludovicianus)	Opp	Grimes (1928)
Prothonotary Warbler (Protonotaria citrea)	Opp	Walkinshaw (1941)
Northern Parula (Parula americana)	Opp	Williams et al. (1958)
Yellow Warbler[d] (Dendroica petechia)	Fac	Ford (unpublished), S. Sealy (personal communication)
Blackpoll Warbler[d] (Dendroica striata)	Fac	B. Eliason (personal communication)

TABLE II (Continued)

Species	Extent of polygyny[a]	Source[b]
Kirtland's Warbler (*Dendroica kirtlandii*)	Fac	Mayfield (1960), Radabaugh (1972)
Prairie Warbler (*Dendroica discolor*)	Fac	Nolan (1978)
Palm Warbler[d] (*Dendroica palmarum*)	Opp	Welsh (1971)
Ovenbird (*Seiurus aurocapillus*)	Opp	Hann (1937)
Northern Waterthrush (*Seiurus noveboracensis*)	Opp	Eaton (1957)
Yellowthroat (*Geothlypis trichas*)	Opp	Stewart (1953), Powell and Jones (1978)
Yellow-breasted Chat[d] (*Icteria virens*)	Opp	Thompson and Nolan (1973)
Wilson's Warbler[d] (*Wilsonia pusilla*)	Fac	Stewart et al. (1978)
Painted Redstart[d] (*Myioborus pictus*)	Opp	Marshall and Balda (1974)
Indigo Bunting[d] (*Passerina cyanea*)	Fac	Carey and Nolan (1975, 1979), Payne (1982)
Painted Bunting[d] (*Passerina ciris*)	Py	C. Thompson (personal communication)
Dickcissel (*Spiza americana*)	Py	Zimmerman (1966)
Rufous-sided Towhee[d] (*Pipilo erythrophthalmus*)	Opp	D. Ewert (personal communication)
Chipping Sparrow (*Spizella passerina*)	Opp	Walkinshaw (1959), Keller (1979)
Lark Sparrow (*Chondestes grammacus*)	Opp	Knowles (1938)
Lark Bunting[d] (*Calamospiza melanocorys*)	Py	Pleszczynska (1978), Pleszczynska and Hansell (1980)
Savannah Sparrow[d] (*Passerculus sandwichensis*)	Fac	McLaren (1972), Welsh (1975)
Song Sparrow (*Melospiza melodia*)	Opp	Nice (1933, 1937), Smith et al. (1982)
Swamp Sparrow (*Melospiza georgiana*)	Fac	Willson (1966b)
White-crowned Sparrow (*Zonotrichia leucophrys*)	Fac	DeWolfe (1968), Petrinovich and Patterson (1978)
Lapland Longspur[d] (*Calcarius lapponicus*)	Opp	Seastedt and MacLean (1979)
Snow Bunting (*Plectrophenax nivalis*)	Opp	Tinbergen (1939)
Bobolink (*Dolichonyx oryzivorus*)	Py	Martin (1974), Wittenberger (1978)
Red-winged Blackbird (*Agelaius phoeniceus*)	Py	Nero (1956), Orians (1961)
Tricolored Blackbird (*Agelaius tricolor*)	Py	Orians (1961)
Eastern Meadowlark (*Sturnella magna*)	Py	Lanyon (1957)
Western Meadowlark (*Sturnella neglecta*)	Py	Lanyon (1957)

(Continued)

TABLE II (Continued)

Species	Extent of polygyny[a]	Source[b]
Yellow-headed Blackbird (*Xanthocephalus xanthocephalus*)	Py	Willson (1966a)
Brewer's Blackbird (*Euphagus cyanocephalus*)	Py	Williams (1952)
Great-tailed Grackle (*Quiscalus mexicanus*)	Py	Selander and Giller (1961)
Boat-tailed Grackle (*Quiscalus major*)	Py	Selander and Giller (1961)
Common Grackle[d] (*Quiscalus quiscula*)	Fac	Howe (1976)
House Finch (*Carpodacus mexicanus*)	Opp	Michener (1925)

[a]Opp = opportunistically polygynous, polygyny rare, often only one case; Fac = facultatively polygynous, polygyny rates up to 20% but considerable spatial or temporal variation in rates; Py = polygynous, polygyny rates usually > 20% in most to all populations sampled.
[b]Additional sources for many polygynous species are not cited.
[c]Overlooked by Verner and Willson (1969).
[d]Polygyny reported since Verner and Willson (1969).
[e]Polygyny known only for European populations.

more records of polygyny, for many share taxonomic affinity and/or behavioral similarity with opportunistically or facultatively polygynous species.

4.2. European Passerines

As further indication of how common opportunistic and facultative polygyny might be among at least north temperate zone passerines, von Haartman (1969) listed 47 European species for which there are records of polygyny. Forty-five of these species are considered part of the mid-European avifauna, which includes 97 species of passerines. Since von Haartman considered only five species to be "commonly" polygynous, at least 40 (41%) mid-European passerine species practice polygyny at levels that might be considered to be opportunistic or facultative. While this is a higher percentage than found among North American passerines, it may only reflect, as von Haartman suggests, that mid-European passerines have been better studied. In fact, in 1969 only 38 species of North American passerines were known to engage in polygyny whereas currently, after more species have been carefully studied, 55 are known to engage in polygyny, at least to a limited extent.

4.3. Non-Passerines

I have not systematically attempted to discover polygynous deviations from monogamy in non-passerines. Nevertheless, I have found some examples of species which appear to practice polygyny at a low rate. For most of these species too little is known to permit labeling them as either opportunistically or facultatively polygynous. The species are listed below by family.

Anatidae. Magpie Goose (*Anseranas semipalmata;* Frith and Davies, 1961), Mute Swan (*Cygnus olor;* Dewar, 1936), and Canvasback (*Aythya valisineria;* M. Anderson, personal communication).

Accipitridae. All records of polygyny for this family are from Newton (1979). Osprey (*Pandion haliaetus*), Sparrow Hawk (*Accipiter nisus*), Shikra (*A. badius*), and Buzzard (*Buteo buteo*). In addition, trios of adults have been observed at nests of the Red-tailed Hawk (*Buteo jamaicensis*), Bald Eagle (*Haliaeteus leucocephalus*), and Bateleur (*Terathopius ecaudatus*), but the mating relationships of the individuals involved are not known.

Falconidae. All records are from Newton (1979). European Kestrel (*Falco tinnunculus*), Peregrine Falcon (*F. peregrinus*), and Merlin (*F. columbarius*).

Phasianidae. White-tailed Ptarmigan (*Lagopus leucurus;* Choate, 1963), Willow Ptarmigan (*L. lagopus;* Watson and Jenkins, 1964), and Rock Ptarmigan (*L. mutus;* Weeden and Theberge, 1972).

Charadriidae. Greater Golden Plover (*Pluvialis apricaria;* Bannerman, 1961), Northern Lapwing (*Vanellus vanellus;* Wilson, 1967), Southern Lapwing (*V. chilensis;* Walters and Walters, 1980), and Snowy Plover (*Charadrius alexandrinus;* Warriner and Warriner, 1978).

Scolopacidae. Common Greenshank (Nethersole-Thompson, 1979).

Laridae. Brown Skua (*Catharacta lonnbergi;* Bonner, 1964; Burton, 1968), Herring Gull (*Larus argentatus;* Shugart and Southern, 1977), and Ring-billed Gull (*L. delawarensis;* Conover et al., 1979).

Nethersole-Thompson (1979) reports the occurrence of "triangles" in three charadriiforms, Oystercatcher (*Haematopus ostralegus*), Eurasian Curlew (*Numenius arquata*), Little Ringed Plover (*Charadrius dubius*), and polygyny in the Black-tailed Godwit (*Limosa limosa*), but details and references are not given. Two females laying clutches in the same nest have been reported for the Lesser Yellowlegs (*Tringa flavipes;* Bannerman, 1961), Little Stint (*Calidris minuta;* Reynolds, 1972), and American Avocet (*Recurvirostra americana;* Yom-Tov, 1980). These records suggest opportunisitic polygyny, but the mating relationships of the individuals are not known.

Since I did not conduct an extensive search of the literature on non-passerines, it is premature to draw conclusions from the few species listed above. Undoubtedly more examples of low rates of polygyny exist, and more will be discovered. However, it seems unlikely that opportunistic and facultative polygyny will prove to be as common among this group as they are among passerines. Mandatory biparental care, including during incubation, is much more prevalent among non-passerine species, which makes it difficult for males of such species to be anything other than monogamous. On the other hand, precocial young that require comparatively little biparental care occur in a number of non-passerine groups, which favors the emancipation of (usually) males and the development of a highly polygynous mating system. Obviously, there are exceptions to these generalizations, but overall it appears that non-passerines are more likely to be either "purely" monogamous or highly polygynous when compared to the passerines in which precociality does not occur and mandatory biparental care, especially during incubation, is uncommon.

5. POLYANDRY

Excluding examples of EPCs involving mated females, which have been called polyandry by some authors (Hann, 1937; von Haartman, 1951), polyandry is decidedly rare among avian species. Collectively, Jenni (1974) and Oring (1982) list 44 species (plus "others" in the Turnicidae) for which polyandry has been "reported." The number of species for which polyandry has been documented is considerably smaller. Jenni (1974) lists the Sanderling (*Calidris alba*) solely on the basis of Parmelee and Payne's (1973) evidence that female Sanderlings are capable of "double-clutching." He also lists seven species of passerines even though he acknowledges that for one of them (House Wren) the record (Kendeigh, 1941) is really an example of sequential monogamy. None of the records for the remaining six passerines are unequivocal cases of polyandry. The Kirtland's Warbler (*Dendroica kirtlandii*; Radabaugh, 1972) and Bay-winged Cowbird (*Molothrus badius*; Fraga, 1972) records are examples of mate-switching between breeding attempts. The Hicks' Seed-eater (*Sporophila aurita*; Gross, 1952) record is probably an EPC. The Eastern Bluebird record (Lasky, 1947) cannot be distinguished from a case of a "helper at the nest" (helpers are known to occur in this species; Skutch, 1961). Hann's (1940) record of polyandry in the Ovenbird is based on his observations of both males that occupied adjacent territories feeding the young in a nest near their common territorial boundary, and polyandry in the Lark Bunting (*Calamospiza melanocorys*) is apparently based on Bent's (1908) statement that each female seemed to have two males in attendance.

Of the remaining 36 species, an examination of the information presented in Oring's (1982) Table IV and a review of some of the original sources cited therein, reveal that there are no documented cases of naturally occurring polyandry for 17 of them. Wilson's Phalarope (*Phalaropus tricolor*) is included in spite of the fact that extensive field study of marked birds has produced only records of monogamous pairings (Johns, 1969; Howe, 1975). The Dunlin (*Calidris alpina*) is included on the basis of three cases of females acquiring new mates after they abandoned their first mates who were caring for young out of the nest (Soikkeli, 1967). The Little Stint is apparently included on the basis of a probable case of double-clutching that is more likely the result of a monogamous mating than a polyandrous mating (Hilden, 1975; Pitelka et al., 1974). The remaining 14 species are included on the basis of polyandrous matings in captivity or such circumstantial evidence as "role reversal," "male parental care," "group of one female, two males," etc. This leaves only 19 of the 44 species for which there are records of polyandrous matings in the wild. Recent records of polyandrous matings in the Wattled Jacana (*Jacana jacana*; Osborne, 1982), Brown Skua (P. Jenkins, paper presented at 1982 meeting of American Ornithologists' Union), and Dunnock (*Prunella modularis*; Birkhead, 1981) bring the total up to 22.

Most of the 22 species for which polyandrous matings are known appear to practice polyandry, or polyandry mixed with polygyny, to such an extent that their mating systems warrant labeling with one (or more) of the terms denoting a form of polygamy. However, four of the species may be basically monogamous species that practice polyandry on a facultative basis. For both the Mountain Plover (*Charadrius montanus*; Graul, 1973) and the Red-legged Partridge (*Alectoris rufa*; Jenkins, 1957) there is but a single documented case of polyandry among a number of known cases of monogamous matings. Similarly, in the Dotterel (*Eudromias morinellus*; Pulliainen, 1971; Nethersole-Thompson, 1973) polyandry appears to be much less frequent than monogamy. Finally, there is a single record of probable polyandry for the Spotted Redshank (*Tringa erythropus*; Raner, 1972), but little else is known about its mating habits. It appears likely that further study of these four species, as well as a number of other species in which double-clutching and/or male incubation occur, will produce additional evidence that facultative (or opportunistic) polyandry is an occasional strategy of some basically monogamous populations.

6. CONCLUSION

It is evident from this review of the mating habits of "monogamous" species that males of many species do not maintain an exclusive mating relationship with only one female per breeding season. Mated males of a taxonomically and behaviorally diverse group of species actively pursue a strategy of seeking copulations with the mates of other males. And males of many species, at least among north temperate region passerines, regularly acquire second mates on an opportunistic or facultative basis. Indeed, as Trivers (1972) hypothesized, monogamously mated males appear to be selected to pursue a mixed reproductive strategy, i.e., one in which they actively seek, or at least capitalize on, opportunities to copulate with additional females. How widespread mixed strategies will ultimately prove to be is currently an open question since so few species have been studied using techniques, e.g., color-banding, that facilitate detection of EPCs and polygamy. However, Trivers' (1972) theoretical arguments, and the frequency with which mixed strategies are encountered among species that have been well studied, suggest that they will prove to be common.

It is hardly surprising that the role of the male during incubation appears to be a major factor in determining which strategy of infidelity a male pursues (EPCs, polygyny, or both). Polygyny is seldom encountered in species in which the male plays a vital role during incubation, i.e., either sharing incubation duties or providing other important aid such as guarding the nest during the female's inattentive periods, e.g., Gray Catbird (Slack, 1976). When polygyny does occur in these species it often has unusual features such as two females using nests that are very close to each other, or two females laying in the same nest, e.g., Herring Gull, Brown Skua, and several falconiforms (see references in Section 4.3). The rarity of polygyny among these species probably reflects the difficulties a male has in finding the time to court and defend additional females, and in keeping his mated status hidden. (Females who require male help with incubation would almost surely have been selected to reject a mated male.) Of course, males may have been selected to not attempt to acquire a second mate if dividing incubation time between two nests results in a lower RS than is achieved by incubating at only one nest.

Polygynous deviations from monogamy are fairly common among species in which the male plays a minor role during incubation, e.g., most of the species listed in Table II. These males are freer to court and defend additional females and, if necessary, are more likely to be able to hide their mated status. They are also less likely to have lower

RS with two females than with one. Even if these males ordinarily feed young but are unable to successfully partition their feeding efforts between two broods, they may be able to acquire a second female in a sequence that avoids extensive overlap between broods. Failing that, they can still selectively feed the young in only one nest. Furthermore, an unmated female who cannot find an unmated male may be willing to accept secondary female status rather than not breed. She can probably hatch the eggs without appreciable male help and may even fledge some young without male help, e.g., Yellow Warbler (Ford, unpublished observations). Even if male help in feeding young is required, she may get that help if her brood does not overlap extensively with that of the primary female. There is the additional possibility that, when nest losses are high, she may have risen to primary female status by the time her eggs hatch. Almost certainly many more examples or opportunistic and facultative polygyny will be discovered in species in which males play a minor role during incubation. This is especially likely when one considers that among the many species whose mating habits have not been well studied are a number that have characteristics often correlated with polygyny, such as sexual dimorphism and male plumage bimaturism, e.g., Thraupinae, Icterini, and Parulinae.

Male ducks (especially *Anas* species) present a conspicuous exception to the generalization about the importance of male incubation. They provide no parental care to either eggs or young and yet (apparently) do not practice polygyny. The male-biased sex ratio typically found in these species (Bellrose et al., 1961) evidently results in males having little opportunity to be anything other than monogamous (Wittenberger and Tilson, 1980). Male ducks, however, appear to practice a strategy of EPC to a greater extent (or perhaps they have merely been better studied) than any other group (McKinney et al., in press). It seems reasonable to expect that males might be more strongly selected to pursue a strategy of EPC when either their ability or the opportunity to acquire second mates is severely limited.

When sufficient data become available, there may prove to be a direct correlation between the degree to which monogamy is enforced on males and the extent to which they practice an EPC strategy. In addition to ducks, EPC appears to be a persistent strategy among species in which the male shares in incubation, e.g., herons and gulls (McKinney et al., in press), but the significance of the number of EPC records for these species is difficult to assess because most are colonial nesters. The combination of the close proximity of many females and the large number of males that an investigator can observe per unit of time may have resulted in an inflated impression of the prevalence of

EPCs in species in which the male shares in incubation. Most of the species in which the male does not incubate are dispersed nesters. Males must travel some distance just to encounter a female, and an investigator can usually watch but one male at a time. EPCs, or evidence suggesting an EPC strategy, do occur in species in which monogamy is not strongly enforced, e.g., Yellow Warbler, Ovenbird, Savannah Sparrow, and Indigo Bunting (see Sections 3.2.2 and 4.1), but only time-based studies of male activities for many species can reveal whether a correlation exists between the degree to which monogamy is enforced and the prevalence of EPCs. Regardless of a correlation, I suspect EPCs will prove to be so widespread among "monogamous" species that one of the more interesting new problems will be the need to identify the factors that have mitigated against the development of an EPC strategy in species in which it appears to be absent.

Some mated females apparently willingly accept, and perhaps even seek, EPCs, e.g., Ovenbird (Hann, 1937), Tree Sparrow (*Spizella arborea*; Weeden, 1975) other references in McKinney et al., (in press), but the information currently available is inadequate to assess either how widespread this behavior might be or its adaptive significance. Mated females of the vast majority of species have little opportunity to practice polyandry since it requires male behavior that is decidedly uncommon, i.e., either two males cooperating in the care of a single clutch (cooperative polyandry), or one male assuming sole care of a clutch (simultaneous or sequential polyandry). I am not aware of any clues that might permit prediction of which additional species might be found to practice cooperative polyandry, but it is reasonable to anticipate that additional examples of polyandrous matings will be found among those species of phasianids, charadriids, scolopacids, and cuculids in which double-clutching and/or male incubation is known to occur.

ACKNOWLEDGMENTS. I am grateful to the many ornithologists who provided me with unpublished information, especially L. Oring, F. McKinney, and D. Buitron for providing me with prepublication copies of their very important papers on topics relevant to this review. I am also indebted to the many graduate students at the University of Minnesota with whom I have had many profitable discussions on monogamous birds, especially J. Burns, D. Bruggers, and B. Eliason. F. McKinney, H. B. Tordoff, D. Bruggers, B. Reaney, and C. Rodell all provided much appreciated and valuable criticisms of the manuscript. B. Prokosch and P. Sinner each provided assistance for one season with the Catbird study. I am especially indebted to T. Bonner for her four summers of invaluable assistance with the Yellow Warblers. Portions of the field work were supported by research grants to J. Poff and to me from St. John's University, and by NSF-URP grants nos. SPI-8026194 (Poff) and EPP75-04606 (Ford).

REFERENCES

American Ornithologists' Union, 1982, Thirty-fourth supplement to the A.O.U. checklist of North American birds, Auk (Suppl.) **99**(3).
Armstrong, E. A., 1955, The Wren, Collins, London.
Bannerman, D. A., 1961, The Birds of the British Isles, Vol. 10, Oliver and Boyd, Edinburgh and London.
Barney, C. C., 1929, Redstart neighbors, Bull. Mass. Aud. Soc. **13(5)**:10.
Bellrose, F. C., Scott, T. G., Hawkins, A. S., and Low, J. B., 1961, Sex ratios and age ratios in North American ducks, Bull. Ill. Nat. Hist. Survey **27**:385–474.
Bent, A. C., 1908, Summer birds of southwestern Saskatchewan, Auk **25**:25–35.
Bent, A. C., 1942, Life histories of North American flycatchers, larks, swallows, and their allies, U. S. Nat. Mus. Bull. **179**:1–435.
Best, L. B., 1977, Territory quality and mating success in the Field Sparrow, Condor **79**:192–204.
Birkhead, M. E., 1981, The social behaviour of the Dunnock Prunella modularis, Ibis **123**:75–84.
Birkhead, T. R., 1979, Mate guarding in the Magpie Pica pica, Anim. Behav. **27**:866–874.
Bonner, W. N., 1964, Polygyny and super-normal clutch size in the Brown Skua, Catharacta skua lonnbergi (Mathews), British Antarctic Survey Bull. **3**:41–47.
Bray, O. E., Kennelly, J. J., and Guarino, J. L., 1975, Fertility of eggs produced on territories of vasectomized Red-winged Blackbirds, Wilson Bull. **87**:187–195.
Brown, C. R., 1975, Polygamy in the Purple Martin, Auk **92**:602–604.
Brown, C. R., 1979, Territoriality in the Purple Martin, Wilson Bull. **91**:583–591.
Bryens, O. M., 1925, Statistics on the House Wren, Wilson Bull. **37**:157–159.
Buitron, D., 1983, Extra-pair courtship in Black-billed Magpies, Anim. Behav. **31**:211–220.
Burns, J. T., 1982, Nests, territories, and reproduction of Sedge Wrens (Cistothorus platensis), Wilson Bull. **94**:338–349.
Burns, J. T., Cheng, K. M., and McKinney, F., 1980, Forced copulation in captive Mallards: I. Fertilization of eggs, Auk **97**:875–879.
Burton, R. W., 1968, Breeding biology of the Brown Skua, Catharacta skua lonnbergi (Mathews), at Signy Island, South Orkney Islands, British Antarctic Survey Bull. **15**:9–28.
Carey, M., and Nolan, V., Jr., 1975, Polygyny in Indigo Buntings: A hypothesis tested, Science **190**:1296–1297.
Carey M., and Nolan, V., Jr., 1979, Population dynamics of Indigo Buntings and the evolution of avian polygyny, Evolution **33**:1180–1192.
Cederholm, G., Flodin, L., Fredriksson, S., Gustafsson, L., Jacobson, S., and Patersson, L., 1974, Ett försök att med nätfangst och ringmärkning bestämma andelen icke häckande faglar i en smafagelpopulation, Fauna och Flora **69(4)**:134–145.

Choate, T. S., 1963, Habitat and population dynamics of White-tailed Ptarmigan in Montana, *J. Wildl Mgmt.* **27**:684–699.
Conover, M. R., Miller, D. E., and Hunt, G. L., Jr., 1979, Female–female pairs and other unusual reproductive associations in Ring-billed and California gulls, *Auk* **96**:6–9.
Crawford, R. D., 1977, Polygynous breeding of Short-billed Marsh Wrens, *Auk* **94**:359–362.
Darley, J. A., Scott, D. M., and Taylor, N. K., 1971, Territorial fidelity of Catbirds, *Can. J. Zool.* **49**:1465–1478.
Dewar, J. M., 1936, Menage á trois in the Mute Swan, *Brit. Birds* **30**:178–179.
DeWolfe, B. D., 1968, Nuttall's White-crowned Sparrow, in: *Life Histories of North American Cardinals, Grosbeaks, Buntings, Towhees, Finches, Sparrows, and Allies* (A. C. Bent, ed.), *U. S. Natl. Mus. Bull.* **23**:1292–1324.
Durango, S., 1956, Territory in the Red-backed Shrike *Lanius collurio*, *Ibis* **98**:476–484.
Eaton, S. W., 1957, A life history study of *Seiurus noveboracensis*, *Sci. Studies of St. Bonaventure Univ.* **19**:7–36.
Eckhardt, R. C., 1976, Polygyny in the Western Wood Pewee, *Condor* **78**:561–562.
Emlen, J. T., Jr., 1954, Territory, nest building, and pair formation in the Cliff Swallow, *Auk* **71**:16–35.
Emlen, S. T., and Oring, L. W., 1977, Ecology, sexual selection and the evolution of mating systems, *Science* **197**:215–223.
Ficken, M. S., 1962, Agonistic behavior and territory in the American Redstart, *Auk* **79**:607–632.
Fraga, R. M., 1972, Cooperative breeding and a case of successive polyandry in the Bay-winged Cowbird, *Auk* **89**:447–449.
Frith, H. J., and Davies, S. J. J. F., 1961, Ecology of the Magpie Goose, *Anseranas semipalmata* Latham (*Anatidae*), *C.S.I.R.O. Wildl. Res.* **6**:91–141.
Gladstone, D. E., 1979, Promiscuity in monogamous colonial birds, *Am. Natural.* **114**:545–557.
Graul, W. D., 1973, Adaptive aspects of the Mountain Plover social system, *Living Bird*, **12**:69–94.
Grimes, S. A., 1928, The Loggerhead Shrike, *Florida Natural.* **1**:48–50.
Gross, A. O., 1952, Nesting of Hicks' Seedeater at Barro Colorado Island, Canal Zone, *Auk* **69**:433–446.
von Haartman, L., 1951, Successive polygamy, *Behaviour* **3**:256–274.
von Haartman, L., 1969, Nest-site and evolution of polygamy in European passerine birds, *Ornis Fenn.* **46**:1–12.
Hann, H. W., 1937, Life history of the Oven-bird in southern Michigan, *Wilson Bull.* **49**:145–237.
Hann, H. W., 1940, Polyandry in the Oven-bird, *Wilson Bull.* **52**:69–72.
Hilden, O., 1975, Breeding system of Temminck's Stint *Calidris temminckii*, *Ornis Fenn.* **52**:117–146.
Howard, H. E., 1929, *An Introduction to the Study of Bird Behavior*, Cambridge University Press, Cambridge.
Howe, H. F., 1976, Egg size, hatching asynchrony, sex, and brood reduction in the Common Grackle, *Ecology* **57**:1195–1207.
Howe, M. A., 1975, Behavioral aspects of the pair bond in Wilson's Phalarope, *Wilson Bull.* **87**:248–270.
Howell, J. C., 1942, Notes on the nesting habits of the American Robin (*Turdus migratorius* L.), *Am. Midl. Natural.* **28**:529–603.
Hunt, G. L., Jr., 1980, Mate selection and mating systems in seabirds, in: *Behavior of Marine Animals*, Volume 4 (H. E. Winn, ed), Plenum, New York, pp. 113–151.

Hunt, G. L., Jr., and Hunt M. W., 1977, Female–female pairing in Western Gulls (*Larus occidentalis*) in southern California, *Science* **196**:1466–1467.
Jenkins, D., 1957, The breeding of the Red-legged Partridge, *Bird Study* **4**:97–100.
Jenni, D. A., 1974, Evolution of polyandry in birds, *Am. Zool.* **14**:129–144.
Johns, J. E., 1969, Field studies of Wilson's Phalarope, *Auk* **86**:660–670.
Keller, M. E., 1979, Breeding behavior and reproductive success of Chipping Sparrows in northwestern Minnesota, M.S. Thesis, University of North Dakota, Grand Forks.
Kendeigh, S. C., 1941, Territorial and mating behavior of the House Wren, *Illinois Biol. Mongr.* **18(3)**:1–120.
Knapton, R. W., 1978, Breeding ecology of the Clay-colored Sparrow, *Living Bird* **17**:137–158.
Knowles, E. H. M., 1938, Polygamy in the western Lark Sparrow, *Auk* **55**:675–676.
Lack, D., 1968, *Ecological Adaptations for Breeding in Birds*, Methuen, London.
Lanyon, W. E., 1957, The comparative biology of the Meadowlarks (Sturnella) in Wisconsin, *Publ. Nuttall Ornithol. Club.* **1**:1–67.
Lasky, A. R., 1946, A nine-year-old Mockingbird and his mates, *Bird-Banding* **17**:36–38.
Lasky, A. R., 1947, Evidence of polyandry at a Bluebird nest, *Auk* **64**:314–315.
Logan, C. A., and Rulli, M., 1981, Bigamy in a male Mockingbird, *Auk* **98**:385–386.
Mader, W. J., 1975, Extra adults at Harris' Hawk nests, *Condor* **77**:482–485.
Marler, P., 1956, Behaviour of the Chaffinch *Fringilla coelebs*, *Behaviour (Suppl.)* **5**:1–184.
Marler, P., and Mundinger, P. C., 1975, Vocalizations, social organization and breeding biology of the Twite *Acanthus flavirostris*, *Ibis* **117**:1–17.
Marshall, J., and Balda, R. P., 1974, The breeding ecology of the Painted Redstart, *Condor* **76**:89–101.
Martin, R. F., 1980, Analysis of hybridization between the hirundinid genera *Hirundo* and *Petrochelidon* in Texas, *Auk* **97**:148–159.
Martin, S. G., 1974, Adaptations for polygynous breeding in the Bobolink, *Dolichonyx oryzivorus*, *Am. Zool.* **14**:109–119.
Mayfield, H., 1960, *The Kirtland's Warbler*, Cranbrook Institute of Science, Bloomfield Hills, Michigan.
McKinney, F., Cheng, K. M., and Bruggers, D. J., Sperm competition in apparently monogamous birds, in: *Sperm Competition and the Evolution of Animal Mating Systems* (R. L. Smith, ed.), Academic Press, New York (in press).
McLaren, I. A., 1972, Polygyny as the adaptive function of breeding territory in birds, *Trans. Conn. Acad. Arts Sci.* **44**:191–210.
Michener, H., 1925, Polygamy practiced by the House Finch, *Condor* **27**:116.
Montagna, W., 1942, The Sharp-tailed Sparrows of the Atlantic coast, *Wilson Bull.* **54**:107–120.
Mumford, R. E., 1964, The breeding biology of the Acadian Flycatcher, *Misc. Publ. Mus. Zool., Univ. Mich.* **125**:1–50.
Nero, R. W., 1956, A behavior study of the Red-winged Blackbird. I. Mating and nesting activities. II. Territoriality, *Wilson Bull.* **68**:5–37, 129–150.
Nethersole-Thompson, D., 1973, *The Dotterel*, Collins, London.
Nethersole-Thompson, D., and Nethersole-Thompson, M., 1979, *Greenshanks*, Buteo Books, Vermillion, S. Dakota.
Newton, I., 1979, *Population Ecology of Raptors*, T. and A. D. Poyser, Berkhamsted.
Nice, M. M., 1933, Relations between the sexes in Song Sparrows, *Wilson Bull.* **45**:51–59.

Nice, M. M., 1937, Studies in the life history of the Song Sparrow, I, *Trans. Linn. Soc. N.Y.* **4**:1–247.
Nice, M. M., 1941, The role of territory in bird life, *Am. Midl. Natural.* **26**:441–487.
Nolan, V., Jr., 1978, The ecology and behavior of the Prairie Warbler, *Dendroica discolor, Ornithol. Monogr.* **26**:i–xxii, 1–595.
Orians, G. H., 1961, The ecology of blackbird (*Agelaius*) social systems, *Ecol. Monogr.* **31**:285–312.
Oring, L. W., 1982, Avian mating systems, in: *Avian Biology*, Volume 6 (D. S. Farner, J. R. King, and K. C. Parkes, eds.), Academic Press, New York, pp. 1–92.
Oring, L. W., and Knudson, M. L., 1972, Monogamy and polyandry in the Spotted Sandpiper, *Living Bird* **11**:59–73.
Osborne, D. R., 1982, Replacement nesting and polyandry in the Wattled Jacana, *Wilson Bull.* **94**:206–208.
Parmelee, D. F., and Payne, R. B., 1973, On multiple broods and the breeding strategy of arctic Sanderlings, *Ibis* **115**:218–226.
Patch, C. L., 1918, A crow polygamist? *Ottawa Natural.* **32**:6.
Payne, R. B., 1982, Ecological consequences of song matching: Breeding success and intraspecific song mimicry in Indigo Buntings, *Ecology* **63**:401–411.
Petrinovich, L., and Patterson, T. L., 1978, Polygyny in the White-crowned Sparrow (*Zonotrichia leucophrys*), *Condor* **80**:99–100.
Pitelka, F. A., Holmes, R. T., and MacLean, S. F., Jr., 1974, Ecology and evolution of social organization in arctic sandpipers, *Am. Zool.* **14**:185–204.
Pleszczynska, W. K., 1978, Microgeographic prediction of polygyny in the Lark Bunting, *Science* **201**:935–937.
Pleszczynska, W, and Hansell, R. I. C., 1980, Polygyny and decision theory: Testing of a model in Lark Buntings, *Am. Natural* **116**:821–830.
Powell, G. V. N., and Jones, H. L., 1978, An observation of polygyny in the Common Yellowthroat, *Wilson Bull.* **90**:656–657.
Power, H. W., and Donner, C. G. P., 1980, Experiments on cuckoldry in the Mountain Bluebird, *Am. Natural.* **116**:689–704.
Power, H. W., Litovich, E., and Lombardo, M. P., 1981, Male Starlings delay incubation to avoid being cuckolded, *Auk* **98**:386–389.
Price, F. E., and Bock, C. E., 1973, Polygyny in the Dipper. *Condor* **75**:457–459.
Pulliainen, E., 1971, Breeding behavior of the Dotterel *Charadrius morinellus*, Rept. 24, Varrio Subarctic Res. Stn., University of Helsinki.
Radabaugh, B. F., 1972, Polygamy in the Kirtland's Warbler, *Jack-Pine Warbler* **50**:48–52.
Raner, L., 1972, Forekommer polyandri hos smalnäbbad simsnäppa (*Phalaropus lobatus*) och svartsnäppa (*Tringa erythropus*)? *Fauna och Flora* **67(3)**:135–138.
Reynolds, J. F., 1972, Little Stint incubating eight eggs, *Brit. Birds* **65**:529.
Richardson, R. A., 1956, Bigamy in swallow, *Brit. Birds* **49**:503.
Roberts, T. A., and Kennelly, J. J., 1980, Variation in promiscuity among Red-winged Blackbirds, *Wilson Bull.* **92**:110–112.
Robinson, A., 1956, The annual reproductory cycle of the Magpie, *Gymnorhina dorsalis* Campbell, in south-western Australia, *Emu* **56**:233–336.
Ryder, J. P. and Somppi, P. L., 1979, Female–female pairing in Ring-billed Gulls, *Auk* **96**:1–5.
Samson, F. B., 1976, Territory, breeding density, and fall departure in Cassin's Finch, *Auk* **93**:477–497.
Seastedt, T. R., and MacLean, S. F., 1979, Territory size and composition in relation to

resource abundance in Lapland Longspurs breeding in arctic Alaska, Auk **96**:131–142.
Selander, R. K., 1972, Sexual selection and dimorphism in birds, in: *Sexual Selection and the Descent of Man: 1871–1971* (B. Campbell, ed.), Aldine, Chicago, pp. 180–230.
Selander, R. K., and Giller, D. R., 1961, Analysis of sympatry of Great-tailed and Boat-tailed grackles, Condor **63**:29–86.
Shelley, L. O., 1935, Notes on the 1934 Tree Swallow breeding-season, Bird-Banding **6**:33–35.
Sherman, A. R., 1952, *Birds of an Iowa Dooryard*, Christopher Publ. House, Boston.
Shugart, G. W., and Southern, W. E., 1977, Close nesting, a result of polygyny in Herring Gulls, Bird-Banding **48**:276–277.
Skutch, A. F., 1961, Helpers among birds, Condor **63**:198–226.
Slack, R. D., 1976, Nest guarding behavior by male Gray Catbirds, Auk **93**:292–300.
Smith, J. N. M., Yom-Tov, Y., and Moses, R., 1982, Polygyny, male parental care, and sex ratio in Song Sparrows: An experimental study, Auk **99**:555–564.
Smith, S., 1950, *The Yellow Wagtail*, Collins, London.
Smith, S. M., 1967, A case of polygamy in the Black-capped Chickadee, Auk **84**:274.
Snow, D. W., 1956, Territory in the Blackbird Turdus merula, Ibis **98**:438–447.
Soikkeli, M., 1967, Breeding cycle and population dynamics in the Dunlin (Calidris alpina), Ann. Zool. Fenn. **4**:158–198.
Southern, W. E., 1959, Foster-feeding and polygamy in the Purple Martin, Wilson Bull **71**:96.
Stewart, R. E., 1953, A life history of the Yellow-throat, Wilson Bull. **65**:99–115.
Stewart, R. M., 1973, Breeding behavior and life history of the Wilson's Warbler, Wilson Bull. **85**:21–30.
Stewart, R. M., Henderson, R. P., and Darling, K., 1978, Breeding ecology of the Wilson's Warbler in the high Sierra Nevada, California, Living Bird **16**:83–102.
Stoner, D., 1942, Bird study through banding, Sci. Monthly **55**:132–138.
Thompson, C. F., and Nolan, V., Jr., 1973, Population biology of the Yellow-breasted Chat (Icteria virens L.) in southern Indiana, Ecol. Monogr. **43**:145–171.
Tinbergen, N., 1939, The behavior of the Snow Bunting in spring, Trans. Linn. Soc. N.Y. **5**:1–95.
Trivers, R. L., 1972, Parental investment and sexual selection, in: *Sexual Selection and the Descent of Man: 1871–1971* (B. Campbell, ed.), Aldine, Chicago, pp. 136–179.
Verbeek, N. A. M. 1973, The exploitation system of the Yellow-billed Magpie, U. Calif. Publ. Zool. **99**:1–58.
Verner, J., 1965, Breeding biology of the Long-billed Marsh Wren, Condor **67**:6–30.
Verner, J., and Engelsen, G. H., 1970, Territories, multiple nest building, and polygyny in the Long-billed Marsh Wren, Auk **87**:557–567.
Verner, J., and Willson, M. F., 1969, Mating systems, sexual dimorphism, and the role of male North American passerine birds in the nesting cycle, Ornithol. Monogr. **9**:1–76.
Walkinshaw, L. H., 1941, The Prothonotary Warbler, a comparison of nesting conditions in Tennessee and Michigan, Wilson Bull. **53**:3–21.
Walkinshaw, L. H., 1959, A Chipping Sparrow nest in which eight eggs were laid and seven young reared, Auk **76**:101–102.
Walters, J., and Walters, B. F., 1980, Co-operative breeding by Southern Lapwings, Vanellus chilensis, Ibis **122**:505–509.

Warriner, J., and Warriner, R., 1978, Pajaro's plovers, *Point Reyes Bird Observatory* **45**:4–5.

Watson, A., and Jenkins, D., 1964, Notes on the behaviour of the Red Grouse, *Brit. Birds* **57**:137–170.

Weeden, J. S., 1965, Territorial behavior of the Tree Sparrow, *Condor* **67**:193–209.

Weeden, R. B., and Theberge, J. B., 1972, The dynamics of a fluctuating population of Rock Ptarmigan in Alaska, *Proc. 15th Int. Ornithol. Congr.* **1970**:90–106.

Welsh, D. A., 1971, Breeding and territoriality of the Palm Warbler in a Nova Scotia bog, *Can. Field Natural.* **85**:31–37.

Welsh, D. A., 1975, Savannah Sparrow breeding and territoriality on a Nova Scotia dune beach, *Auk* **92**:235–251.

Welter, W. A., 1935, The natural history of the Long-billed Marsh Wren, *Wilson Bull.* **47**:3–34.

Wetherbee, K. B., 1933, Some complicated Bluebird family history, *Bird-Banding* **4**:114–115.

Williams, L., 1952, Breeding behavior of the Brewer Blackbird, *Condor* **54**:3–47.

Williams, L., Legg, K., and Williamson, F. S. L., 1958, Breeding of the Parula Warbler at Point Lobos, California, *Condor* **60**:345–354.

Willson, M. F., 1966a, Breeding ecology of the Yellow-headed Blackbird, *Ecol.Monogr.* **36**:51–77.

Willson, M. F., 1966b, Polygamy among Swamp Sparrows, *Auk* **83**:666.

Wilson, J., 1967, Trigamy in Lapwing, *Brit. Birds*, **60**:217.

Wittenberger, J. F., 1978, The breeding biology of an isolated Bobolink population in Oregon, *Condor* **80**:355–371.

Wittenberger, J. F., 1979, The evolution of mating systems in birds and mammals, in: *Handbook of Behavioral Neurobiology, Volume 3, Social Behavior and Communication* (P. Marler and J. Vandenbergh eds.), Plenum, New York, pp. 271–349.

Wittenberger, J. F., and Tilson, R. L., 1980, The evolution of monogamy: Hypotheses and evidence, *Annu. Rev. Ecol. Syst.* **11**:197–232.

Wolf, L. L., and Wolf, J. S., 1976, Mating system and reproductive biology of Malachite Sunbirds, *Condor* **78**:27–39.

Woolfenden, G. E., 1976, A case of bigamy in the Florida Scrub Jay, *Auk* **93**:443–450.

Yom-Tov, Y., 1980, Intraspecific nest parasitism in birds, *Biol. Rev.* **55**:93–108.

Zenone, P. G., Sims, M. E., and Erickson, C. J., 1979, Male Ring Dove behavior and the defense of genetic paternity, *Am. Natural.* **114**:615–626.

Zimmerman, J. L., 1966, Polygyny in the Dickcissel, *Auk* **83**:534–546.

CHAPTER 12

THE EVOLUTION OF DIFFERENTIAL BIRD MIGRATION

ELLEN D. KETTERSON and VAL NOLAN JR.

1. INTRODUCTION

The evolution of bird migration and the role of migration in life history have long been matters of general interest, and the volume of recent literature on these subjects (Baker, 1978; Dingle, 1980; Gauthreaux, 1978, 1979, 1982; Fretwell, 1980; Greenberg, 1980; Greenwood, 1980; Myers, 1981a; Ketterson and Nolan, 1982) reflects their continuing importance to students of avian ecology and evolutionary biology.

In the effort to understand why some birds make long migrations while others do not migrate or travel only short distances, analysis of intraspecific variation in migratory behavior seems likely to be especially fruitful (Morton, 1980). Focus on differences among individuals from a common gene pool minimizes confounding variables and offers a system more amenable to a quantitative approach. In this paper, we review hypotheses to account for differential migration, i.e., the situation in which all individuals of a population migrate but distance traveled varies according to sex and/or age. The same hypotheses can be applied to partial migration, in which some classes of a population migrate while others do not. In testing hypotheses against data from a single short-distance migrant, the Dark-eyed Junco (*Junco h. hyemalis*),

---

ELLEN D. KETTERSON and VAL NOLAN JR. • Department of Biology, Indiana University, Bloomington, Indiana 47405.

which spends its life in the north temperate zone, we acknowledge the risk that our view may be narrower than would be ideal.

In 1976, we reported latitudinal clinal variation in the winter sex ratio of Dark-eyed Juncos in the eastern United States (Ketterson and Nolan, 1976). Male juncos were found to predominate in the northern parts of the winter range, females in the southern (see Fig. 1). We considered what factors might have led to the evolution of the differential migration that produces this distribution and suggested that among the selective pressures that might have been responsible were (1) intrasexual competition for breeding resources, which might have caused members of the sex that defends territories to winter nearer the breeding ground; (2) winter climate, which might have caused members of the smaller-bodied sex to migrate farther toward the south; (3) intersexual competition for resources during the non-breeding season, which might have forced members of the subordinate sex to segregate themselves and; (4) risk of mortality in transit, which might have varied according to sex and led one sex to abbreviate its migrations. Two or more of these factors could have operated simultaneously, as we later proposed (Ketterson and Nolan, 1979), but assessment of their relative importance is complicated by the fact that in the junco the predicted effects of the first three are the same. Regions of more severe climate are closer to the breeding ground, and males, the territorial sex, are larger than females. Thus, both factors 1 and 2 predict shorter migrations by males. Factor 3 also predicts shorter male migrations: males are socially dominant to females in winter.

Because the same conditions hold true for many, if not most, temperate-zone migrant bird species, the validity of any general hypothesis designed to account for the evolution of differential bird migration on the basis of only one of these factors becomes extremely difficult to evaluate (Myers 1981a). Strong arguments have been made, nevertheless, that one or another of these four factors has been the factor of primary importance (Gauthreaux, 1978, 1982; Myers, 1981a). In marked contrast is a multifactor model proposed by Baker (1978).

In this paper, we first report unpublished findings on the winter distribution of the age classes of migratory juncos, according to sex. We then discuss and evaluate the single-factor hypotheses for differential migration and the winter distribution that results, drawing both on general considerations and on various data from juncos. Finally, having concluded that none of these hypotheses is sufficient to explain the junco's distribution, we turn to Baker's model and find that it comes closest to dealing adequately with the complexities of differential migration, but that it lacks predictive power.

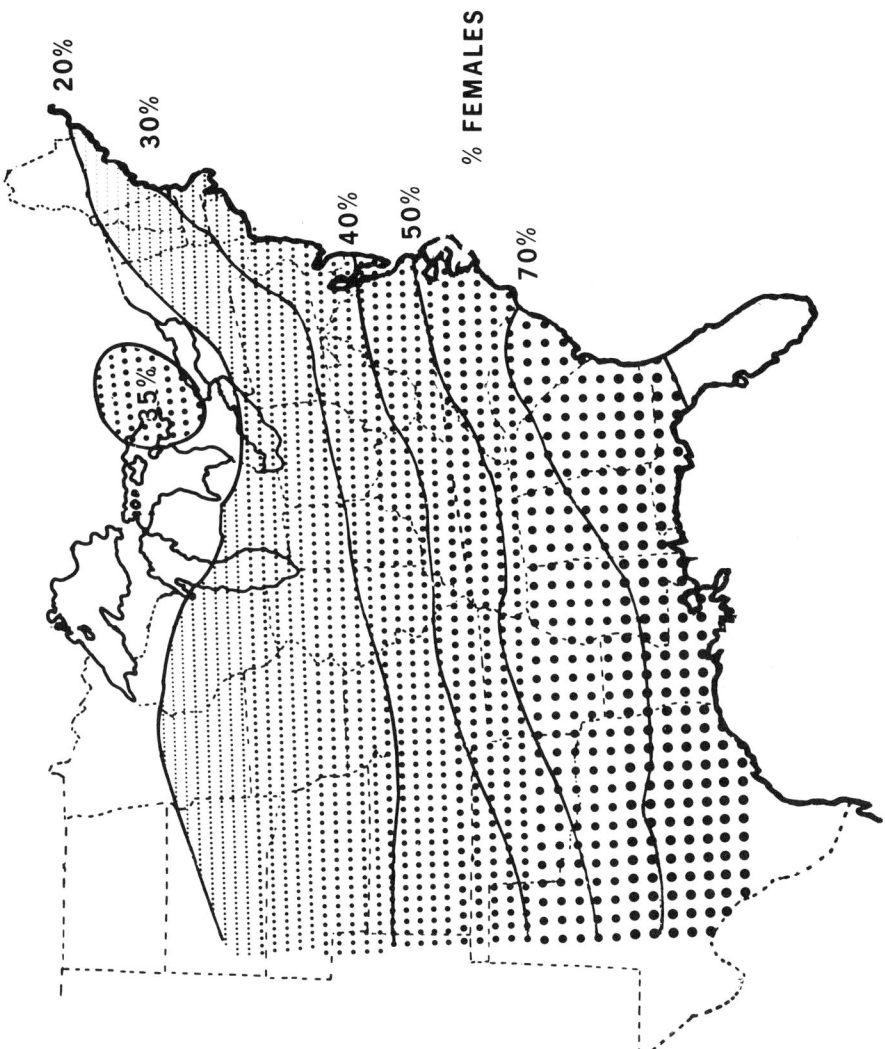

FIGURE 1. Clinal variation in the winter sex ratio of the Dark-eyed Junco (from Ketterson and Nolan, 1976, reprinted by permission from *Ecology*; copyright 1976, The Ecological Society of America).

2. WINTER DISTRIBUTION OF EASTERN MIGRATORY JUNCOS

Figure 2 shows in the upper curve the relative abundance of juncos at the various latitudes of the winter range at the end of December (see Appendix II for methods and other details about the figure). At this date, post-migratory winter populations have recently been established, and most of the severe weather of winter is still ahead. The series of lower curves indicates the relative abundance of each sex–age class at the various latitudes, again in late December. In Fig. 2 and hereafter, young juncos are those produced in the preceding breeding season and adults are all others.

We draw the following conclusions from Fig. 2: (1) The pattern of abundance at middle latitudes is trimodal. This pattern was apparent in four of six years analyzed (1974–1979), and we suspect it is real. Despite the three small peaks, however, the overall abundance from north to south is strikingly invariant. (2) Except for adult males, the distribution of each sex–age class exhibits a pronounced peak, with that for adult females farthest south, for young females at mid-range, and for young males farthest north. Adult males seem to be distributed bimodally, with a northern peak at the latitude of greatest abundance of young males and a second peak south of this. Because these patterns showed considerable stability in the years analyzed and the unimodal peaks approximately correspond to the upper trimodal pattern of overall abundance, we believe that the winter distribution of junco sex–age classes is fairly estimated by the figure. (3) Adults of each sex winter somewhat south of the young of that sex, although the difference is not as clear in males as in females. Using a Keuffel and Esser compensating polar planimeter to measure areas under the respective sex–age curves, we calculated for each class the proportion found south of 38.5°N latitude, approximately the mid-point on the north-south axis of the winter range. These percentages for adult males, young males, adult females, and young females were 49%, 44%, 80%, and 68%, respectively.

Two other points should be made about Fig. 2. First, the sex–age curves yield estimates of relative abundance of the four classes after autumn migration and before the major toll of overwinter mortality. A planimeter reveals that the areas under the four curves bear the following relations to one another: adult males to young males 1:1.20; adult females to young females 1:1.17; adult males to adult females 1:0.72; young males to young females, 1:0.71. These calculations imply that in early winter of the years investigated young birds constituted about 54% of the population and that in both age classes males constituted

DIFFERENTIAL BIRD MIGRATION

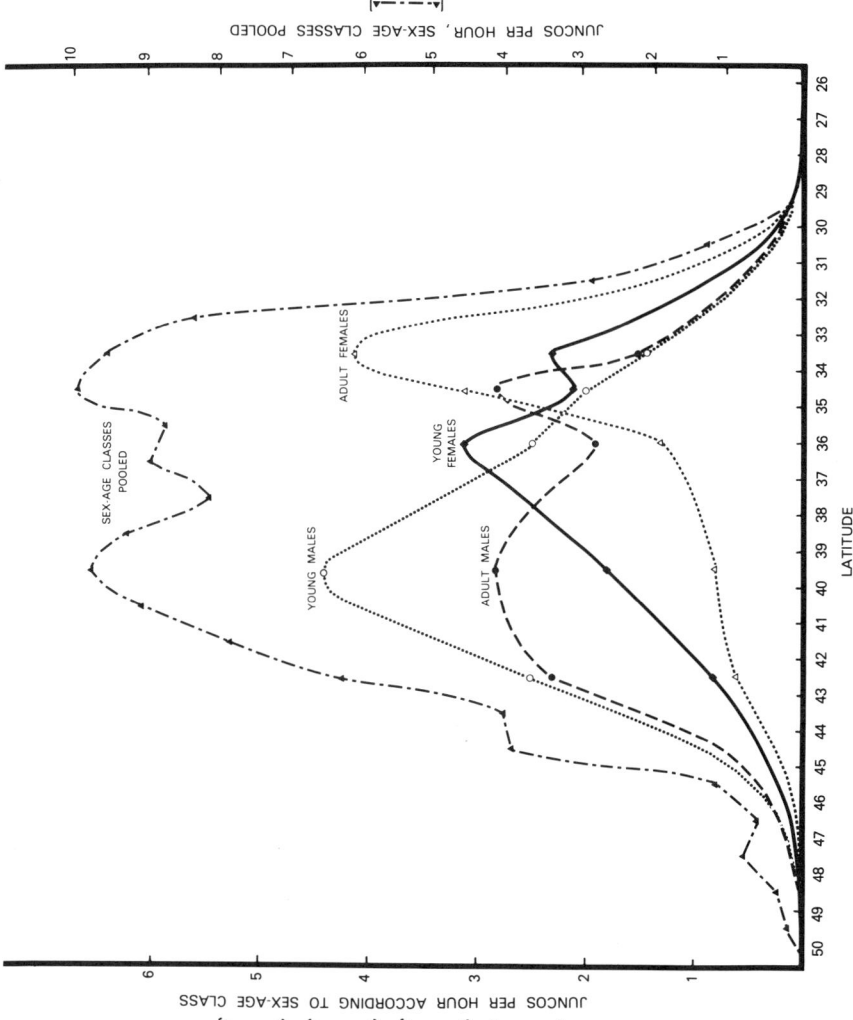

FIGURE 2. Estimated relative abundance of juncos in the winter range (top curve) and estimated relative abundance of the sex and age classes (lower series of curves), all according to degree of latitude (see Appendix II for methods).

some 59%. Second, the distributions as drawn could be maintained from year to year only if the age classes differ either in annual survivorship from December to December, or in year-to-year fidelity to the winter site, or in both. This is most easily demonstrated by the female distributions. If young and adult females at latitudes 36°N and 33.5°N had the same survival rate and survivors tended equally to return to the sites in which they had spent the previous winter, the age structure at the more northerly latitude would shift toward a higher proportion of adult females, and the age ratios at the two latitudes would soon become the same. To account for the maintenance of the age distributions in Fig. 2, we have argued elsewhere (Ketterson and Nolan, 1982) (1) that December-to-December survival of northern populations and southern populations is probably equal and that it is the same for adults and young, (2) that northern juncos show less fidelity to the winter site occupied when young than do southern juncos, and (3) that juncos that do not show winter site fidelity tend to shift southward when they are adult. Discussion of the data supporting these conclusions appears in Ketterson and Nolan (1982) and is summarized in Sections 3.2.2, 4.1.3.a, b, and c.

3. SINGLE-FACTOR HYPOTHESES FOR THE EVOLUTION OF DIFFERENTIAL MIGRATION

3.1. The Body-Size Hypothesis

3.1.1. The Body-Size Hypothesis Stated

If smaller-bodied individuals were less likely than larger-bodied conspecifics to survive winter at higher latitudes, then sex–age classes (or races) having smaller bodies might be expected to evolve toward longer migrations that would take them into milder climates (Ketterson and Nolan, 1976).

The mechanism proposed (Ketterson and Nolan, 1976) to account for this putative size-related variation in probability of overwinter survival is differential fasting endurance (Calder, 1974). On the assumption that energy stores are proportional to body mass (i.e., 1:1), bigger individuals should have greater reserves relative to their basal metabolic rate, because of the less than proportional relationship between body size and metabolism (Calder, 1974). As a result, during severe winter storms when food is temporarily unavailable, they should be able to survive for longer periods of time, drawing on their fat stores to support their energy needs. The ultimate selective factor under this hypothesis would not operate until after migration is over; therefore, the proximate mechanism causing variation in migratory distance would be either some form of endogenous control or a differential response to one or more environmental variables encountered before final choice of the wintering site.

3.1.2. The Body-Size Hypothesis Evaluated Generally

Field evidence for the hypothesis was sought by Dolbeer (1982) in a comparative study of certain icterids and of the Starling (*Sturnus vulgaris*). Dolbeer predicted that if body size has been important in the evolution of differential migration of the sexes, then species with a higher degree of sexual size dimorphism would show a greater intersexual difference in distance migrated. A comparison of the distance separating banding and recovery locations of individuals banded during the summer months and recovered during the winter months showed that the winter distributions of both the dimorphic Common Grackle (*Quisculus quiscula*) and Red-winged Blackbird (*Agelaius phoeniceus*) fulfill Dolbeer's prediction. That the sexes of the monomorphic Starling do not separate in winter also supports his prediction. However, female Brown-headed Cowbirds (*Molothrus ater*), although considerably smaller than males, migrate no farther than males.

The hypothesis would receive experimental support if northern-wintering, larger individuals were found to have greater fasting endurance than their southern-wintering conspecifics when both were held under identical conditions. Ketterson and King (1977) reported that among White-crowned Sparrows (*Zonotrichia leucophrys gambelii*), a species in which males are larger and females migrate farther, males can fast for longer periods than females. In contrast, among juncos and Tree Sparrows (*Spizella arborea*) no significant sexual difference in fasting endurance was found (Stuebe and Ketterson, 1982), although in both species the trend favored males. In none of these experiments were the fat stores of the subjects at the time food was withdrawn from them known, and the assumption that stores were proportional to body size may be questioned. In fact, we know of no demonstration among conspecific birds either that winter fat stores are proportional to lean body mass or that size-related differences in metabolic rate are other than negligible. Clearly both these points are testable; but in species whose fat stores vary in response to recent environmental conditions, an adequate test of the proportionality point will require a large sample of individuals collected at the same time.

3.1.3. The Body-Size Hypothesis Applied to the Junco

Among juncos, males exceed females in wing length and in lean—i.e., metabolizing—mass (Helms et al., 1967). Adults of each sex are slightly heavier (wet weight) and have longer wings than young (Fig. 3; Nolan and Ketterson 1983); whether lean mass varies with age class is unknown, but it seems safe to assume that lean mass of adults is at least as great as that of young. Turning to size variation within each of the four sex and age classes, wet weight and wing length are significantly correlated (Nolan and Ketterson, 1983); but here too the relationship between wing length and lean mass is not known.

On the basis of the foregoing facts, the Body-Size Hypothesis predicts that males should winter farther north than females. It does not, however, predict a distribution in which young settle north of adults of their sex (Fig. 2). Nor, if we are willing to assume that within a sex–

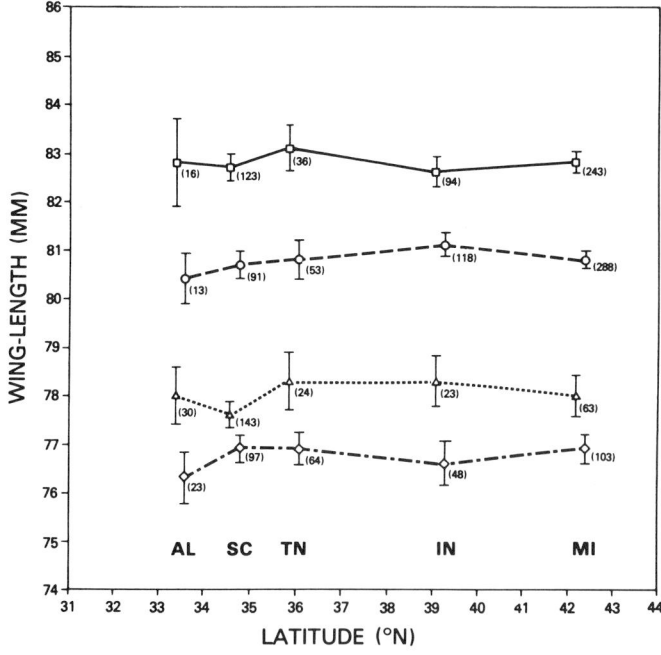

FIGURE 3. Mean wing-length of the sex and age classes of juncos at five latitudes in the winter range. Curves (top to bottom) apply to old males, young males, old females, and young females. In no class did wing length vary with latitude. (From Nolan and Ketterson, 1983, reprinted by permission from *The Wilson Bulletin*; copyright 1983, The Wilson Ornithological Society.)

age class wing length and lean mass co-vary (James, 1970), does it predict the absence of within-class latitudinal variation shown in Fig. 3.

3.2. The Dominance Hypothesis

3.2.1. The Dominance Hypothesis Stated

According to Gauthreaux (1978, 1982), a single underlying force drives all forms of intraspecific segregation in winter. These forms of segregation are habitat apportionment among sedentary populations, differential dispersal, partial migration, differential migration, and intraspecific variation in irruptive movements; and the single driving force is dominance. In the model, when competition for food or some other nonbreeding resource is intense, socially dominant individuals will be more likely to obtain an adequate supply, subordinates will then depart, and dominance-based winter segregation will result. If all individuals can survive winter within the breeding range, there will be no migration in the usual sense; subordinates will simply be found in the poorer habitats. If the breeding range can support only a portion of the population, subordinates will be those that are forced to leave. If the entire population is required to emigrate, dominants will migrate only so far as necessary to reach a suitable habitat. Subordinates will either migrate farther or, if they do not go farther, will occupy habitat of lesser quality.

In a more recent statement Gauthreaux (1982) reiterates and develops his views. He notes that a corollary of wintering on the breeding ground, or nearer to it, or (if no nearer than subordinates) in habitat that is of higher quality, is that dominants can begin to breed at an earlier date. This is true either because they do not have to migrate or, if they do, because their spring migrations are shorter or initiated sooner than those of subordinates. Gauthreaux's emphasis is not on mechanisms, but he states that dominance probably exerts its influence on migratory behavior proximately.

3.2.2. The Dominance Hypothesis Evaluated Generally

Gauthreaux's contribution has without question advanced the general understanding of the importance of social behavior in the evolution of migration. We nevertheless have three specific reservations about this hypothesis, and it is on these that we necessarily focus.

First, the Dominance Hypothesis relies to a considerable extent on the numerous species of birds in which young and/or females tend to be both subordinate and more likely to disperse, or to migrate under circumstances in which dominants remain sedentary, or to travel farther when all migrate. Recent papers have reported similar data and interpreted them as supporting Gauthreaux (Lundberg et al., 1981). Nevertheless, as pointed out in the Introduction, in most of these examples males are also larger than females and the first to initiate breeding activities, and adults are larger than young. Consequently, the examples could be cited equally well in support of the Body-Size Hypothesis or the Arrival-Time Hypothesis (see below). Further, as Gauthreaux (1982) recognizes, there are exceptions. White-crowned Sparrows (King et al., 1965, M. L. Morton, personal communication) and American Goldfinches (Carduelis tristis; A. L. A. Middleton, personal communication) apparently migrate farther when adult, and Starlings in at least some parts of their range evidently do the same (Spaans, 1977). The behavior of female Sharp-shinned Hawks (Accipiter striatus) is inconsistent with the hypothesis. They are much larger than males and almost certainly dominant over them (Smith, 1982) but tend to migrate farther (Evans and Rosenfield, in press). (For other exceptions, see Section 3.3.2.)

How might we test Gauthreaux's model? Herein lies our second reservation. The hypothesis supposes that the gradients of proximity to the breeding grounds and of habitat quality either always covary and produce an orderly geographic separation of dominance classes or, if they do not, that dominants and subordinates in the same geographic area segregate according to habitat quality. Distance is a readily measurable variable, but it is not clear that variation in habitat quality can be identified independently of the very distribution for which the model seeks to account.

Our final objection refers not to the empirical support for the hypothesis or to its falsifiability, but rather arises from its assumptions of how dominants and subordinates behave and survive when resources are limiting in winter. The Dominance Hypothesis assumes that food is the resource most likely to be limiting in winter, an assumption that is often granted for the north temperate zone (Lack, 1954; Smith et al., 1980; Jansson et al., 1981; Pulliam and Millikan, 1982) and one we do not dispute. The hypothesis then assumes that during food shortages subordinates invariably suffer more and accordingly are the first to disperse, and it is these last assumptions that we believe should be examined. There is certainly evidence to support them. An early instance is Fretwell's (1969) much-cited report, based on a small sample of banded juncos, that subordinates were less likely than dominants to be recaptured at winter's end. Kikkawa's (1980) more convincing overwinter-survival estimates of dominant and subordinate Silvereyes (*Zosterops lateralis*) agree with Fretwell's results. Similarly, Smith et al. (1980) found that subordinate Song Sparrows were more likely to be alive at winter's end if food supplies were supplemented. Finally, in an aviary experiment, Baker and Fox (1978) gradually restricted the food of captive juncos and noted the social status of individuals whose body mass fell below 17 g, which they took as a stage of emaciation equivalent to death. More subordinates fell below the critical value than dominants. [In evaluating this result it should be noted that female juncos tend to be both subordinate to males and to have a smaller lean body mass (Helms et al., 1967). Therefore, reduction of mass below 17 g is less likely to indicate imminent starvation in females than in males, and it may be questioned whether Baker and Fox's results are entirely convincing in view of the confounding effects of the sexual difference in body size.]

Opposed to the view that subordinates always suffer disproportionately and are the first to emigrate are our own data on juncos (Ketterson and Nolan, 1982). Male juncos are dominant over females and, within the sex classes, adults are dominant over young (Balph, 1977;

TABLE I

Recapture in Late Winter of Juncos Marked in Early Winter According to Sex, Age, and Location[a,b,c,d]

	Male		Female		Classes combined
	Adult	Young	Adult	Young	
Indiana	10/68 (15%)	8/78 (10%)	1/13 (8%)	3/29 (12%)	24/198 (12%)
South Carolina	15/46 (33%)	10/22 (41%)	23/43 (53%)	13/25 (52%)	60/136 (44%)

[a]After Ketterson and Nolan, 1982.
[b]Data are from the two winters 1977–1978 and 1978–1979 and are pooled. Denominators of fractions show the numbers of juncos marked and released in good condition in early winter, and numerators show numbers of marked juncos recaptured at the same sites in late winter. A few marked individuals were of unknown sex or age, causing the size of the "classes combined" fraction to exceed the totals of the pooled fractions for the sex and age classes.
[c]In Indiana, the sex and age classes were equally likely to be recaptured [$\chi^2 = 0.94$, $df = 2$, n.s. (female age classes combined because samples are small)]. The same was true of South Carolina ($\chi^2 = 4.72$, $df = 3$, n.s.).
[d]Late-winter recapture frequencies for each sex and age class were compared for the two locations. Each class was significantly more likely to be recaptured in South Carolina than in Indiana (adult males: adj. $\chi^2 = 4.15$, $df = 1$, $p < 0.05$; young males: adj. $\chi^2 = 9.36$, $df = 1$, $p < 0.01$; adult females: adj. $\chi^2 = 6.77$, $df = 1$, $p < 0.01$; young females: adj. $\chi^2 = 9.26$, $df = 1$, $p < 0.01$).

Ketterson, 1979a). Table I shows that among juncos captured and banded in Indiana in early winter and recaptured in late winter, frequencies of recapture of the sex and age classes were statistically indistinguishable. The same was true in South Carolina, although it is notable that the South Carolina overwinter recapture rate of all classes pooled was 3.7 times higher than that rate in Indiana. In an independent and rather large Indiana sample (Table II) obtained in two winters, the sex and age structure of the junco population at winter's end was the same as it had been in early winter, i.e., before unusually severe weather had periodically restricted food. Essentially, the same was true in South Carolina with its milder climate (Table II), although this sample was

TABLE II
Early- and Late-Winter Comparisons of Sex and Age Ratios of Winter Junco Populations at Three Latitudes[a,b]

	n	Male		Female	
		Adult	Young	Adult	Young
Indiana[c]					
December 1978	395	28%	42%	10%	21%
February 1979	400	24%	45%	9%	22%
December 1979	346	17%	52%	10%	21%
February 1980	320	22%	44%	13%	22%
Tennessee[d]					
December 1979	37	27%	27%	14%	32%
February 1980	84	18%	30%	17%	36%
South Carolina[e]					
December 1977	87	37%	17%	28%	18%
February 1978	170	23%	15%	34%	28%
December 1978	132	26%	17%	33%	24%
February 1979	104	34%	10%	41%	15%

[a]After Ketterson and Nolan, 1982.
[b]The Indiana site was at 39°N, the Tennessee at 36°N, and the South Carolina at 34.5°N.
[c]Indiana seasonal comparisons of sex and age ratios, December vs. February, 1978–1979 ($\chi^2 = 1.90$, $df = 3$, n.s.); December vs. February, 1979–1980 ($\chi^2 = 5.10$, $df = 3$, n.s.).
[d]Tennessee seasonal comparison of sex and age ratio, December vs. February, 1979–1980 ($\chi^2 = 1.35$, $df = 3$, n.s.).
[e]South Carolina seasonal comparisons of sex and age ratios, December vs. February, 1977–1978 ($\chi^2 = 7.96$, $df = 3$, $p < 0.05$); December vs. February, 1978–1979 ($\chi^2 = 5.95$, $df = 3$, n.s.).

not fully independent of the South Carolina recapture data in Table I. Finally, early- and late-winter sampling of a population near Nashville, Tennessee, in a single winter produced this same result. During the course of the winter the proportions of dominant and subordinate Tennessee juncos remained unchanged (Table II). Accordingly, sex, age, and dominance status appear to be unrelated to the probability of death and dispersal during winter in this species.

Can the apparent conflict between our work and that of other investigators be reconciled? The situations studied by Baker and Fox (caged flocks) and by Kikkawa and Smith et al. (nonmigratory populations on very small islands; see Tompa, 1964, for details on the population and the characteristics of the island on which Smith et al. worked) had in common the fact that dispersal or emigration was prevented or severely limited by physical barriers, whereas the juncos we studied were migratory and free to disperse both in advance of winter and during that season. We propose that in this latter situation it cannot always be assumed that subordinates will be the only class to emigrate or that they will die in greater numbers. In fact, it seems likely that there are circumstances in which middle- to high-ranking birds should benefit more from dispersal than subordinates, as the next paragraph describes.

The critical question to the individual that confronts the possibility of scarce resources is what alternatives are available to it. It is a truism, but it seems worth emphasizing, that dominance rank can be expressed only in a social context and is likely to vary with context. Individuals of equally high *expected* rank, i.e., rank resulting from the inherent traits of size, sex, and age, will match their expectation in varying degrees, depending upon the characteristics of their associates. The *realized* rank of an average male junco would almost surely be higher in a flock composed largely of females than in a group in which males predominated. Conversely, in a location in which females predominate, an individual female's realized position may greatly exceed her expectation. Further, Rohwer and Ewald (1981) have suggested that birds at the top of a dominance hierarchy may prefer association with subordinates and may attempt to drive away their social equals or near equals by directing more aggressive behavior toward them (see the similar conclusion of Ketterson, 1979b). In such a situation the more-often attacked individuals of intermediate status might suffer more than subordinates and thus be the group more likely to disperse. If both the rank an individual achieves and the impact of that rank on its access to resources vary as a function of the relative frequency of individuals with high or low expected rank among its flockmates, then the relationships of dominance status, aggressive behavior, order of dispersal, and survival are likely to be more complex than is usually recognized.

3.2.3. The Dominance Hypothesis Applied to the Junco

The relevant data have been given in the preceding subsection.

The prediction of the hypothesis is that dominant classes will settle closer than subordinates to the breeding grounds or, if no closer, in higher-quality habitat. In juncos, dominant and subordinate sex and age classes intermingle throughout the winter range, although in proportions that vary with distance from the breeding ground. Nothing suggests that subordinates are being selected against when they overwinter with dominants. On average, male juncos select winter sites nearer the breeding grounds than do females, and this could be taken as evidence that supports the hypothesis; but the distribution of age classes, with adults of each sex farther south than young, indicates that dominance rank cannot predict the winter distribution of the junco (Fig. 2).

3.3. The Arrival-Time Hypothesis

3.3.1. The Arrival-Time Hypothesis Stated

Myers (1981a) reasons that if members of one sex (or of some other class) experience more intense competition for breeding resources than do members of the other, then individuals of the more competitive sex should benefit by returning earlier to the breeding ground. The social system sets the relative levels of intrasexual competition. Thus, where one sex establishes territory, its members should arrive first in order to gain priority of access to territories. Although early return could result either from migrating a shorter distance, and therefore wintering nearer the breeding ground, or from earlier departure in spring from a common wintering ground, Myers speculates that intense intrasexual competition may often lead members of the more competitive sex to do both. Based on his review of seven species for which the necessary information was available, including a number of shorebirds with reverse size dimorphism and unusual mating systems, Myers concluded "that to predict latitudinal segregation of the sexes, information about arrival schedules is both necessary and sufficient." A comparable statement could not be made about either the Body-Size or the Dominance

Hypothesis. Other time-related selective pressures can modify the effects expected from interclass variation in intensity of intraclass competition, as Myers recognizes when he notes that breeding at high latitudes may abbreviate the breeding season so greatly that differences in competition would have no detectable influence on arrival schedules.

3.3.2. The Arrival-Time Hypothesis Evaluated Generally

Myers has made an important contribution to our understanding of the evolution of differential migration by calling attention to the importance of social systems as a selective factor. However, as before, we concentrate on what we believe to be the limitations of the hypothesis.

First, it seems unlikely that priority of arrival should always be linked to shorter migrations. Southern-wintering White-Crowned Sparrows, for example, may arrive on their breeding grounds no later than northern-wintering conspecifics; at least, they are known to initiate premigratory fattening at an earlier date (King and Mewaldt, 1981). It remains to be seen whether knowledge of arrival schedules will prove a sufficient general predictor of winter distributions.

Second, Myers does not consider the impact that the winter distribution of one class may have on the distribution of the others. He proposes that for each class some ideal location or range of locations exists, presumably determined by the greater probability of survival there between breeding seasons. For the more competitive class the effect of intraclass competition for breeding resources is superimposed upon this survivorship-based ideal and may shift the class' distribution toward the breeding range. That the distribution—or the redistribution—of one class may affect the distribution of the others through density-dependent feedback is not considered, and we suggest that a comprehensive hypothesis should speak to this complication.

Third, we think it worth questioning whether males are more competitive than females, which is the reason proposed to account for their greater proximity to and earlier arrival on the breeding ground. Do the sexes differ in intensity of competition, as many (e.g., Greenwood, 1980) have concluded, or is the critical difference between them one of the seasonal timing of their competition? It is probably true that in most migratory bird species males return first, compete for territories, and court females when they appear. [In an interesting exception that supports Myers' views, female Spotted Sandpipers (*Tringa macularia*) arrive earlier than males (Oring and Lank, 1982), set up territories, and are the active sex in courtship.] Other evidence of the importance, and perhaps greater intensity, of male–male competition comes from numerous removal experiments (Krebs, 1971; Samson, 1976; Thompson, 1977) in which the elimination of territorial males has increased the breeding opportunities of others of that sex. However, the literature also reports examples of female–female competition. The presence of unpaired females during the breeding season (von Haartman, 1971; Saether and Fonstad, 1981) and of females paired to other females (Hunt et al., 1980), as well as the occurrence of delayed breeding by juvenile females as the result of aggressive behavior by adult females (Hannon et al., 1982) have all been reported. Thus whether males usually are more competitive than females seems to us a question that may still be open. What does seem clear is that in territorial species, the sex that establishes territory engages in its intrasexual competition at an earlier date than does the sex whose role is to choose territory owners as mates. Apparently, for a full understanding of differential migration we must have a better understanding of what it is that selects for territoriality.

Finally, we note that the cause–effect relationship between arrival time and success in competition for breeding resources is likely to have exceptions. Late arrivals do not necessarily lose out in competition for territories. In both Song Sparrows (*Melospiza melodia*; Nice, 1943, p. 154) and Prairie Warblers (*Dendroica discolor*; Nolan, 1978, p. 40), returning males that find their previous year's territories preempted by earlier arrivals very rarely have difficulty in ejecting the usurpers; and this result may be widespread (but see the contrary results of Catchpole, 1972, and the interesting experimental findings of Krebs, 1982). In many species the percentage of breeding sites reoccupied by former owners is so high that the reoccupancy rate is probably also the survival rate. We doubt that all survivors arrive earlier than all individuals seeking sites for the first time in their lives, and we therefore suggest that priority of arrival is not sufficient to establish an indefeasible claim when the claim is contested by a former owner. One could respond that intraclass competition would then be greatest among inexperienced breeders, and we would agree. Our point is only that the most competitive class, even if it arrives first, does not always have its choice of resources. We regard this not as inconsistent with Myers' view but as a refinement of it.

3.3.3. The Arrival-Time Hypothesis Applied to the Junco

The hypothesis would predict that classes of juncos that arrive first in spring also migrate the shortest distances in autumn and/or depart earliest in spring, and that they do so because for them competition at the start of the breeding season is most intense.

We have no observations about the order of arrival and can find nothing in the literature. Thus we cannot test directly Myers' thesis that knowledge of arrival sequence is necessary and sufficient to predict winter distribution. We do have extensive data on the spring migration through Indiana, and these indicate that the peak of male passage precedes that of females by several days (Nolan and Ketterson, unpublished data). We therefore expect that males arrive first, a result that does not require, however, that they depart earlier. If all juncos from all wintering sites began to move northward at about the same time and rate, the earlier peak of male passage through Indiana and the assumed earlier male arrival on the breeding ground could result entirely from the differential sexual distribution in winter. Is there evidence of such simultaneous initiation of spring migration? We have considered this possibility for a single Indiana population by comparing (Table III) final capture dates of banded juncos skull-aged (in December or earlier) and known to have wintered at the study site. The median dates of last capture of adult males, young males, adult females, and young females were March 22, March 25, March 26, and March 29, respectively. Differences were nonsignificant, although young juncos did show greater variability than adults. If departure also does not differ among latitudes, then winter distribution alone would determine arrival time, and the predicted order of arrival would be young males, adult males, young

TABLE III
Latest Spring Capture Dates of Migratory Juncos Known to Have Wintered Near Bloomington, Indiana: An Approximation of Their Departure Time[a,b]

	Males		Females	
	Adult	Young	Adult	Young
n	32	93	12	34
Median date	March 22	March 25	March 26	March 29
Extremes	March 6–April 8	March 2–April 13	March 12–April 8	March 1–April 14

[a]Juncos in the sample met the following criteria: first captured on or before January 2 and aged by skull ossification; captured at least two times during December, January, or February; last captured on or after March 1. Capture efforts were made on a near-daily basis from October 1–May 1 near Bloomington, Indiana during the winters of 1973–1974, 1974–1975, 1975–1976, 1976–1977, 1977–1978, and 1978–1979 and the data were pooled across years.
[b]The classes did not differ significantly in date of last capture (Kruskal–Wallis one-way analysis of variance, $H = 2.70$, $df = 3$, n.s.). The median date regardless of class was March 25. Females were no more likely than males to be among those captured after March 25 (28 females, 56.5 males, $\chi^2 = 3.14$, $df = 1$, 2-tailed $p < 0.10$), nor were young more likely than adults (66 young, 20.5 adults, $\chi^2 = 0.19$, $df = 1$, n.s.).

females, adult females. If this prediction were borne out, Myers would be supported.

We also have no information on the relative intensity of intrasexual competition among juncos, but competition probably occurs earlier among males than females, given the territoriality of males. Further, it seems likely that at the beginning of the breeding season young males are more competitive among themselves than are males that have bred previously. We base this statement on the degree of male site fidelity we have observed to breeding locations, applying the argument made above (Section 3.3.2) about the inference to be drawn when all survivors reoccupy their former breeding territories. In field studies at Wawa, Ontario, Canada, 50% of males banded in the preceding year reoccupied their former territories, and in the only exception the former territory had been flooded. This 50% reoccupancy rate is the same as three independent estimates that put the annual survival rate of juncos at about 50% (Ketterson and Nolan, 1982; Sections 4.1.3.a and 4.1.3.b), indicating that surviving males regularly are able to take over their territories of the year before. Accordingly, competition among them should be less intense than among young males, which must contest for the habitat left unoccupied by the death of former owners. In this competition we could expect priority of occupation to confer a considerable and perhaps decisive advantage (see Balph, 1979; Yasukawa and Bick, 1983), putting young males under strong pressure to arrive early.

In contrast to males, few banded females have returned to nest on our breeding-study areas, although females are fully capable of showing site fidelity: they home to their former winter sites in the same proportions as males (Section 4.1.3.a). Because we have no reason to suspect a sexual difference in survivorship, we attribute the lower fidelity of females to the breeding site either to weaker motivation to reoccupy former sites or to lesser ability to retake former sites from competitors that arrive earlier. The first alternative would be expected if competition were slight and if reoccupation of the former site conferred little reproductive advantage derived from experience there. The second alternative might be true if there were strong intrasexual competition for sites but if prior residents had no psychological or other advantage in such contests. Whatever the level of competition among females may be, the age classes appear to be on an equal footing.

We conclude that the fact that young males winter somewhat closer to the breeding site than older males fulfills the prediction and rationale of the Arrival-Time Hypothesis. But if our argument based on the low site fidelity of females is sound, that is, if female age classes are equally competitive, then the tendency of these classes to separate in winter does not conform to expectation. Therefore, while we suspect that advantages associated with arrival time on the breeding range are important to the junco's differential winter distribution, we think that arrival time is only one of several important pressures and that a multifactor model is necessary to explain the data.

4. A MULTIFACTOR HYPOTHESIS FOR THE EVOLUTION OF DIFFERENTIAL MIGRATION

4.1. The Migration-Threshold Hypothesis

4.1.1. The Migration-Threshold Hypothesis Stated

Baker's (1978) model appears to remain unfamiliar to most, perhaps because few choose to devote the necessary time to its complex presentation. (The present paper arose out of an invitation to apply Baker's model to birds.) Two reviews of current knowledge of the evolution of migration have paid it scant attention and/or have ignored what we view as its essential points (Keast and Morton, 1980; Gauthreaux, 1982). Therefore, in Appendix I, we summarize those elements of the model, its symbols and its terminology, that we believe most interesting to avian biologists and here assume that the reader will consult this appendix, if interested. Baker uses the word "migration" to include any non-accidental change of location by any metazoan, but we confine our statement here to migration as it is usually defined for birds, i.e., to cyclic to-and-fro movements between the breeding and the non-breeding ranges. (For general reviews of Baker's book see, e.g., Krebs, 1979, and Dingle, 1979.)

According to Baker, birds migrate when their migration thresholds have been exceeded. Each individual has an inherited threshold that has been shaped by natural selection in such a way that it will be exceeded, and the individual will initiate migration, at the point at which the advantages of remaining at a site are just outweighed by the advantages of leaving it. This point is described by relating the suitability of the currently occupied habitat, h_1, to the suitability of habitats attainable by migration ($\bar{h}$, e.g., h_2, h_3, etc.), corrected for the cost of making a round trip to one of those latter habitats and back again. Habitat suitability is measured in terms of potential reproductive success (p.r.s.), and the suitability of any particular habitat is the ratio of the individual's p.r.s. at the time it departs from that habitat to what its p.r.s. was at the time it arrived. Because p.r.s. declines throughout life, h is always equal to or less than 1.0. The relative suitability of two alternative habitats is Baker's habitat quotient, h_q.

The migration factor, M_R, which is also measured in terms of potential reproductive success, expresses p.r.s. at the termination of a migration as a proportion of what it would have been at that same time had the animal not undertaken the migration. Because it is typically more costly to be moving than to remain in a familiar location, M_R is also usually a number less than 1.0.

These ideas are expressed in two equations used repeatedly by Baker. First, it becomes advantageous for an animal to migrate at the point

$$h_1 < h_2 M_R,$$

where h_1 is the suitability of the breeding or natal site, h_2 is the suitability of the wintering site, and M_R is the migration factor for the round trip. Once the migration threshold has been exceeded, selection should favor behavior that acts to maximize the quantity

$$h_2/h_1 \, M_R = h_q M_R.$$

In one of his several contradictions, Baker suggests that in order to maximize $h_q M_R$ birds will sometimes initiate segments of the migration under conditions where $h_q M_R$ is less than 1.0 *for that segment*, because by doing so they are able ultimately to reach and spend the winter in regions of very high suitability.

If classes of individuals differ in the incidence of migration, the necessary implication is that they also differ in the average value per class of h_1, h_2, M_R, or some combination of these. Further, in those classes that migrate, if distance migrated differs, the classes must vary with respect to the location of habitats where $h_q M_R$ approaches a maximum. We now consider briefly how $h_q M_R$ might differ according to sex and age, beginning with the habitat quotient.

Suitability of the winter habitat is a composite variable, the value of which for a given individual is a function of (1) the physical attributes of the habitat (h_p), (2) the individual's prior experience (if any) in the habitat, and (3) the distance of the winter habitat from the breeding site. Also important are (4) the density of the population in the habitat (h_d) and (5) the individual's resource holding power relative to the power of the other occupants of the habitat (h_{rhp}). When suitability is determined by several factors, e.g., h_p, h_d, h_{rhp}, the overall habitat suitability is the product of the separate suitabilities and thus remains a number less than 1.0. Suitability for sex and age classes will differ only if the value of the component(s) is sex- or age-dependent.

For example, h_p, the habitat's physical and biological suitability (excluding competition), is determined by its climate, absolute abundance of predators and of food, availability of roosting sites, and many similar factors. Because climate generally improves for birds that move away from the earth's poles in autumn, habitat suitability and thus the habitat quotient should improve accordingly. However, unless climatic conditions are more important to the fitness of members of one class than of another (as in the Body-Size Hypothesis), climate can play no role in explaining interclass variation in the incidence of migration. Prior experience at a non-breeding location can also influence the suitability of that location, and several authors have suggested that in a species for which the attributes of a non-breeding habitat are stable from year to year, experience gained there in earlier years should be sufficiently beneficial to select for high site fidelity. Fretwell (1980) has gone so far as to suggest that prior residency provides so great an advantage in winter (because it confers dominance at the site) that selection has caused some species or populations to become sedentary simply because to do so preserves the advantage of familiarity with the winter site. In any case, since young of migratory species can have had no prior experience in the winter range, that aspect of the non-breeding habitat quotient will necessarily vary with age. Finally, if occupation of more distant habitats delays return to breed in the spring in a way that diminishes fitness, then the habitat quotient will also be reduced. If that delay is more important to one class than to another, suitability will be correspondingly reduced for that class. (Baker is inconsistent in his treatment of distance. On p. 678, distance is treated as a component of habitat suitability, but in his definition of the migration factor he includes loss of copulations as a cost that may be associated with migration. Because in the former treatment he refers specifically to the evolution of seasonal return migration in birds, we infer that to be his view and take distance to be a component of h_q.)

The components of habitat suitability considered in the preceding paragraph are density-independent in their effects. Variation in population density also has important effects on mean habitat suitability, which will be depressed if density is high in relation to resource levels. If individuals do not differ in resource holding power (and they differ only when they have unequal access to a necessary resource that is in short supply), then each will have an equally depressing effect on the mean habitat suitability of its associates. In the case where alternate habitats of greater suitability are available, some will respond to high density by emigrating to one of these habitats. The effect of this on those that remain (and do not incur the cost of migration) will be to reduce density and improve habitat suitability, and the cost and benefits to those that emigrate and those that remain will be balanced. This situation is said to be *free* (*sensu* Fretwell, 1972), and the fitness of migrants and of non-migrants is equal. On the other hand, when individuals differ in resource holding power and those with lower power suffer from reduced access to resources, a *despotic* situation prevails. In this case, for individuals or sex and age classes of low power the habitat is less suitable than for conspecific individuals or classes of higher power. These latter will be expected not to migrate (or, assuming that the entire population has migrated and has reached the most suitable site in the winter range, they will be expected not to migrate any farther); they will settle and restrict access to resources by those with less power, which by definition will have lower reproductive success. Whether these less favored birds nevertheless remain in the habitat or migrate (or continue to migrate) depends upon whether, for them, the suitability of the currently occupied habitat, despite the presence there of individuals of greater resource holding power, is greater or less than the suitability of other available habitats, corrected for the cost of getting there. Only if it is less will they initiate migration. [Although Gauthreaux (1982) equates dominance with resource holding power, we suggest that the concept of h_{rhp} will be most useful if it is defined not strictly in terms of rank but in terms of rank-associated gain or loss of potential reproductive success. For example, if subordinates are not at a disadvantage relative to high-ranking birds so long as they co-occur in low relative frequencies, their resource holding power may be equal to that of birds of higher rank. We stress again our view (Section 3.2.2) that predicting the behavior or relative fitness of individuals of subordinate rank is no simple matter.]

We turn now to possible sex- and age-related variation in the migration factor, M_R, a variable whose value rises as the risk of mortality during migration falls. That value probably varies with age in two opposing ways. First, all other things being equal, a given migration cost (m; see Appendix I) is more likely to be assumed by young individuals than by old. The reason is that potential reproductive success decays throughout life, and therefore the p.r.s. remaining to younger individuals at any particular time is higher than that of older individuals. Thus the impact of any given value of m will be proportionately smaller in young birds than in adults. On the other hand, young animals are inexperienced, and any variation in migration cost associated with experience will clearly favor older individuals. In small birds that do not travel in organized social groups, we regard the higher probability of death confronted by first-time migrants as greatly outweighing any advantage associated with their higher potential reproductive success (see Ralph, 1971; Nolan, 1978, pp. 448–451, 472–473; Greenberg, 1980). Turning from age to sex, sex-associated differences in migratory cost per unit distance migrated have rarely been described. In small-bodied, essentially size-monomorphic birds that do not store energy for reproduction in advance of migration, no sex differences are anticipated; but waterfowl, shorebirds, hawks, and gallinaceous birds may provide interesting exceptions.

In summary, individuals tend to initiate migration when the cost is low relative to the gain in habitat suitability. To the extent that members of sex–age classes differ in assessing these variables because over evolutionary time the variables have exerted different selective pressures according to sex or age, classes will differ in the frequency with which they initiate migration and thus in the distance they travel.

4.1.2. The Migration-Threshold Hypothesis Evaluated Generally

An indication of the all-encompassing scope of Baker's model is the fact that all of the premises and predictions of the single-factor hypotheses can be comfortably accommodated within it. Thus the model's treatment of the despotic situation makes it broad enough to incorporate the Dominance Hypothesis: when sex-age classes differ in dominance rank, they may (but they need not) also differ in resource holding power. Migration distances will then be greater in those with lower power if (but only if) the alternative habitats available to them are sufficiently suitable to offset the cost of reaching those habitats. Note also, however, that if dominants do not have greater resource holding power, Baker's model would not predict dominance rank-associated differences in migratory behavior. The concept of the habitat quotient can also incorporate the Arrival-Time Hypothesis. If early arrival on the breeding ground is advantageous and is correlated with wintering nearby, the relative suitability of the more distant habitats is reduced by a measure that reflects the loss of potential reproductive success resulting from delayed return. Finally, as already noted (Section 4.1.1) the model includes a counterpart of the Body-Size Hypothesis.

While the comprehensiveness of the model makes it admirable in the abstract, in practice it may render predictions untestable. Thus, in order to apply the migration equation to differential migration by sex and age classes, numerous, detailed, species-specific data are required. Assuming these are obtainable for a particular species, predictions based on them would probably no longer be necessary for that species and would be unlikely to be generalizable to others.

At a different level of criticism, the physiological reality of the proposed critical mechanism, the migration threshold, seems debatable and largely beyond reach of investigation. When it is noted that the seasonal migration threshold is only one of a supposed large family ("hierarchy") of inherited thresholds on the basis of which Baker would account for every change of location ("migration")—a threshold for leaving the nest to forage, for leaving the foraging site to roost, for flying from roost to song post, etc.—it seems that we are dealing more with a convention for describing bird behavior than with real mechanisms.

This is not to suggest that Baker's model has no value, but to us its utility lies in providing an organizing and heuristic scheme for *a posteriori* analysis of data and in emphasizing how numerous and varied the relevant data are likely to be.

4.1.3. The Migration-Threshold Hypothesis Applied to the Junco

As will become obvious when we attempt to use Baker's ideas in relation to the winter distribution of the junco, we are not testing the model in the scientific sense. Rather, we are accepting it more or less at face value for the purpose of argument and asking whether its predictions are consistent with what we already believe to be true about the junco. These beliefs are based to a large extent on information on population dynamics, and we emphasize that the data are imperfect. Nevertheless, in spite of insufficiencies, a considerable number of independent data sets (from free-living populations at several latitudes in the winter range, from two locations in the breeding range, and from United States Fish and Wildlife Service recovery records) all converge to provide an interpretation that is at least internally consistent. One especially important gap (Myers, 1981b), however, probably cannot be filled. We do not know the breeding-range origins of the populations that we follow in winter, nor do we know the wintering sites of the individuals composing our breeding populations.

Baker's model predicts that the sex–age classes should concentrate in those regions in which, for them, $h_q M_R$ reaches its maximum. If $h_q M_R$ is at a maximum for young males at high latitudes, whereas for adult females the maximum lies at lower latitudes, these facts would support the model (see Fig. 2). Rough approximations of the relative mean values of certain components of h_q and M_R can be obtained if we initially grant that overwinter survival approximates h_2 and that the product of autumn and spring migration mortality is inversely proportional to M_R.

If we further make the simplifying assumption that on the breeding ground (h_1) habitat suitability in winter falls to some unknown but geographically invariant value, then h_q will vary as a function of h_2. We now examine data on population dynamics and interpret them in terms of h_q and M_R.

4.1.3.a. *Estimates of Junco Population Parameters, Adults.* Overwinter survival appears to be lower among male and female adults of northern-wintering junco populations than among male and female adults of southern-wintering populations (see Section 3.2.2, Ketterson and Nolan, 1982). Despite this latitudinal difference in survivorship during winter, we believe that when the full year is considered members of northern-wintering populations survive as well as southern. This conclusion is based on two observations: (1) Annual recapture rates of marked adults in the north and south in the year subsequent to capture were equal (Table IV, statistical comparison restricted to males). (2) After the first return to the north by marked individuals, returns by these same birds in subsequent years produced an estimated survival rate of 53%, a conclusion based on an independent and much larger sample (Ketterson and Nolan, 1982). That percentage is not lower than the expected annual survivorship of many temperate-wintering species

TABLE IV
Recapture in December of Juncos Caught in a Previous December According to Sex, Age, and Location[a,b,c,d]

	Male		Female		Classes combined
	Adult	Young	Adult	Young	
North	25/220	9/260	1/59	3/100	38/639
	(11%)	(3%)	(2%)	(3%)	(6%)
South	6/71	8/46	8/100	8/58	30/275
	(8%)	(17%)	(8%)	(14%)	(11%)

[a] After Ketterson and Nolan, 1982.
[b] Sites treated as "North" were in Michigan and Indiana, those regarded as "South" were in South Carolina and Alabama. Denominators of fractions are the numbers of juncos marked and released in good condition in December, and numerators are the numbers of those marked juncos recaptured at the same sites in a subsequent December. Recapture efforts were made in three Decembers in Michigan, Indiana, and South Carolina and in two Decembers in Alabama.
[c] Comparisons of return of three sex and age classes to northern and southern locations follow: adult males: adj. $\chi^2 = 0.23$, $df = 1$, n.s.; young males: adj. $\chi^2 = 11.28$, $df = 1$, $p < 0.001$; young females: adj. $\chi^2 = 4.85$, $df = 1$, $p < 0.05$. For all classes pooled, rate of return to the South was significantly higher (adj. $\chi^2 = 7.07$, $df = 1$, $p < 0.01$). Returns of adult females were too few to be compared.
[d] Frequency of return of the sex and age classes to northern sites differed ($\chi^2 = 17.87$, $df = 3$, $p < 0.001$). Returns to southern sites did not differ ($\chi^2 = 3.80$, $df = 3$, n.s.).

(Greenberg, 1980), nor is it lower than the return rate of male juncos to Canadian breeding sites (50%; Section 3.3.3). These facts make it unlikely that southern-wintering juncos have a higher annual survival rate than the 53% found for northern-wintering juncos. If, then, annual survival is equal among winter populations, juncos from the southern part of the winter range must suffer more heavily than do northern juncos in seasons other than winter; and for several reasons (Ketterson and Nolan, 1982), it is more likely that this compensating heavier mortality occurs during the longer migrations of southern winterers.

Expressing the foregoing in Baker's terms, h_q of adult juncos increases with distance traveled in migration, but because M_R decreases, the product, $h_q M_R$, does not vary with latitude of the wintering site.

4.1.3.b. Estimates of Junco Population Parameters, Young. Whether annual survivorship is independent of latitude of the wintering site in young juncos as well as in adults depends on the date selected as the start of the annual period. If survivorship is measured forward for a 12-month period beginning at the onset of winter, say from December 1, young at a site probably survive at the same rate as those adults that winter at that same site, for the reasons that follow. Based on the argument in the preceding paragraph, it appears that young survive the 12 months equally well whether they winter in the north or the south. For two reasons we believe in this geographic equality despite the capture–recapture evidence (Table IV) that young exhibit a lower rate of return to the north than do adults (and also than do southern-wintering young to the south): (1) Recapture rates in late winter of young first caught and banded in early winter at northern and southern stations do not differ from recapture rates of adults at those same locations (Section 3.2.2.). Any youth-related disadvantage in survivorship during the period December 1–December 1 would be expected to be most pronounced early in that period, when the individuals are youngest and the weather most severe. (2) United States Fish and Wildlife Service data indicate that juncos banded in northern localities are more likely than those banded in southern localities to be recovered in subsequent winters away from the banding site (Ketterson and Nolan, 1982). That is, northern juncos are less site-faithful than southern, and most individuals that change wintering sites move southward in the second or subsequent winter (Fig. 4). The sex and age of the group that moves is unknown. But because the rates of return of adults to northern and southern sites are equal, in contrast to the rates of young (see above), the non-site-faithful element among northern winterers probably consists largely of young. When these move to more southerly locations in the second winter of life, their migrations as adults are longer. The

FIGURE 4. Initial capture location and recovery location of juncos shown by United States Fish and Wildlife Service records to have been captured and recovered in different winters and at places separated by at least 30 min of latitude. Each line represents an individual. The arrow point is at the recovery location (from Ketterson and Nolan, 1982, reprinted by permission from *The Auk*; copyright 1982, the American Ornithologists' Union).

population structure in Fig. 2, showing adults to be more common at lower than at higher latitudes, is an entirely independent finding that is consistent with this conclusion.

If we tentatively accept our crude estimate of the ratio of young to adults—54:46 (Section 2)—on December 1 and look not forward from that date but backward, it seems probable that the mortality of young juncos exceeds that of adults in the months between the end of the breeding season and December 1. Clutch size in juncos is often five, and six-egg clutches are reported (Godfrey, 1979). Further, second broods are common, at least in some years in parts of the range (personal observations). If we take as a possible indicator of the junco's productivity the productivity of other temperate-zone passerines like the Red-winged Blackbird (4.2 fledglings per female) and the Song Sparrow (6.4 fledglings per female; see Greenberg, 1980; Table II), it seems that in the junco the ratio of young to adults in August could be as high as 75:25. Thus it is likely that young suffer a greater loss than adults during autumn migration, their first migration, even though they tend to travel shorter distances. If the migrations made by young were prolonged, this age differential could presumably be even greater.

If we express these points in Baker's terms, we conclude that h_q increases with distance migrated and the increase is the same as that described for adults. M_R declines with distance as it did in adults but, because the autumn death rate is higher among young, $h_q M_R$ is lower in young than in adults.

4.1.3.c. Estimates of Junco Population Parameters, Males and Females. Because females as a whole are concentrated toward the southern part of the winter range, their overwinter survival will exceed that of males as a whole. However, female survivorship during autumn and spring will be lower than male survivorship, because females make longer migrations. In early winter the population sex ratio favors males (59%; Section 2); by the following breeding season, it should approach 50:50. In terms of $h_q M_R$, this *product* does not vary within any age class according to sex, but h_q tends to be greater for females and M_R greater for males.

4.1.3.d. Predicting Sex and Age Distribution in Terms of $h_q M_R$. According to Baker's model, the latitude of its maximal $h_q M_R$ represents the ideal wintering location for each class, and the upper and lower latitudes of the region within which $h_q M_R$ is greater than 1.0 define the limits of its winter range. When our data on seasonal survival and inferences from those data are used to estimate h_q and M_R, the model predicts that young juncos should winter north of adult juncos: M_R is lower for young than for adults, whereas h_q as estimated by overwinter survival is independent of age. On the other hand, we have no data that would predict a sex bias in $h_q M_R$, and in the absence of such information the model does not predict the fact that male juncos winter north of females. In an effort to obtain such information we would, if we could, examine the relationships within each sex (1) between latitude of the wintering site and time of arrival at the breeding site and (2) between this arrival time and reproductive success. But data bearing on the first point would require that we track individual juncos between nonbreeding and breeding sites, an impossible feat in the current state of technology; thus testing of this aspect of the model cannot be completed.

Despite the conclusion just reached, it seems useful to continue in our original objective of applying Baker's hypothesis to a bird species, making reasonable assumptions where data are lacking. Figure 5A–E does this, graphing what we consider to be the critical elements of $h_q M_R$ separately; Figure 5F then combines these elements to show for each sex–age class a north–south range of values of $h_q M_R$ that would produce a distribution like that presented in Fig. 2.

To explain Fig. 5: Our data indicate that overwinter survival for all sex and age classes improves with distance migrated, and we suspect this is attributable to the north–south winter climatic gradient. For each class then, h_p should increase toward the south, probably reaching an asymptote (Fig. 5A). We expect this asymptote because (1) variation in snow cover is probably the prime determinant of h_p, and snowfall becomes infrequent well to the north of the southern limits of the winter range; and (2) prolongation of southward migration could contribute to delayed return to the breeding ground and loss of time for breeding, even in the absence of competition for breeding resources. Assuming that to some extent delayed return lowers the reproductive success of males more than of females, h_{at} (habitat suitability as a function of arrival time), and thus h, at any latitude should be lower for males (Fig. 5B, C; for the sake of simplicity, sexes not subdivided into age classes). The southward increase of h_q should therefore differ according to sex; and for males h_q might ultimately begin to decrease (Fig. 5D), if in that sex competition for breeding resources is more intense (or earlier) and the outcome of the competition more dependent on time of arrival on the breeding ground. As distance migrated southward increases M_R should decrease, with good reason to believe that the decrease is greater for young birds because of their higher probability of death during their first migration (Fig. 5E). Combining these considerations, Fig. 5F locates hypothetical maxima of $h_q M_R$ of the four sex–age classes on the north-south axis of the winter range. As was intended, their relative positions correspond to those derived from field data and shown in Fig. 2. Thus, if the necessary information could be obtained, Baker's model is capable of predicting the distribution reported herein. But its very flexibility in allowing new and alternate terms to be inserted at will as components of h and therefore h_q (as we inserted h_{at} above) may render its predictions uninteresting.

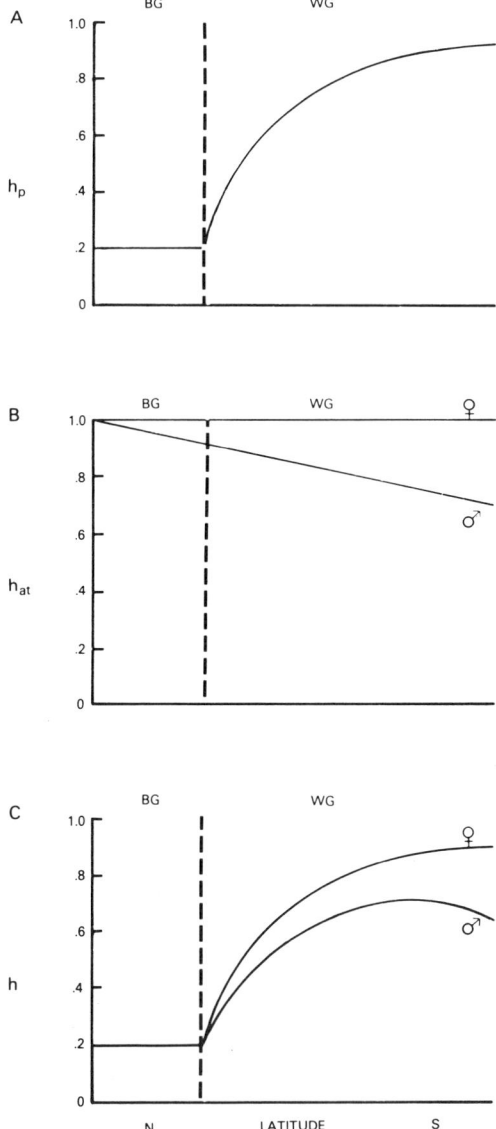

FIGURE 5. Habitat suitability and/or the migration factor in relation to latitude. BG signifies breeding ground and WG signifies wintering ground. (A) Physical habitat suitability (h_p) in relation to latitude: h_p improves north to south. The rate of improvement with latitude is presumed greater at high than at low latitudes because of snow and non-class-specific impact of added distance on time available for breeding. (B) (Spring) arrival-time component of habitat suitability (h_{at}) which estimates the impact of delayed arrival on intraclass competition for breeding resources. Because male juncos are believed to arrive sooner than females and to benefit more from early arrival, added distance has greater impact on h_{at} for males. (C) Sex-specific habitat suitability, the product of 5A and

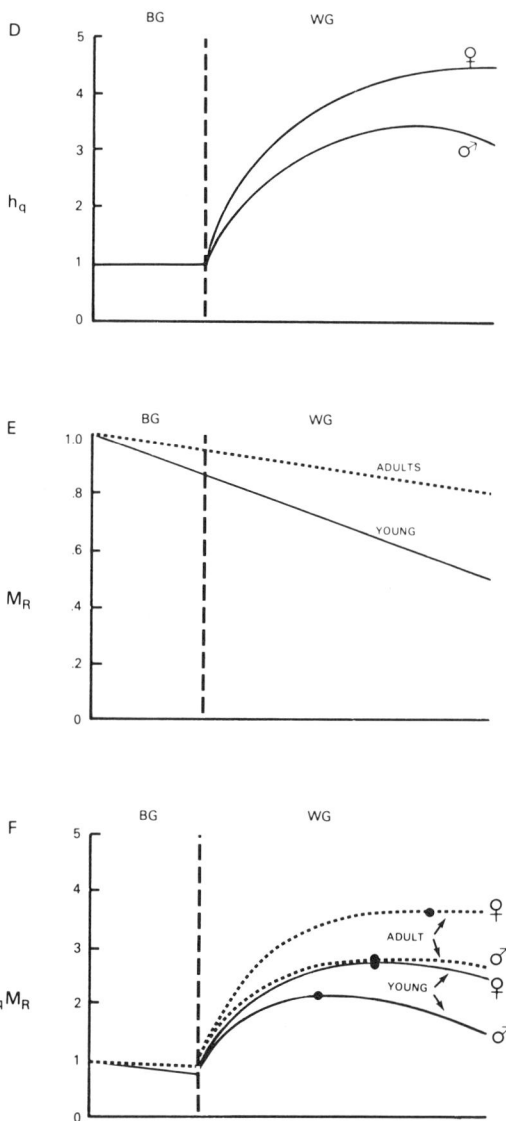

5B. (D) Habitat quotient (h_q), the ratio of winter habitat suitability to breeding (or natal) habitat suitability, in relation to latitude. We assumed that the suitability of the breeding ground in winter is (1) invariant with latitude and sex, and (2) is some non-zero number, arbitrarily 0.2. (E) Migration factor (M_R) in relation to latitude. We assumed that risk of mortality in transit is distance-dependent and greater per unit distance in young during their first migration than in adults. (F) Product of habitat quotient and migration factor ($h_q M_R$) in relation to latitude: migration is advantageous only if $h_q M_R > 1.0$ and most advantageous where $h_q M_R$ is a maximum. The points in the curves indicate the respective latitudes where $h_q M_R$ achieves its maximum for each sex–age class.

5. CONCLUSIONS

In closing, we ask whether this review brings us closer to understanding how or why differential migration has evolved in the junco or in any other bird species? Based on the data now available, no single-factor hypothesis predicts the winter distribution we have described here. As we see it, each sex–age class of the junco tends to settle where for it an optimal balance of several selective pressures—migration mortality, overwinter survival, and reproductive success as a function of time of return to the breeding ground—may be achieved. Males presumably set a higher premium on early return than do females. The behavior of young birds is probably shaped more strongly by advantages of minimizing risk of death en route and perhaps of early arrival to breed than is the behavior of adults. Adults may also tend to avoid regions where young are most abundant, because for adults the risks inherent in prolonged migration are balanced by the increased probability of overwinter survival. These views are summarized in Fig. 6.

Attempting now to generalize to other species, we tend to agree with Myers' (1981a) views as stated in Section 3.3 and to reach the following much-qualified restatement of his conclusion: Where priority in time of arrival on the breeding range permits control of limiting resources there and a consequent gain in productivity, and where members of one class have more to gain by early arrival than do members of another, then the class with more to gain would be expected to evolve a migration schedule and/or a nonbreeding distribution that promotes priority of arrival, provided the gain is great enough to counteract any costs associated with that schedule or distribution.

Whether there are interclass differences in potential gain in productivity as the result of early arrival, and also the magnitude of such gain, will be strongly affected (1) by the species' mating system, (2) by the degree of spatial and temporal variability or stability of breeding-season resources, insofar as these are independent of mating system, and (3) by the duration of the period available for breeding. Polygynous males and polyandrous females that defend territories should have more to gain than their prospective mates (Myers, 1981a), as should the sex (usually male) that acquires and defends the nest site when sites are limiting, as in cavity nesting species (von Haartman, 1968; Lundberg, 1979). In the many monogamous species in which males are territorial but nest sites are not limiting, males with previous experience in breeding should gain less by arriving early than should first breeders, provided the breeding habitat remains suitable from year to year. Given temporal stability, experienced breeders typically show site fidelity

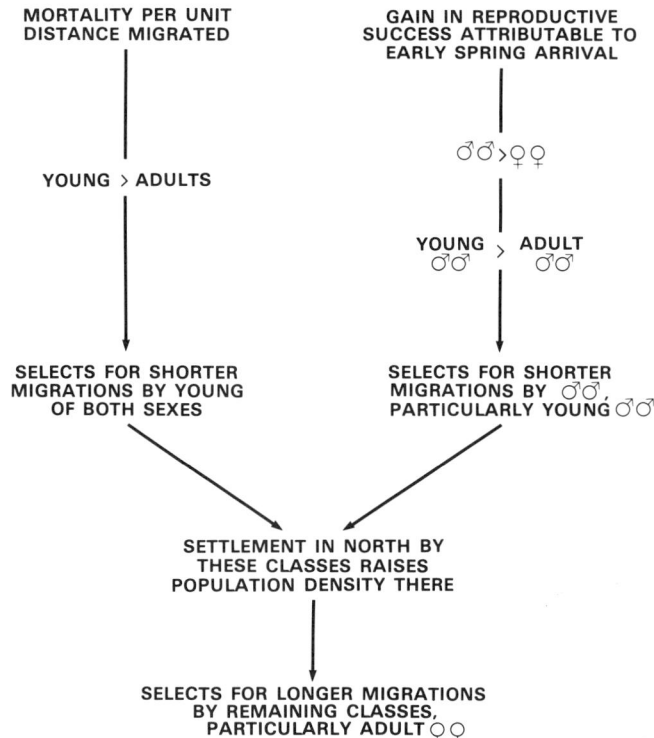

FIGURE 6. Selective factors proposed to account for the differential winter distribution of juncos.

(Greenwood, 1980) and apparently are often able to reclaim their former territories even if they are not the first to arrive on them (Section 3.3.2). If under these circumstances, breeding habitat is also homogeneous over large areas, experienced breeders would profit little from moving to new locations. They would also lose whatever benefits are associated with familiarity with the territory, its boundaries, and its neighbors of the previous season that have survived and returned (see Nolan, 1978; p. 41–42); site fidelity should then be even stronger and the gain from arriving early even less. Finally, when time for breeding is short, any interclass differences in potential gain from early arrival may be swamped by the uniformly shared necessity to arrive as early as conditions permit or risk losing any opportunity to reproduce. Time may be short in an absolute sense, as at high latitudes (Myers 1981a; Greenberg, 1980), or it may in effect be constraining because high rates of nest predation put a premium on the ability to make repeated renesting attempts (Nolan, 1978; pp. 92–93).

Longer migrations by younger birds, rather than the reverse, have been most often reported in the literature (Gauthreaux, 1978, 1982). We suggest that in fully migratory species shorter migrations by young may be more common than is realized and that information to date may be biased by emphasis on species that are partial migrants or species whose young delay maturation. Partial migration may be one attribute of species in which breeding resources are in such short supply that they must be defended on a year-round basis, thus causing experienced breeders to remain on or near the resources that they possess (again, cavity nesters come to mind as an example). In these cases, young birds, not yet the possessors of breeding resources, may have more to gain by migrating and thereby raising their probability of survival. In extreme cases the expected reproductive success of young may be so low that postponement of breeding has been selected for and migratory behavior so modified that younger individuals do not return to the breeding range until they attain reproductive age.

Assuming that the migratory behavior of members of one class has been shaped by the greater significance that arrival time on the breeding range holds for their productivity, how might the winter distribution of that class interact with the distribution of the other classes? The answer depends in part on whether the gain in productivity associated with early arrival causes the class with more to gain to winter where survivorship is less than maximal. If it does, the areas where winter survivorship is highest will be open to settlement by members of the other classes, which would be expected to occupy them. A sex or age bias in distribution would result. If, however, the gain in productivity by members of the class most influenced by priority of arrival is not sufficient to cause its members to sacrifice overwinter survivorship (if they as well as members of the other classes attempt to settle where survival is maximal) and if habitat suitability at those most favorable locations is density-dependent, then the distributions of the other classes may or may not be affected. That is, if more distant habitat is available and is of sufficient suitability to offset the associated cost of migrating there, differential migration should result and the class with least to gain by early arrival should migrate farther with no resulting loss of fitness. If equally suitable habitat is not available, then the magnitude of the inequality of habitats and the relative resource holding power of the classes will determine whether there are differences in migratory behavior and what any differences will be.

In the junco we envision a free distribution (Fretwell, 1972), resulting from the availability of extensive suitable winter habitat that can be reached at bearable cost; but the junco's solution is obviously only one possibility. In some species, migration cost will be too high or the geographic extent of the winter range too narrow to permit one or more classes to prolong migration and settle where their potential reproductive success is equal to that of the other classes. We conclude that predicting the factors accounting for a species' differential migration and estimating the relative values of these factors for the population classes involved will never be easy and may prove impossible. But even *a posteriori* explanations may serve a useful purpose in attempts to understand the evolution of migration.

ACKNOWLEDGMENTS. We thank the many people who have helped make our capture efforts possible, particularly the following: Ray Adams, Pat Adams, Thomas S. H. Baxter, Michael L. Bierly, Sidney Gauthreaux, Jr., Paul Hamel, Carl Helms, James V. Peavy, Jr., Anna Ross, Ruth Schatz, and Paul Schatz. Members of our own group who accompanied us and whose field work was indispensable were Sue Braatz, L. Jane Clay, Jeff Hill, Mike Kowalski, Mary Latham-Weeks, Margaret Londergan, Tim Londergan, Dorothy Mammon, Catherine Meyer, Cindy B. Patterson, Toy S. Poole, Ann W. Richmond, Richard Rowlands, Michael Shelton, Robert Steel, and Ken Yasukawa. John M. Emlen made several very helpful suggestions about an earlier version of this paper, as did J. P. Myers, Robert Prŷs-Jones, Sievert Rohwer, and R. R. Baker. Bowling Green State University, Indiana University, Clemson University, the Kalamazoo Nature Center, and the Banding Laboratory of the U.S. Fish and Wildlife Service supported us in important ways. This research was funded in part by NSF DEB-78-11982 and DEB-81-10457.

APPENDIX I

Abstract of Baker's View of Avian Migration

Baker (1978) defines migration to include any non-accidental movement by a metazoan from one spatial unit to another. He assumes that migrations have evolved because they are, or historically have been, adaptive at the level of individual selection. (As he puts it, chickens cross the road because they find conditions sufficiently better on the other side to have made worthwhile the risk involved in getting there.) Although we do not agree with all of his ideas, we summarize them here because we think they will stimulate considerable discussion and are not readily accessible in their original form.

The individual begins life with a certain maximum *potential reproductive success*, S_{max}, that is measured in terms of number of offspring that individual and its descendents are capable of producing before some arbitrarily set date, assuming all of them, ancestor and progeny, live in a perfect environment. Because the environment is not perfect, potential reproductive success (S) declines throughout life, even before the individual in question reaches reproductive age, but at any particular time has a specific value, S_p. Spending time in any spatial unit, a *habitat* (H), will consequently result in some "measurable" loss of potential reproductive success. S_p will be lower at departure than arrival, and this loss is a measure of the habitat's suitability and is some number less than 1.0. Focusing on the reproductive potential that remains and that is capable of being affected by the individual itself, i.e., its *action-dependent potential reproductive success*, S_d, rather than its *action-independent reproductive success* (which will be realized through the success of any offspring that are already independent), it is possible to determine relative *habitat suitabilities* (h) of two habitats. This requires comparison of the relative diminution in S_d that would be expected as the result of residence in one habitat or the other.

Presented with two habitats, H_1 and H_2, with respective expected habitat suitabilities of h_1 and h_2, under what conditions would a bird gain a selective advantage by migrating from H_1 to H_2? The answer depends on how superior H_2 is to H_1, as measured by the *habitat quotient* (h_2/h_1), and on the cost of getting from H_1 to H_2. Obviously, in order for the migration to prove advantageous, h_2 must be greater than h_1, but how much greater? From the time an animal departs from H_1 and arrives in H_2, or, if it does not depart, from the time it would have departed and would have arrived, its S_d will undergo the usual process of diminution. Typically, this diminution will be greater during time spent in migration than it would be if the animal spends that same time period in H_1. This *migration cost* (m) of moving is written as (S_1-S_2), where S_2 is the potential reproductive success remaining at the moment of arrival at H_2, and S_1 is the potential reproductive success it would have had at that same moment if it had not migrated from H_1. The *migration factor*, M, refers not to the absolute difference between S_2 and S_1, but to S_2/S_1, the individual's (action-dependent) potential reproductive success, again at the time it arrives in H_2, relative to what it would have been at that time in H_1 if it had not migrated. [An equivalent expression for M is $1 - (m/S_1)$]. Note that M varies from 0–1 and large values of M are associated with small values of m. After arriving (or not arriving because it remained in H_1), the animal's potential reproductive success will continue to decay at one rate or the other until some specified time, the difference in rate of decay depending upon the habitat quotient, i.e., the relative suitabilities of H_1 and H_2.

The *migration equation* states that if h_1 is less than h_2M, an animal should initiate migration. In Baker's words, "migration is an advantage when the realization of potential reproductive success on the way to and in a spatial unit to which an animal migrates is greater than the realization of potential reproductive success that would have been achieved during the same period if the animal had remained in the spatial unit vacated" (Baker, 1978, p. 37).

Migrations are *calculated* or *non-calculated*, depending upon whether the migrant has information about its destination (H_f) and the habitat suitability there (h_f). Information is acquired by prior experience, current sensory contact, or social communication. For such a migration the equation can be written $h_1:h_fM_f$. Calculated migrations are more likely than non-calculated migrations to prove advantageous, and therefore they are selected for. The migrant making a non-calculated migration can expect only to settle in a destination with average habitat suitability $(\bar{h})$ after a migration of average cost $(\bar{m})$. Thus, its mean expectation of migration $(\bar{E})$ is $\bar{h}M$, and the migration equation for non-calculated migration compares the ratio $h_1:\bar{E}$. When birds show fidelity to a seasonal home range occupied in a previous year, their migrations obviously are calculated and presumably have been selected for by the advantages of familiarity with the site to which the migration is made, i.e., its food sources, refuges, in many cases its conspecific occupants, and sometimes by the advantages of having made structures there in previous years. Depletion or exhaustion of non-renewable resources at a previously occupied site would, of course, select against return to it.

The most important mechanism by which selection is expected to minimize the ratio of non-calculated to calculated migrations is *exploratory migration*. All birds have a *familiar area* within which they are able to move from any point to any other point. An exploratory migrant is one that leaves its familiar area but retains the ability to return there, i.e., to the starting point. Thus, the act of exploring enlarges the familiar area and presents the explorer with a series of new habitats in which to settle. If one of these proves superior to H_1 and the explorer does settle, the exploratory migration is considered to have become a calculated migration.

In birds making *seasonal return migrations* in which individuals depart and return to a breeding (or wintering) area, the return trip, at least, is often calculated. The migration equation for a return migration must take into account the migration factor for both the to and fro components of the journey. If M has the same value on both legs of the journey, then the animal should initiate the first leg of a return migration when $h_1 < h_2 M^2$ (or more generally $h_2 M_R$). This concept is elusive, and we present an example. A migrant departing from its breeding grounds in Canada and arriving in the northern United States has the choice of settling (in H_1) or continuing southward (to H_2). Climate will improve if it continues, but because the bird must return to Canada to breed it must compare not only the suitability of its present location (h_1) with that of the one farther south (h_2); it must also account for the relative loss of potential reproductive success in transit both during movement from H_1 to H_2 and from H_2 back to H_1 in the spring. It must compare that loss in both autumn and spring migrations to the loss that would be experienced during the same time periods had it remained in H_1.

To relate seasonal return migrations to the familiar area concept, Baker believes that the individuals of many, probably most, bird species migrate over a familiar area that was thoroughly explored during the first migration of life. Each stopping place along the way is also a familiar area.

How are h_1 and h_2 and M evaluated? In every "decision" to migrate or not, at least two of three different classes of environmental variables are monitored in H_1. *Habitat variables* are those whose fluctuating values are correlated directly with the probability of surviving and of reproducing in H_1. Food availability and nest-site abundance are examples. *Indirect variables* are those that change predictably on some cyclical basis but whose changes have no immediate and direct impact on chances of survival and reproducing. Photoperiodic change is a common indirect variable. Birds whose habitat suitability varies unpredictably (is not correlated with fluctuations in indirect variables) must of necessity actively monitor habitat variables and initiate migration when the value of these falls below the threshold described by the migration equation. Such birds are *facultative* migrants (roughly, what others have called weather or irruptive migrants). But such monitoring requires energy and takes time, and the very fact that habitat suitability is deteriorating means that attainment of the physiological state necessary for migration may be difficult at the time h_1 falls below $h_2 M$. Accordingly, when the variation in habitat suitability is predictably correlated with an indirect variable, selection favors incorporation of a response to that variable. Migrants that respond to indirect variables are *obligatory*. A field observer noting the departure of obligatory migrants (roughly, what others have called instinct migrants) when an indirect variable reaches a certain level would probably not consider that there had been any deterioration in habitat suitability.

A third category of environmental variables must be evaluated by both obligatory and facultative migrants. This group consists of *migration-cost variables*, those factors that impinge on the bird while migrating and affect the probability that it will reach its destination at the optimal time and in optimal condition. Wind direction and wind speed are examples.

A key concept is the *migration threshold*, which is proposed to be the physiological mechanism that suppresses or leads to the initiation of natural selection, the ability, albeit imperfect, to assess h_1 and $\overline{h_f M_f}$ (or $\overline{hM}$) through (1) perception of habitat variables, migration cost variables, or indirect variables, and (2) comparison of the perceived levels of those variables with its own internal state. This state can be expected to vary, particularly according to age and season. The migration threshold is then that inherited value, v_t, of habitat, migration cost, or indirect variable (or composite of all three) that is perceived by an animal having a particular internal state and above which the animal initiates migration (Baker, 1978, p. 346). Translating, birds may initiate migration as the result of wholly endogenous events, as the result of endogenous cycles entrained by predictive external variables such as daylength, or wholly as the result of events in their immediate environment, such as diminishing food supply, a severe winter storm, favorable winds, or changing photoperiod.

Baker then refines the model to account for the differences in migratory behavior found among individuals of the same sex and age and also among classes of individuals that differ in sex and age. Two factors are said to account for within-sex-age-class variation: individual differences in (1) experience, or (2) migration threshold. As an example of the first, consider two adult females having identical inherited thresholds and with the same migration "decision" to resolve. Suppose the first female were still in molt because molt had been arrested while she had been successful in raising a second brood. She might initiate migration later than the the other female, which had produced no fledglings and had molted early. On the date the unsuccessful female departed, the molting female's migration cost would still be high, M low; for her, $h_1 > h_2 M$.

Intraclass variation in the inherited migration threshold will exist to the extent that, all other things (including resource holding power; see below) being equal, h_1 is a function of population density. In a *free distribution* (Fretwell, 1972) of the sort described, selection should favor the evolution of a within-class frequency distribution of migration thresholds wherein, for any particular population density, some individuals perceive h_1 to be less than h_2M, while others do not. Departure of members of the former group reduces population density, thus increasing h_1 for those that remain. The migrants, in turn, benefit from having responded to an $\overline{E}$ that is relatively high because of the presence of suitable but underoccupied habitats, and both non-migrants and migrants have equal potential reproductive success. Each time h_1 deteriorates, a new exodus occurs by the fraction of the population whose migration threshold is exceeded. Selection should stabilize at the frequency distribution of thresholds at which potential reproductive success is the same for all members of the population, whatever the setting of their individual thresholds. When these conditions hold, the *migratory fraction* of the population, f_m, is determined by the current size of the population, N_m, and that population size, N_e, which corresponds to $h_1 = \overline{E}$, i.e., the habitat suitability for those that remain is equal to the mean expectation of migration for those that depart; $f_m = 1-(N_e/N_m)$. The greater the excess of N_m over N_e, the higher the proportion of individuals that migrate.

Predicting the incidence of migration becomes more complicated if some individuals are better able to defend resources than others and if in the entire range of the population resources are limiting. In that case, the *despotic* situation, potential reproductive success of individuals with high resource holding power will be greater than that of low-power individuals. Higher power individuals will settle (or remain) in areas of greatest habitat suitability (H_1). Individuals of lower power will then either find $h_1 < h_fM_f$ (or $\overline{E}$) or $h_1 > h_fM_f$. In the former case they will migrate because their potential reproductive success will be greater than if they stay behind, although it will not be as high as that of the high-power non-migrants in H_1. On the other hand, if $h_1 > h_fM_f$ (or $\overline{E}$), they will not migrate despite the greater resource holding power of their co-occupants in H_1 and their own resultant lower potential reproductive success. Thus, in order to predict whether an individual will initiate migration one must know, in addition to everything else, whether the situation is free or despotic, and if it is despotic, what that individual's relative resource holding power is.

Turning now to sexual and ontogenetic variation in the incidence of migration (and to the subject of this paper), Baker defines for each sex and/or age class an *initiation factor*. Recall from the basic migration equation that an animal should migrate when h_1 falls below h_fM_f if the migration is calculated, or below $\overline{hM}$ if it is non-calculated. Extrapolating to a group of potential migrants of a particular class occupying a habitat, H_1, the proportion that experiences the condition $h_1 < h_f M_f$ (or $\overline{hM}$) depends on the average perception within the group of the suitability of the habitat, $\overline{h}_1$, in relation to both the suitability of other available habitats and also to the migration factor, i.e., in relation to $\overline{h_f M_f}$ (or $\overline{hM}$). The initiation factor, i', is defined as $\overline{h_f M_f}/\overline{h}_1$; and the incidence of migration, I, which may be measured as the percentage of migrants (or the proportion initiating migration at a particular time), should be some positive function of i'. Baker rationalizes as follows: As long as members of a group remain in H_1, we may presume that for them $\overline{h}_1 > \overline{h_f M_f}$ (or $\overline{hM}$). As suitability of H_1 deteriorates (or migration risk decreases, or alternate habitats become more suitable) $\overline{h}_1$ will approach $\overline{h_f M_f}$ (or $\overline{hM}$) in value, and i' will approach unity (from below unity). As i' increases, so does I, and the biologist observes an increase in the incidence of migration. For reasons we shall not explore, Baker chooses to rewrite his expression as follows:

$$i = \overline{h}_q (1 - \overline{m}/\overline{S}_d),$$

or, in the case of seasonal return migration, as $i = h_q M_R$. From these formulations it follows that the incidence of migration should be higher in groups that (1) experience a greater habitat quotient, (2) a lower cost of migration, or (3) will have remaining to them at the conclusion of migration a higher portion of their potential reproductive success.

Obviously, the value of $\overline{m}$ that results in $\overline{h}_1 < \overline{h_f M_f}$ (or $\overline{hM}$) depends on the simultaneous values of $\overline{S}_d$ and $\overline{h}_q$, and this is true for each of the other variables. Just as obviously, the values of $\overline{m}$, $\overline{h}_q$, and $\overline{S}_d$ may be expected to vary with sex and age, e.g., $\overline{S}_d$ will typically be higher in young animals. Consequently, the values of the habitat variables, migration-cost variables, or indirect variables that combine to set the value of v_t are expected to vary in a corresponding manner, and the incidence of migration of the sex–age classes will vary under any particular set of environmental conditions.

Once the migration threshold is exceeded, Baker contends that selection should act to maximize for each individual the quantity $h_q M_R$ (Baker, 1978, p. 678). That is, individuals should seek the maximum gain in habitat suitability compared with the minimal cost of a round trip migration, and members of sex–age classes should concentrate in regions in which for them $h_q M_R$ is greatest. Baker grants that maximization of $h_q M_R$ may entail the initiation of segments of a migratory journey at values of $h_1 > h_2 M$ if rapid travel serves to increase the ultimate value of $h_q M_R$.

APPENDIX II

Methods Relative to Data Presented in Fig. 2 and Table V

As described in detail elsewhere (Ketterson and Nolan, 1982; Nolan and Ketterson, 1983) we conducted early-winter (December 1–January 10) capture and banding operations during 2–4 years at five sites (Table V). The efforts, each several days long, were at Kalamazoo, Michigan (42°N), Bloomington, Indiana (39°N), Nashville, Tennessee (36°N), Clemson, South Carolina (34.5°N), and Birmingham, Alabama (33.5°N). Similar operations were also conducted in late winter (February 18–March 3) at Bloomington and Clemson in 2 years and at Nashville and Birmingham in 1 year. Birds were sexed and aged as described previously (Ketterson and Nolan 1976, 1982).

To obtain a single percentage for each of the four sex–age classes in the winter population at the five sites mentioned, we first calculated annual percentages of each class (Table V). For Tennessee, this process involved averaging the two sets of numbers from early and late winter, 1979–1980. We then obtained the mean of the annual percentages in order to produce Fig. 2. Significant annual variation existed only in Michigan and Indiana and age ratios were considerably more variable than sex ratios.

In the other process that gave rise to Fig. 2, we analyzed Christmas Bird Counts published by the National Audubon Society for counts made between 70°W and 100°W in six consecutive winters beginning 1974–1975. So-called censuses at about 750 sites per winter yielded the number of juncos seen at each site in the period we regard as early winter. We divided each count by the total party hours devoted to that count (juncos/party hour), adjusting the number of party hours to take account of the fact that in most counts some stated percentage of time was spent in habitats not occupied by juncos, e.g., pelagic habitats. We then grouped counts according to degree of latitude and determined the mean adjusted number of juncos per party hour for each latitudinal group of counts, considering the mean per degree to be a measure of relative junco abundance at that latitude in early winter of the year analyzed. We next calculated for each latitude the mean of annual means for the years for which we had data on sex–age structure, i.e., 1976–1977, 1977–1978, 1978–1979, 1979–1980 and plotted these (Fig. 2). Finally, we multiplied the relevant latitudinal mean of means by the percentages of the four sex–age classes represented in the samples captured in Michigan, Indiana, Tennessee, South Carolina, and Alabama (Table V). The products gave the points on the lower curves seen in Fig. 2, (curves drawn by hand).

TABLE V
Sex–Age Ratios According To Year and Location[a]

	n	Male Adult	Male Young	Female Adult	Female Young
Michigan (42°N)					
1976–1977	119	42%	34%	15%	8%
1977–1978	105	44%	39%	7%	10%
1978–1979	240	29%	46%	6%	19%
1979–1980	114	31%	41%	11%	17%
		37%	40%	10%	14%
Indiana (39°N)					
1976–1977	33	33%	39%	9%	18%
1977–1978	80	35%	46%	5%	14%
1978–1979	395	28%	42%	10%	21%
1979–1980	346	17%	52%	10%	21%
		28%	45%	8%	18%
Tennessee (36°N)					
1978–1979	62	21%	29%	15%	35%
1979–1980	121	22%	28%	15%	34%
		22%	29%	15%	34%
South Carolina (34.5°N)					
1976–1977	123	25%	19%	37%	19%
1977–1978	87	37%	17%	27%	18%
1978–1979	132	27%	17%	33%	24%
1979–1980	113	22%	28%	26%	23%
		28%	20%	31%	21%
Alabama (33.5°N)					
1976–1977	59	22%	14%	36%	29%
1977–1978	23	13%	22%	39%	26%
1979–1980	37	11%	19%	54%	16%
		15%	18%	43%	24%

[a] Locations were sampled in early winter except that the Tennessee sample from 1979–1980 was sampled in December and February and the data combined, and the Alabama 1979–1980 sample was taken in February 1980. Except for Michigan and Indiana, there were no significant annual differences in sex and age structure at a locality (Michigan: $\chi^2 = 25.85$, $df = 9$, $p < 0.01$; Indiana: $\chi^2 = 22.04$, $df = 9$, $p < 0.01$; Tennessee: $\chi^2 = 0.07$, $df = 3$, n.s.; South Carolina: $\chi^2 = 13.21$, $df = 9$, n.s.; Alabama: $\chi^2 = 8.29$, $df = 6$, n.s.). There was no significant annual variation in sex ratio at any location.

REFERENCES

Baker, R. R., 1978, The Evolutionary Ecology of Animal Migration, Holmes and Meier, New York.

Baker, M. C., and Fox, S. F., 1978, Dominance, survival, and enzyme polymorphism in dark-eyed juncos, Junco hyemalis, Evolution **32**:697–711.

Balph, M. H., 1977, Winter social behaviour of dark-eyed juncos: Communication, social organization, and ecological implications, *Anim. Behav.* **25:**859–884.

Balph, M. H., 1979, Flock stability in relation to social dominance and agonistic behavior in dark-eyed juncos, *Auk* **96:**714–722.

Calder, W. A., 1974, Consequences of body size for avian energetics, in: *Avian Energetics* (R. A. Paynter, ed.), Nuttall Ornithol. Club Publ. 15, Cambridge, Massachusetts, pp. 86–144.

Catchpole, C. K., 1972, A comparative study of territory in the reed warbler (*Acrocephalus scirpaceus*) and sedge warbler (*A. schoenobaenus*), *J. Zool., Lond.* **166:**213–231.

Dingle, H., 1979, Migration (book review), *Science* **204:**609–610.

Dingle, H., 1980, Ecology and evolution of migration, in: *Animal Migration, Orientation, and Navigation* (S. A. Gauthreaux, ed.), Academic Press, New York, pp. 1–101.

Dolbeer, R. A., 1982, Migration patterns for age and sex classes of blackbirds and starlings, *J. Field Ornithol.* **53:**28–46.

Evans, D. L., and Rosenfield, R. N., Migration mortality of sharp-shinned hawks ringed at Duluth, Minnesota, USA, *World Conference on Birds of Prey*, April 26–29, 1982, Thessaloniki, Greece (in press).

Fretwell, S. D., 1969, Dominance behavior and winter habitat distribution in juncos (*Junco hyemalis*), *Bird-Banding* **40:**1–25.

Fretwell, S. D., 1972, *Populations in a Seasonal Environment*, Princeton University Press, Princeton, New Jersey.

Fretwell, S., 1980, Evolution of migration in relation to factors regulating bird numbers, in: *Migrant Birds in the Neotropics: Ecology, Behavior, Distribution, and Conservation* (A. Keast and E. S. Morton, eds.), Smithsonian Inst. Press, Washington, D.C., pp. 517–527.

Gauthreaux, S. A., Jr., 1978, The ecological significance of behavioral dominance, in: *Perspectives in Ethology*, Volume 3 (P. P. G. Bateson and P. H. Klopfer, eds.), Plenum Press, New York, pp. 17–54.

Gauthreaux, S. A., Jr., 1979, Priorities in bird migration studies, *Auk* **96:**813–815.

Gauthreaux, S. A., Jr., 1982, The ecology and evolution of avian migration systems, in: *Avian Biology*, Volume 6 (D. S. Farner and J. R. King, eds.), Academic Press, New York and London, pp. 93–167.

Godfrey, W. E., 1979, *The Birds of Canada*, National Museum of Natural Sciences, Ottowa, Canada.

Greenberg, R., 1980, Demographic aspects of long-distance migration, in: *Migrant Birds in the Neotropics: Ecology, Behavior, Distribution, and Conservation* (A. Keast and E. S. Morton, eds.), Smithsonian Inst. Press, Washington, D.C., pp. 493–504.

Greenwood, P. J., 1980, Mating systems, philopatry, and dispersal in birds and mammals, *Anim. Behav.* **28:**1140–1162.

von Haartman, L., 1968, The evolution of resident versus migratory habit in birds. Some considerations, *Ornis Fennica* **45:**1–6.

von Haartman, L., 1971, Population dynamics, in: *Avian Biology*, Volume 1 (D. S. Farner and J. R. King, eds.), Academic Press, New York, pp. 391–459.

Hannon, S. J., Soperck, L. G., and Sopuck, F. C., 1982, Spring movements of female blue grouse; evidence of socially induced delayed breeding in yearlings, *Auk* **99:**687–694.

Helms, C. W., Aussiker, W. H., Bower, E. B., and Fretwell, S. D., 1967, A biometric study of major body components of the slate-colored junco, *Junco hyemalis*, *Condor* **69:**560–578.

Hunt, G. L., Jr., Wingfield, J. C., Newman, A., and Farner, D. S., 1980, Sex ratio of western gulls on Santa Barbara Island, California, *Auk* **97:**473–479.

James, F. C., 1970, Geographic size variation in birds and its relationship to climate, Ecology **51**:365–390.
Jansson, C., Ekman, J., von Bromssen, A., 1981, Winter mortality and food supply in tits Parus spp., Oikos **37**:313–322.
Keast, A., and Morton, E. S. (eds.), 1980, Migrant Birds in the Neotropics: Ecology, Behavior, Distribution, and Conservation, Smithsonian Inst. Press, Washington, D.C.
Ketterson, E. D., 1979a, Aggressive behavior in wintering dark-eyed juncos: Determinants of dominance and their possible relation to geographic variation in sex ratio, Wilson Bull. **91**:371–383.
Ketterson, E. D., 1979b, Status signaling in juncos, Auk **96**:94–99.
Ketterson, E. D., and King, J. R., 1977, Metabolic and behavioral responses to fasting in the White-crowned Sparrow (Zonotrichia leucophrys gambelii), Phys. Zool. **50**:115–129.
Ketterson, E. D., and Nolan, V., Jr., 1976, Geographic variation and its climatic correlates in the sex ratio of eastern-wintering dark-eyed juncos (Junco hyemalis hyemalis), Ecology **57**:679–693.
Ketterson, E. D., and Nolan, V., Jr., 1979, Seasonal, annual, and geographic variation in sex ratio of wintering populations of dark-eyed juncos (Junco hyemalis), Auk **96**:532–536.
Ketterson, E. D., and Nolan, V., Jr., 1982, The role of migration and winter mortality in the life history of a temperate-zone migrant, the dark-eyed junco, as determined from demographic analyses of winter populations, Auk **99**:243–259.
Kikkawa, J., 1980, Winter survival in relation to dominance classes among silvereyes Zosterops lateralis chlorocephala of Heron Island, Great Barrier Reef, Ibis **122**:437–446.
King, J. R., and Mewaldt, L. R., 1981, Variation in body weight in Gambel's white-crowned sparrows in winter and spring: Latitudinal and photoperiodic correlates, Auk **98**:752–764.
King, J. R., Farner, D. S., and Mewaldt, L. R., 1965, Seasonal sex and age ratios in populations of the white-crowned sparrow of the race gambelii, Condor **67**:489–504.
Krebs, J. R., 1971, Territory and breeding density in the great tit, Parus major L., Ecology **52**:2–22.
Krebs, J. R., 1979, Survival value of migration (book review), Nature **278**:671–672.
Krebs, J. R., 1982, Territorial defence in the Great Tit (Parus major): Do residents always win?, Behav. Ecol. Sociobiol. **11**:185–194.
Lack, D., 1954, The Natural Regulation of Animal Numbers, Oxford Press, London.
Lundberg, A., 1979, Residency, migration and a compromise: Adaptations to nest-site scarcity and food specialization in three Fennoscandian owl species, Oecologia **41**:273–281.
Lundberg, P., Bergman, A., and Olsson, H., 1981, On the ecology of wintering dippers (Cinclus cinclus) in northern Sweden, J. Ornithol. **122**:163–172.
Morton, E. S., 1980, The importance of migrant birds to the advancement of evolutionary theory, in: Migrant birds in the Neotropics: Ecology, Behavior, Distribution, and Conservation (A. Keast and E. S. Morton, eds.), Smithsonian Inst. Press, Washington, D.C., pp. 555–557.
Myers, J. P., 1981a, A test of three hypotheses for latitudinal segregation of the sexes in wintering birds, Can. J. Zool. **59**:1527–1534.
Myers, J. P., 1981b, Cross-seasonal interactions in the evolution of sandpiper social systems, Behav. Ecol. Sociobiol. **8**:195–202.
Nice, M. M., 1943, Studies in the Life History of the Song Sparrow, Volume 2, Trans. Linnaean Soc. N.Y. **6**:xiv + 1–328.

Nolan, V., Jr., 1978, The Ecology and Behavior of the Prairie Warbler *Dendroica discolor*, *Ornithol. Monogr.* **26**:1–595.

Nolan, V., Jr., and Ketterson, E. D., 1983, Geographic variation in body mass of wintering dark-eyed juncos (*Junco h. hyemalis*), *Wilson Bull.* (in press).

Oring, L. W., and Lank, D. B., 1982, Sexual selection, arrival times, philopatry and site fidelity in the polyandrous spotted sandpiper, *Behav. Ecol. Sociobiol.* **10**:185–191.

Pulliam, H. R., and Millikan, G. C., 1982, Social organization in the nonreproductive season, in: *Avian Biology*, Volume 6 (D. S. Farner and J. R. King, eds.), Academic Press, New York, pp. 169–197.

Ralph, C. J., 1971, An age differential of migrants in coastal California, *Condor* **73**:243–246.

Rohwer, S., and Ewald, P. W., 1981, The cost of dominance and the advantage of subordination in a badge signaling system, *Evolution* **35**:441–454.

Saether, B.-E., and Fonstad, T., 1981, A removal experiment showing unmated females in a breeding population of Chaffinches, *Anim. Behav.* **29**:637–639.

Samson, F. B., 1976, Territory, breeding density, and fall departure in Cassin's finch, *Auk* **93**:477–497.

Smith, J. N. M., Mongomerie, R. D., Taitt, M. J., and Yom-Tov, Y., 1980, A winter feeding experiment on an island song sparrow population, *Oecologia* **47**:164–170.

Smith, S., 1982, Raptor "reverse" dimorphism revisited: A new hypothesis, *Oikos* **39**:118–122.

Spaans, A. L., 1977, Are startlings faithful to their winter quarters? *Ardea* **64**:83–87.

Stuebe, M. M., and Ketterson, E. D., 1982, A study of fasting in tree sparrows (*Spizella arborea*) and dark-eyed juncos (*Junco hyemalis*): Ecological implications, *Auk* **99**:299–308.

Thompson, C. F., 1977, Experimental removal and replacement of territorial male yellow-breasted chats, *Auk* **94**:107–113.

Tompa, F. S., 1964, Factors determining the numbers of song sparrows, *Melospiza melodia* (Wilson), on Mandarte Island, B.C., Canada, *Acta Zool. Fenn.* **109**:4–73.

Yasukawa, K., and Bick, E. I., 1983, Dominance hierarchies in dark-eyed juncos (*Junco hyemalis*): A test of a game theory model, *Anim. Behav.* **31**:439–448.

AUTHOR INDEX

Ackerman, R. A., 77, 83, 85, 98, 100, 101
Ahlquist, J. E., 245–292; 246, 249, 253, 257, 261, 268, 274, 277, 281–83, 285, 286
Ainley, D. G., 14, 29
Almquist, H. J., 95, 98
Amadon, D., 168, 184
Amprino, R., 295, 303, 321
Anderson, B., 135–37, 140, 142, 145–47, 151, 155
Andersson, M., 52
Ankney, C. D., 14, 29, 38, 43, 61
Ansari, H. A., 195, 209
Aquadro, C. F., 243
Ar, A., 43, 66, 75–80, 83, 84, 86, 94, 95, 97, 98, 100, 102
Armstrong, E. A., 331, 342, 351
Arrighi, F. E., 195, 198, 209
Ashmole, N. P., 3, 22, 27, 29, 53, 61
Ashkenazie, S., 48, 61
Askenmo, C., 14, 18, 27, 29
Audubon, M. R., 135, 151, 156
Aussiker, W. H., 400
Avise, J. C., 212, 215, 217–219, 226, 231, 232, 238, 243

Bachmann, K., 249, 255, 288
Baerends, G. P., 52
Baker, A., 197, 208
Baker, J. R., 2, 30, 34, 61
Baker, M. C., 212, 218, 242, 367, 399
Baker, R. R., 357, 358, 375–385, 399, 391–397
Balda, R. P., 343, 353
Balls, M., 322
Balph, M. A., 367, 374, 400
Bammi, R. K., 209
Bannerman, D. A., 345, 351
Barbour, E. H., 123, 126
Barker, W. C., 269, 288
Barney, C. C., 341, 351
Barrowclough, G. F., 151, 156, 168, 174, 184, 212, 215–218, 221, 226, 229, 231, 233, 238, 242
Barry, W. T., 46, 61
Barsbold, R., 128
Bateson, P. P. G., 46, 61
Batt, B. D. J., 40, 61
Bauer, G., 119, 120, 126
Becking, J. H., 88, 89, 93, 98
Beer, C. G., 52, 62
Bell, G., 1, 5, 25, 29
Bellrose, F. C., 349, 351
Bengtson, S. A., 43, 55
Bennet, M. R., 325

Benoist, C., 208, 209
Bent, A. C., 347, 351
Benveniste, R. A., 257, 289
Bergman, A., 401
Berger, A. J., 46, 68
Bertram, B. C. R., 56
Best, L. B., 341, 351
Bick, E. I., 374, 402
Biederman, B. M., 195, 205, 208
Birchard, G. F., 77, 79, 86, 98, 101
Birkhead, T. R., 334, 336, 339, 347, 351
Bledsoe, A. H., 288
Blem, C. R., 34, 63
Bloom, G., 16, 29
Bloom, S. E., 206, 208
Blount, J. E., 99
Board, R. G., 82, 88–91, 93, 94, 97–99, 102
Boardman, 99, 101
Bock, C. E., 342, 354
Bock, W. J., 111, 127, 127, 267, 287, 289
Bonner, T. I., 247, 257, 275, 279, 289
Bonner, W. N., 351
Bossert, 4, 6, 7, 32
Bower, E. B., 400
Bramwell, C. D., 109, 127

403

Bray, D. E., 351
Breathnach, R., 208, 209
Breitenbach, R. P., 14, 29
Brenner, J. D., 289
Brewer, R., 139, 156
Britten, R. J., 246, 248, 242, 253, 255, 266, 289, 290
Brockleman, W. K., 39, 61
Brodkorb, P., 111, 127, 282, 289
Broom, R., 106, 127
Brown, C. R., 351
Brown, W. L., 166, 171, 187
Bruggers, D. J., 353
Brunning, D. F., 56, 59, 61
Bryant, P. J., 321, 322
Bryant, S. V., 310, 321, 322
Bryens, O. M., 342, 351
Buitron, D., 333, 335, 338, 339, 351
Burns, J. T., 330, 333, 339, 342, 351
Burton, R. W., 351
Buckland, R. A., 209
Burleigh, T. H., 236, 243
Bush, G. L., 160, 176, 185

Cain, A. J., 163, 176, 185
Calandra, A. J., 305, 321
Calder, W. A., 72, 98, 362, 400
Call, A. B., 61
Cameron, J., 325
Cami, B., 209
Camosso, M., 303, 321
Caplan, A. I., 297, 315, 322
Carey, C., 69–103; 70, 72–76, 80, 85–87, 96, 98, 99
Carey, M., 330, 343, 351
Carlenius, C., 106, 108, 198
Carrington, J. L., 322
Carter, N. W., 90, 91, 94, 100
Caswell, H., 4, 25, 30
Catchpole, C. K., 372, 400
Catterall, J. F., 209
Cederholm, G., 339, 351
Cetta, A., 289
Chamberlain, M. E., 289
Chambon, P., 208, 209
Chan, H.-C., 255, 289

Charig, A. J., 108, 118, 120, 127
Charlesworth, B., 4, 6, 7, 19, 24, 29
Charnov, E. L., 38, 61
Cheng, K. M., 353
Chevallier, A., 314, 321, 324
Chiscon, J. A., 290
Choate, T. S., 345, 352
Christ, B., 314, 315, 321, 324
Čihak, R., 319, 321
Clark, G. A., 48, 61
Cochet, M., 208, 209
Cockerham, C. C., 224, 242
Cody, M. L., 4, 5, 22, 29, 33, 62
Colbert, E. H., 109, 121, 127
Colless, D. H., 287–289
Commorford, S. L., 255, 289
Conover, M., 56, 62, 352
Cooke, A. S., 88, 96, 99
Cooke, J., 306, 312, 327
Corbin, K. W., 140, 142, 145, 151, 156, 211–244, 212, 215–218, 220, 221, 224–226, 228, 229, 231, 233, 238–243
Coulson, J. C., 38, 41, 48, 54, 62
Cowen, R., 123, 127
Cracraft, J., 159–187; 35, 62, 110, 127, 170, 173, 178, 179, 184, 185
Craig, J. P., 288
Cramp, S., 40, 59, 62
Crawford, R. D., 324, 352
Crompton, A. W., 111, 114, 127
Cronin, J. E., 279, 291
Cronquist, A., 162, 174, 185
Crooks, R. J., 99
Crosby, G. M., 305, 309, 322
Crossner, K. A., 62

Daan, S., 39, 62
Daly, E., 124, 127
Darley, J. A., 341, 352
Darling, K., 355
Darwin, C., 105, 127

Daugherty, R. J., 145–147, 155
Davies, S. J. F., 56, 59, 62, 345, 352
Davidson, E. H., 253, 289, 29
Davidson, M. F., 103
Davidson, N., 292
Davis, D. E., 46, 62
Dawes, C. M., 99
Dawkins, R., 36, 62
Dayhoff, M. O., 269, 288
de Beer, 111, 126
de Boer, L. E. M., 190, 209
De Master, D. P., 14, 18, 29
De Steven, D., 18, 27, 29
De Wolfe, B. D., 343, 352
Dewar, J. M., 345, 352
Dhouailly, D., 314, 321
Diamond, J., 168, 185
Dingle, H., 357, 375, 400
Dixon, K. L., 132, 139, 156
Dolbeer, R. A., 363, 400
Donner, C. G. P., 335, 354
Doolittle, R. F., 269, 289
Draper, M. H., 103
Drent, R. H., 39, 62, 70, 74, 95, 99
Drets, M. E., 197, 198, 209
Drobny, R. D., 43, 62
Dunn, E. H., 36, 62
Durango, S., 352
Doskocil, M., 321
Doutroupas, S., 315, 321
Dutrillaux, B., 205, 209
Dykhuizen, D., 280, 290

Eastin, W. C., 91, 99
Eastlick, H. L., 319, 321
Eaton, S. W., 343, 352
Eckhardt, R. C., 342, 352
Ede, D. A., 293, 297, 320, 322
Eden, F. C., 289
Ehrlich, P. R., 162, 185
Ekman, J., 401
Eldredge, N., 170, 173, 179, 185
Ellenberger, P. P., 111, 127
Emlen, J. T., Jr., 341, 352
Emlen, S. T., 48, 62, 146, 148, 152, 156, 331, 352

Endler, J. A., 151, 152, 156
Engelson, G. H., 342, 355
Erben, H. K., 88–91, 99, 100
Erickson, C. J., 356
Ericksson, M. O. G., 56, 61
Evans, D. L., 366, 400
Evans, H. J., 209
Ewald, P. H., 369, 402

Fallon, J. F., 297, 298, 305, 307, 309, 310, 318, 320, 322, 325, 326
Farner, D. S., 400, 401
Farquhar, M. N., 248, 290
Farris, J. S., 221, 238, 243
Faust, R., 100
Feduccia, A., 109, 111, 124, 127
Ferguson, A., 156, 243
Fiala, V., 62
Ficken, M. S., 336, 352
Fischel, A., 314, 322
Fisher, C., 270, 290
Fitch, W. M., 264, 266, 267, 269, 277, 289
Florence, D., 208
Fonstad, T., 372, 402
Ford, J., 183, 185
Ford, N. L., 329–356, 335, 341, 342, 349
Foster, M. S., 4, 29, 38, 62
Fouvet, B., 319, 324
Fowler, S., 88, 94, 95, 102
Fox, S. F., 367, 399
Fraga, R. M., 330, 346, 352
Frederick, J. M., 322
Fredrickson, L. H., 52, 62
French, V., 310, 321, 322
Fretwell, S. D., 39, 64, 357, 366, 377, 378, 390, 396, 400
Frisch, H. L., 297, 320, 325
Frith, H. J., 48, 62, 345, 352
Fugle, G. W., 47, 62
Fuller, R., 98

Gadgil, M., 4, 29
Galau, G. A., 277, 287
Galton, P. M., 110, 127
Gannon, F., 208, 209

Garber, S. D., 99
Garcia-Porrero, J. A., 324
Gardiner, B. G., 105, 127
Garrapin, A., 209
Gasseling, M. T., 295, 303, 326
Gauthreaux, S. A., Jr., 357, 358, 365, 366, 375, 378, 390, 400
Geake, F. G., 80, 102
Ghiselin, M. T., 169, 185
Gibson, F., 48, 52, 62
Gill, F., 168, 185
Giller, D. R., 137, 157, 344, 355
Gillespie, D., 251, 289
Gingerich, P. D., 117, 127
Gladstone, D. E., 329, 334, 352
Glutz von Blotzheim, U. N., 40, 41, 63
Godfrey, W. E., 146, 148, 156, 384, 400
Goodman, D., 3–5, 29, 38, 63
Goodpasture, C., 206, 208
Goodwin, D., 179, 185
Goryainova, G. P., 95, 99
Gottlieb, S. S., 289
Gould, S. J., 297, 322
Graham, D. E., 289
Graham, G. J., 289
Grant, C. H. B., 59, 64
Grant, G. S., 81, 96, 99, 100
Grant, V., 176, 185
Graul, W. D., 347, 352
Greenberg, R., 357, 379, 382, 384, 400
Greene, D. G., 98
Greenwood, P. J., 357, 389, 400
Grimes, S. A., 342, 352
Griffiths, G. C. D., 286, 287, 289
Griffiths, P. J., 297, 307, 310, 324
Grim, M., 314, 322
Gross, A. O., 346, 352
Grula, J. W., 254, 290
Guigna, T. D., 287

Gumpel-Pinot, M., 293–327, 301, 307, 309, 312, 314, 323, 324
Gutiérrez, R. J., 212, 216, 221, 222, 231, 235, 238, 243

von Haartman, L., 33, 39, 43, 53, 63, 332, 342, 344, 346, 352, 372, 388, 400
Haeckel, E. H. P. A., 127
Haffer, J., 175, 185
Hahn, D. C., 101
Haiden, G. J., 209
Hall, T. J., 282, 289, 290
Hamburger, V., 313, 318, 319, 322, 324
Hamilton, H. L., 322
Hamilton, W. D., 10, 30
Hampe, A., 301, 322
Handford, P., 218, 243
Hanka, L. P., 76, 95, 99
Hann, H. W., 336, 343, 346, 350, 352
Hannon, S. J., 54, 63, 372, 400
Hansell, R. I. C., 343, 354
Harrington, B. A., 288
Harris, E. D., 88, 89
Harris, M. P., 47, 48, 52, 63
Hartl, D., 280, 290
Hastings, A., 4, 30
Haymes, G. T., 39, 64
Hecht, M. K., 110, 112, 113, 129
Heilmann, G., 106, 122, 123, 127
Heinemann, R., 289
Heinroth, O., 38, 63
Helms, C. W., 364, 367, 400
Hen, R., 208
Henderson, R. P., 355
Hendrick, J. P., 289
Hennig, W., 254, 285, 287, 290
Henning, U., 213, 243
Henny, C. J., 8, 31
Hesse, R., 53, 63
Heusmann, H. W., 52, 53, 63
Hicks, M. J., 297, 322
Hilborn, R., 38, 54, 63

Hilden, O., 52, 63, 347, 352
Hilfer, S. R., 325
Hills, S., 63
Hinchliffe, J. R., 293, 293–327, 297–299, 301–304, 307–309, 312, 322, 323, 327
Hinegardner, R., 255, 290
Hirschfield, M. F., 4, 30
Högstedt, G., 38, 41, 43, 63
Hollyday, M., 318, 323, 324
Holmes, R. T., 27, 30, 354
Holmgren, N., 106, 119, 127, 297, 323
Holst, W. F., 95, 98
Honig, L. S., 304–306, 312, 323, 324, 326
Horder, T. J., 318, 323
Hornbruch, A., 301, 308, 315, 323, 325
Horobin, J., 41, 62
Hough, B. R., 289
Houston, C. S., 140, 146, 156
Howard, H. E., 336, 352
Howe, H. F., 38, 63, 72, 73, 99, 344, 352
Howe, M. A., 347, 353
Howell, J. C., 342, 352
Howell, T. R., 52, 63, 85, 99
Howey, P. W., 80
Hoyt, D. F., 76, 82, 83, 88, 94, 97, 99, 102
Hsu, T. C., 194, 209
Huber, P. J., 262, 290
Hull, D. L., 163, 169, 185
Hunt, C. B., 133, 156
Hunt, E. A., 313, 323
Hunt, G. L., Jr., 56, 63, 333, 352, 372, 400
Hunt, J. A., 289
Hunt, M. W., 56, 63, 333, 352
Huxley, T. H., 105, 127

Inglis, I. P., 43, 50, 51, 63
Iten, L. E., 310–312, 323, 324

Jacob, H. T., 314, 315, 321, 323
Jacob, M., 314, 321, 323
Jacobson, M., 318, 324

James, F. C., 99, 365, 401
Janners, M. Y., 315, 326
Jansson, C., 366, 401
Jansson, H., 208, 209
Jarrell, G. H., 195, 206, 209
Jarvik, E., 197, 324
Javois, L. C., 312, 324
Jenkins, D., 345, 347, 355
Jenni, D. A., 56, 63, 331, 346, 354
Johns, J. E., 347, 355
Johnsgard, P. A., 59, 64
Johnson, A. W., 64
Johnson, D. R., 293, 297, 299, 324
Johnson, M. S., 275, 290
Johnson, N. K., 168, 185
Johnson, S. R., 14, 30
Johnston, D. W., 52, 64
Johnston, H. S., 103
Jones, H. L., 343, 354
Jones, P. J., 14, 30, 34, 38, 64
Jorquera, L. C., 300, 324
Jukes, T. H., 224, 243

Kaul, D., 195, 209
Kear, J., 99
Keast, A., 165, 168, 175, 179, 185, 375, 401
Keefe, E. L., 318, 322
Keeton, W. T., 96, 100
Keller, M. E., 334–336, 339, 343, 353
Kendeigh, S. C., 48, 64, 336, 342, 346, 353
Kennelly, J. J., 333, 354
Kenyon, D. W., 47, 66
Ketterson, E. D., 357–402, 358, 359, 362–364, 367–369, 373, 374, 381–383, 398
Kieny, M., 300, 301, 303, 313–315, 319, 321, 325
Kikkawa, J., 366, 369, 401
Kilgore, D. L., 77, 86, 98
Kimura, M., 225, 243, 275, 276, 290
King, C. E., 26, 31
King, J. L., 224, 243
King, J. R., 363, 366, 371, 401

Kleinschmidt, O., 162, 185
Klomp, H., 19, 30, 33, 43, 47, 56, 64
Knapton, R. W., 146, 147, 149, 156, 341, 353
Knowles, E. H. M., 343, 353
Knudson, M. L., 332, 354
Kohne, D. E., 246, 248, 249, 252, 255, 257, 275, 279, 280, 287, 290
Kolata, G., 269, 290
Kollar, E. J., 270, 290
Korshgen, C. E., 43, 63
Krampitz, G., 88, 91, 94, 100
Krapu, G. L., 14, 30, 43, 64
Krebs, J., 38, 61, 372, 375, 401
Kreithen, M. L., 96, 100
Kriesten, K., 91, 99, 100
Krishan, A., 209
Krivonosov, G. A., 41, 66
Kroodsma, R. L., 145, 147, 148, 156
Kutchai, H., 82, 100

Lack, D., 2–5, 19, 30, 33, 38, 39, 41, 46, 48, 52, 53, 56–59, 64, 329, 330, 332, 353, 366, 401
Lack, E., 38, 64
Lai, E. C., 208, 209
Lambrecht, K., 127
Landmesser, L., 316, 324
Lane, J., 140, 156
Langham, N. P. E., 41, 64
Lank, D. B., 371, 402
Lanser, M. E., 322
Lanyon, W. E., 168, 185, 343, 353
Lasky, A. R., 342, 346, 353
Lazarus, J., 50, 51, 64
Leal, C. W., 324
Leal, K. W., 324
LeDouarin, N., 298, 314, 324
Ledoux, T., 102
Leech, R. M., 99
Legg, K., 356
Lejeune, J., 205, 209
Lemmetyinen, R., 41, 46, 64
Leon, J. A., 4, 5, 24, 30

Leopold, F., 52, 64
Le Pennic, J. P., 209
Levin, D. A., 162, 185
Lewis, J., 316, 318, 319, 324
Lewis, J.H., 300, 326
Lewontin, R. C., 163, 185
Lin, C. C., 208
Lindeman, R., 325
Lipps, J. H., 123, 127
Litovich, E., 354
Logan, C. A., 342, 353
Lombardo, M. P., 354
Lomholt, J. P., 82, 85, 100
Long, R. A., 120, 129
Lotka, A. J., 8, 30
Love, G., 88, 90, 91, 98, 102
Low, B. S., 36, 64
Lowe, P. R., 106, 109, 125, 127, 128
Lu, M. R., 190, 209
Lubs, H. A., 209
Lundberg, P., 366, 388, 401
Lundy, H., 79, 100

MacArthur, R. H., 4, 23, 30, 38
MacCabe, J. A., 299, 305, 306, 308, 309, 311, 320, 321, 324
Mace, M. L., 209
MacInnes, C. D., 14, 29, 38, 43, 61
Mackworth-Praed, C. W., 59, 64
Maclean, G. L., 38, 59, 64
MacLean, S. F., Jr., 343, 354
Mader, W. J., 332, 353
Maderson, P. F. A., 297, 324
Marler, P., 335, 336, 339, 353
March, O. C., 105, 122, 128
Marchalonis, J. J., 269, 290
Marmur, J., 255, 290
Marshall, A. G., 168, 185
Marshall, J., 343, 353
Marshall, J. T., Jr., 134, 156
Martin, L. D., 105–129, 110–112, 114, 119, 120, 124, 128
Martin, R. F., 333, 343, 353
Mauger, A., 314, 321, 325

Maxson, L. R., 292
Maxson, S. J., 48, 65
Mayfield, H., 343, 353
Mayr, E., 133, 155, 156, 160–168, 173–176, 185, 225, 227, 285, 290
McBride, W. G., 319, 325
McCarthy, B. J., 248, 290
McCredie, J., 319, 325
McFarlane, R. N., 41, 64
McKinney, F., 329, 332–336, 349, 350, 353
McLachlan, J. C., 315, 325
McLaren, I. A., 335, 339, 343, 353
Means, A. R., 209
Mednikov, B. M., 251, 290
Mengden, G. A., 199, 205
Mengel, R. M., 131, 151, 156, 177, 186
Merker, H.-J., 297, 325
Merrick, S., 209
Metcalfe, J., 84, 102
Mewaldt, L. R., 371, 401
Meyer, R. K., 14, 29
Michener, H., 344, 353
Michod, R. E., 4, 24, 30
Mickevich, M. F., 279, 290
Mikami, T., 102
Miller, D. E., 352
Miller, E. H., 65
Miller, R. S., 47, 65
Millikan, G. C., 366, 402
Mills, J. A., 46, 54, 65
Mongin, P., 90, 91, 94, 100, 101
Monroe, B. L., Jr., 168, 186
Montagna, W., 297, 325, 333, 353
Montevecchi, W. A., 72, 101
Montgomerie, R. D., 402
Mook, C. C., 121, 127
Moreay, R. E., 2, 4, 30, 38, 41, 65
Morris, D. G., 316, 324
Morris, R. D., 39, 64
Morton, E. S., 357, 375, 401
Moses, R., 355
Moss, R., 39, 43, 65
Mumford, R. E., 342, 353

Mundinger, P. C., 335, 339, 353
Murdoch, W. W., 26, 30
Murrillo-Ferrol, N. L., 313, 325
Murphy, D. J., 310, 311
Murphy, G. I., 4, 25, 30, 39, 65
Myers, J. P., 357, 358, 370, 380, 388, 389, 401

Nathans, D., 255, 290
Nathanson, M. A., 315, 325
Nau, H., 325
Nei, M., 212, 220–224, 226, 232, 239, 242, 243
Nelson, G. J., 170, 172, 178, 179, 183, 186, 286, 290
Nelson, J. B., 3, 30
Nero, R. W., 343, 353
Nethersole-Thompson, D., 334, 345, 347, 353
Nethersole-Thompson, M., 345, 353
Neubert, D., 325
Neuchterlein, G. L., 266, 290
Neufeld, B. R., 289
Newman, A., 400
Newman, S. H., 297, 320, 325
Newton, I., 14, 19, 30, 345, 353
Nice, M. M., 34, 65, 73, 100, 336, 343, 353, 372, 401
Niethammer, G., 40, 65
Nisbet, I. C. T., 41, 46, 65, 72, 100
Niswander, J. D., 221, 244
Nolan, V. Jr., 330, 335, 338, 343, 351, 355, 357–402, 358, 359, 362, 364, 367, 368, 372–374, 379, 381–383, 389, 390, 398, 402
Nopsca, F., 122, 128
Nottebohm, F., 218, 243

Oberholser, H. C., 137, 139, 143, 156
Occam, W., 54

O'Connor, R. J., 19, 30, 34, 65
O'Hare, K., 208, 209
Ohta, I., 224, 243
Ohta, T., 276, 290
Olson, S. H., 109, 124, 128
Olsson, H., 401
O'Malley, B. W., 209
O'Neill, J. P., 168, 186
Orians, G. H., 343, 354
Oring, L. W., 48, 62, 329–332, 346, 347, 352, 354, 371, 402
Osborn, H. F., 128
Osmolka, H., 115, 121, 125, 128
Ostrom, J. H., 106, 108, 110, 111, 115, 117–119, 121, 122, 128
Otin, J. S., 209
Owen, D. F., 4, 30
Owen, R., 105, 128

Packard, G. C., 95, 99, 101
Packard, M. J., 99
Padien, K., 115, 118, 122, 125, 128
Paganelli, C. V., 43, 66, 69, 76, 81, 82, 86, 97–101
Palmer, R. S., 40, 47, 65
Paludan, K., 46, 47, 65
Parisi, P., 99
Parker, H., 47, 67
Parkes, K. C., 122, 124, 128
Parmelee, D. F., 48, 65, 346, 354
Parsons, J., 47, 65, 72, 100
Patch, C. L., 343, 354
Patil, S. R., 197, 209
Patnaik, A. K., 195, 209
Patterson, C., 106, 128, 162, 178, 182, 186
Patterson, I. J., 19, 30
Patterson, T. L., 343, 354
Patton, J. C., 243
Payne, R. B., 38, 48, 65, 334, 336, 346, 354
Pearson, T. H., 48, 65
Pellis, S. M., 50, 51, 65
Pellis, V. C., 50, 51, 65
Perrin, F., 209

Perrins, C. M., 4, 27, 31, 38, 39, 52, 54, 66
Perrot, H. R., 90, 93, 98, 102
Petronievics, B., 105, 129
Petronovich, L., 343, 354
Pettigrew, A. G., 316, 318, 235
Pettit, T. N., 86, 98–100
Phelps, W. H., Jr., 168, 186
Pianka, E. R., 23, 31, 39, 66
Pierotti, R., 41, 66
Piiper, J., 102
Pinot, M., 313, 325
Pitelka, F. A., 347, 353
Pizzy, G., 59, 106
Platnick, N. I., 163, 170, 172, 178, 179, 183, 186
Platter-Rieger, M., 85, 98
Pleszcynska, W., 343, 354
Plowright, R. C., 46, 61
Popper, K. R., 161, 186
Powell, G. V. W., 343, 354
Power, H. W., 335, 354
Pozlavskii, A. N., 41, 66
Prasad, R., 209
Prensky, W., 255, 290
Prestige, M. C., 318, 325
Price, F. E., 342, 354
Prince, H. H., 40, 65
Pugesek, B. H., 38, 66
Pugin, E., 300, 324
Pullainen, E., 347, 354
Pulliam, H. R., 366, 402

Radabaugh, B. F., 330, 346, 354
Rahn, H., 43, 66, 69, 71, 72, 75–84, 86, 95–102
Ralph, C. J., 379, 402
Rand, A. L., 146, 148, 156, 284, 290
Raner, L., 347, 354
Ratti, J. T., 266, 288, 290
Raveling, D. G., 38, 43, 66
Raven, P., 162, 172, 185
Reeves, R. B., 98
Remington, C. L., 131, 132, 151, 144, 156
Reynolds, J. F., 345, 354
Reznick, D. N., 26, 31
Rice, D. W., 47, 66

Rice, N. R., 248, 290
Richardson, R. A., 342, 354
Richardson, K. E. Y., 305, 320, 324
Ricklefs, R. E., 1–32, 4, 5, 10, 14, 16, 17, 19, 31, 34, 36, 38, 48, 53, 66, 72, 74, 101
Rinkel, G. L., 47, 66
Ripley, S. D., 41, 66
Rising, J. D., 131–157, 135, 137, 140–142, 146–150, 152, 153, 156, 157
Roberts, T. A., 333, 354
Robinson, A., 336, 354
Rogers, S. J., 221, 243
Rohwer, S. A., 140, 157, 369, 402
Romanoff, A. K., 69, 72, 88, 101
Romanoff, A. L., 69, 70, 72, 88, 101
Roncali, L., 316, 318, 325
Roniewicz, E., 128
Rooney, P., 321
Rosen, D., 160, 162–164, 170, 174, 178, 186
Rosenfield, R. N., 366, 400
Ross, H. A., 43, 66
Rothstein, S. I., 47, 62
Roughgarden, J., 6, 19, 31
Rowe, D. A., 307, 309, 310, 322, 325
Royal, A., 209
Royama, T., 4, 31, 38, 39, 66, 67
Roychoudhury, A. K., 212, 243
Rubin, L., 300, 325
Rulli, M., 342, 353
Ruyechan, W. T., 289
Ryder, J. P., 43, 53, 56, 67, 333, 354
Ryttman, H., 205, 208, 209

Saether, B.-F, 372, 402
Safriel, U., 38, 48, 50, 67
Samanta, M., 209
Samson, F. B., 335, 354, 372, 402

Sarich, V. M., 270, 279, 292
Sasaki, M., 199, 205, 209
Sauer, E. G. F., 56, 59, 67
Sauer, E. M., 56, 59, 67
Saunders, J. W., Jr., 295, 298–300, 303, 309, 313, 314, 326
Saunders, L. C., 298, 320
Sauveur, B., 91, 101
Savile, D. B., 122, 129
Savory, C. J., 43, 67
Schaffer, W. M., 4, 9, 31, 39, 67
Schifferli, L., 39, 67
Schueler, F. W., 145, 149, 152, 157
Schmekel, L., 101
Schmidt, W. J., 89, 101
Schönwetter, M., 71, 101
Scott, D. M., 352
Scott, V. D., 88, 89, 93, 94, 98, 102
Scudder, G. G. E., 186
Sealy, S. G., 41, 67
Searls, R. L., 298, 306, 314, 325, 326
Seastedt, T. R., 335, 339, 343, 354
Seeley, H. G., 105, 129
Seichert, V., 321
Selander, R. K., 137, 157, 160, 175, 186, 341, 344, 355
Seymour, R. S., 85, 101
Shapiro, H. S., 254, 291
Sharov, A. G., 111, 129
Shaw, M. W., 197, 209
Shelley, L. O., 342, 355
Shellsworth, G. B., 303, 319, 326
Sherman, A. R., 342, 355
Shields, G. F., 189, 189–210, 191, 193, 195, 198, 199, 209, 255, 291
Shoffner, R. N., 190, 210
Shoobridge, R., 325
Short, L. L., Jr., 131, 132, 135, 136, 140, 142, 146, 148, 151, 157, 161, 165, 175, 186
Shugart, G. W., 355

Sibley, C. G., 132, 140, 142, 146, 148–150, 152, 156, 157, 161, 186, 243, 245–292, 246, 249, 253, 257, 261, 268, 272, 274, 277, 281–283, 285, 286, 288, 291, 292
Simandl, B. K., 322
Simkiss, K., 79, 84, 88, 90, 91, 93, 94, 99, 101, 102
Simons, P. C. K., 88–90, 101
Simpson, G. G., 122, 129, 287, 292
Sims, M. E., 356
Skutch, A. F., 3–5, 31, 346, 355
Slack, R. D., 348, 355
Slagsvold, T., 38, 52, 67
Smith, A. H., 101
Smith, C.C., 39, 67
Smith, H. O., 255, 290
Smith, J. C., 306, 307, 326
Smith, J. N. M., 335, 355, 366, 367, 369, 402
Smith, K. K., 111, 114, 127
Smith, S., 335, 355, 366, 402
Smith, S. M., 342, 355
Sneath, P. M., 238, 243, 287, 292
Snell, T. W., 26, 31
Snow, D. W., 339, 355
Snow, B. K., 4, 31
Snyder, G. K., 79, 101
Soikkeli, M., 347, 355
Sokal, R. R., 162, 164, 186, 238, 243, 287, 292
Somppi, P. L., 56, 67, 333, 354
Soperk, F. C., 400
Soperk, L. G., 400
Sotherland, P. R., 76, 95, 99, 101
Southern, W. E., 342, 355
Spaans, A. L., 366, 402
Spaziani, E., 91, 99
Stark, R. J., 298, 306, 326
Stearns, S. C., 2, 4, 23, 31, 39, 54, 67
Steel, R. G. D., 226, 243
Steen, J. B., 47, 67, 82, 100
Stefos, K., 195, 198, 209

Stein, J. P., 209
Steube, M. M., 363, 402
Stewart, J. D., 124, 128
Stewart, R. E., 139, 157
Stewart, R. M., 339, 343, 355
Stirling, R. V., 316, 236
Stock, A. D., 195, 197, 199, 205, 209
Stokes, P. A., 325
Stoner, D., 342, 355
Straus, N. A., 248, 254, 291, 292
Straznicky, K., 319, 326
Street, M. G., 140, 146, 156
Stresemann, E., 160, 162, 186
Summerbell, D., 299, 300, 304–307, 309, 312, 316, 321, 326, 327
Sumner, A. T., 193, 197, 209
Sutton, G. M., 136, 137, 139, 140, 143, 146, 148, 157
Swinton, W. E., 122, 129

Tagaki, N., 199, 205, 209
Taigen, T. L., 77, 99, 101
Taitt, M. J., 402
Tanner, J. T., 27, 31
Tarnovskaya, T. V., 95, 99
Tarsitano, S., 110, 112, 113, 129
Tasker, C. R., 46, 67
Tazawa, H., 79, 82–84, 101, 102
Tegelstrom, H., 208, 209
Temple, G. F., 84, 102
Templeton, A. R., 159, 186
Theberge, J. B., 345, 356
Thomson, A. L., 59, 67
Thompson, C. F., 343, 355, 372, 402
Thompson, E. L., 99
Thompson, W. L., 156
Thorneycroft, A. B., 193, 198, 209
Thorogood, P. V., 293, 326
Tickle, C., 296, 307, 308, 327
Tilson, R. L., 329, 332, 349, 356
Tinbergen, L., 335, 343, 355
Tinkle, D. W., 4, 30
Todaro, G. J., 257, 289

Tompa, F. S., 369, 402
Tordoff, H. B., 109, 124, 127
Torrie, J. H., 226, 243
Traylor, M. A., Jr., 284, 290
Trivers, R. L., 36, 67, 348, 355
Tullett, S. G., 82, 85, 88, 90, 93–95, 97, 98, 102
Tyler, C., 80, 88–91, 93, 94, 96, 101, 102

Uzzell, T., 225, 239, 243

Vady, P.H., 325
Van Camp, L. F., 8, 31
Van Tyne, J., 46, 67
Van Valen, L., 275, 277, 292
Veen, J., 41, 67
Vehrencamp, S. L., 56, 67
Verbeek, N. A. M., 336, 355
Verner, J., 331, 332, 335, 341, 342, 355
Visschedijk, A. H. J., 84, 86, 102
Vleck, C. M., 82, 99, 102
Vleck, D., 82, 99, 102
Vuilleumier, F., 166, 175, 186

Waddington, C. H., 298, 327
Wahlund, S., 214, 244
Walker, A. D., 110–112, 114, 129
Walkinshaw, L. H., 342, 343, 355
Wallace, D. G., 277, 292
Walsberg, G. E., 80, 81, 86, 102
Walters, B. F., 345, 355
Walters, J., 345, 355
Walters, J. R., 33–68, 40, 41, 45, 48–50, 59, 67

Wangensteen, O. D., 69, 75, 76, 83, 100, 102, 103
Ward, J. G., 48, 67
Ward, P., 14, 30, 38, 64
Warren, A. E., 301, 327
Warriner, J., 345, 355
Warriner, R., 345, 355
Watson, A., 27, 31, 345, 355
Weeden, J. B., 350, 355
Weeden, R. B., 345, 356
Weidmann, U., 47, 68
Weller, M. W., 48, 56, 68
Welles, S. P., 120, 129
Wellnhofer, P., 117, 129
Welsh, D. A., 336, 343, 356
Welter, W. A., 342, 356
Welty, J. C., 48, 68
West, D. A., 132, 147, 149, 150, 157
West, G. C., 14, 30
Wetherbee, K. B., 342, 356
Wetmore, A., 246, 272, 289, 292
Wetmur, J. C., 253, 292
Wheat, P., 157
Whetstone, K. N., 110–112, 114, 128, 129
White, E., 38, 49, 54, 62
White, M. J. D., 176, 186
Whittlow, G. C., 98–100
Whybrow, P. J., 117, 129
Wiley, E. O., 171, 178, 179, 186
Williams, A. J., 47, 52, 68
Williams, G. C., 4, 32, 38, 68
Williams, G. R., 292
Williams, L., 342, 356
Williams, O., 157
Williamson, F. S. L., 356
Williston, S. W., 122, 129
Willson, M. F., 331, 332, 335, 341, 343, 344, 355, 356

Wilson, A. C., 260, 279, 291, 292
Wilson, D., 103
Wilson, E. O., 4, 6, 7, 23, 32, 38, 65, 166, 171, 187
Wilson, J., 345, 356
Wingfield, J. C., 400
Winkler, D. S., 33–68, 44, 46, 59, 68
Witt, W., 88, 91, 100
Wittenberger, J. F., 330, 332, 343, 349, 356
Wyburn, G. M., 90, 91, 103
Woese, C. R., 285, 292
Wold, L. L., 335, 356
Wolf, J. S., 335, 356
Wolpert, L., 295, 298, 299, 303, 307, 308, 321, 326, 327
Woo, S. L. C., 207
Woolfenden, G. E., 342, 356
Workman, P. L., 221, 244
Wright, S., 223, 224, 242, 244

Yalden, D. W., 109, 129
Yallup, B. L., 301–303, 312, 327
Yang, S. Y., 243
Yanofsky, C., 213, 243
Yasukawa, K., 374, 402
Yom-Tov, Y., 38, 53, 75, 78, 345, 355, 356, 402
Yoshimoto, C., 102

Zenone, D. G., 335, 356
Zink, R. M., 212, 216, 219, 221, 226, 228, 231, 234, 238, 242–244
Zimmerman, J. L., 343, 356
Zusi, R. L., 168, 187
Zwilling, E., 295, 314, 327

BIRD NAME INDEX

Acanthis flavirostris, 335, 339
Acanthisitta chloris, 272
Acanthisittidae, 272–74
Acanthiza
  chrysorrhoa, 264–65
  sp., 275
Acanthizidae, 276
Acanthorhynchus, 265
Accipiter
  badius, 345
  nisus, 345
  striatus, 345
Aechmophorus
  clarkii, 266
  occidentalis, 266
Aepyornis, 71, 93
Agelaius
  phoeniceus, 79, 333, 343, 363
  tricolor, 343
Albatross, Galapagos, see *Diomedea irrorata*
Alcedinidae, 259, 284
Alcidae, 34, 35, 59
Alectoris rufa, 347
Alectura lathami, 85
Amblyospiza albifrons, 258–59
Ammodramus
  caudacutus, 333
  maritima, 333
Ammomanes, 265
Ampeliceps coronatus, 259

Amytornis textilis, 274–75
Anas platyrhynchos, 195, 333
Anatidae, 58, 89
Anatini, 60
Anhimidae, 58
Anser brachyrhynchus, 50, 51
Anseranas semipalmata, 345
Anseriformes, 35, 36, 40, 41, 58
Anserini, 60
Anthreptes, 265
Anthochaera carunculata, 264–65
Apapane, see *Himatione sanguinea*
Aphelocephala, 275
Aphelocoma coerulescens, 342
Aplonis
  cantoroides, 259
  spp., 212
Apodiformes, 35
Apterygidae, 58
Apteryx australis, 263
Aramidae, 59
Archaeopteryx, 105–15, 124
Ardea herodias, 280
Asio
  flammeus, 205
  otus, 205
Automolus rufipileatus, 259
Avocet, American see *Recurvirostra americana*
Aythya valisineria, 345

Bateleur, see *Terathopius ecaudatus*
Blackbird, see *Turdus merula*
Blackbird, Red-winged, see *Agelaius phoeniceus*
Blackbird, Tricolored, see *Agelaius tricolor*
Blackbird, Yellowheaded, see *Xanthocephalus xanthocephalus*
Blackbird, Brewer, see *Euphagus cyanocephalus*
Bluebird, see *Sialia*
Bluethroat, see *Luscinia svecica*
Bobolink, see *Dolichonyx oryzivorus*
Bombycillidae, 284
Brambling, see *Fringilla montifringilla*
Bubo virginianus, 195
Bunting
 Indigo, see *Passerina cyanea*
 Lark, see *Calamospiza melanocorys*
 Snow, see *Plectrophenax nivalis*
 Yellow, see *Emberiza citrinella*
Burhinidae, 59
Bush-Shrike, Black-headed, see *Laniarius barbarus*
Buteo
 buteo, 345
 jamaicensis, 345
Buzzard, see *Buteo buteo*

Cacomantis merulinus, 93
Calcarius lapponicus, 335, 339, 343
Cairinini, 60
Calamospiza melanocorys, 343, 347
Calidris
 alba, 346
 alpina, 347
 minuta, 345
 pusilla, 50
Canvasback, see *Aythya valisineria*
Caprimulgidae, 59
Caprimulgiformes, 35, 59
Carduelis tristis, 366
Cariamidae, 59
Carpodacus
 cassinii, 345
 mexicanus, 81, 344
Cassowary, see *Casuarius casuarius*
Casuariidae, 58
Casuarius casuarius, 72
Catamblyrhynchus, 283
Catbird, Gray, see *Dumetella carolinensis*

Catharacta lonnbergi, 345, 348
Centurus
 aurifrons, 137
 carolinus, 137
Cereopsis novaehollandiae, 50, 51
Certhia, 284
Certhionyx, 265
Ceryle alcyon, 259, 284
Charadriidae, 58
Chaffinch, see *Fringilla coelebs*
Charadriiformes, 40, 46, 47, 58, 59
Charadrius
 alexandrinus, 81, 345
 dubius, 345
 hiaticula, 279
 montanus, 347
Chat, Yellow-breasted, see *Icteria virens*
Chen caerulescens, 45
Chick, see *Gallus gallus*
Chicken, Lesser Prairie, see *Tympanuchus pallidicinctus*
Chionididae, 59
Chlamydera nuchalis, 268
Chloroceryle americana, 259, 284
Chondestes grammacus, 343
Ciconiiformes, 35, 59
Cinclidae, 284
Cinclodes excelsior, 259
Cinclodes, Stout-billed, see *Cinclodes excelsior*
Cincloramphus cruralis, 268
Cinclus mexicanus, 342
Cistothorus
 palustris, 342
 platensis, 339, 342
Colaptes
 auratus, 132, 134–37
 auratus X *cafer*, 134–37
Coliiformes, 35
Colinus virginianus, 195
Columba livia, 195
Columbiformes, 35
Conopophaga castaneiceps, 271, 273
Conopophila, 265
Contopus sordidulus, 342
Coot, American, see *Fulica atra*
Copsychus, 265
Coraciiformes, 35
Cormorant
 Great, see *Phalacrocorax carbo*

BIRD NAME INDEX

Cormorant (cont.)
    Little, see *Phalacrocorax niger*
Corvidae, 259
Corvus brachyrhynchos, 259, 342
Coturnix coturnix, 96, 294ff
Cowbird
    Bay-winged, see *Molothrus badius*
    Brown-headed, see *Molothrus ater*
Cracidae, 58
Crow, Common, see *Corvus brachyrhynchos*
Cuckoo, Plaintive, see *Cacomantis merulinus*
Cuculiformes, 35
Curlew, Eurasian, see *Numenius arquata*
Cyanocitta cristata, 259
Cygnus olor, 345

Daphoenositta
    chrysoptera, 268
    spp., 277
Delichon urbica, 334
Dendrocitta occipitalis, 259
Dendrocygninae, 60
Dendrocopos minor, 195
Dendroica
    coronata, 133, 212
    discolor, 343, 372
    kirtlandii, 343, 346
    palmarum, 339, 343
    petechia, 335, 337–38, 339, 342, 349
Dickcissel, see *Spiza americana*
Diomedea irrorata, 279–80
Diomedeidae, 59
Dipper, see *Cinclus mexicanus*
Dolichonyx oryzivorus, 343
Dotterel, see *Eudromius morinellus*
Dove, Ring, see *Streptopelia risoria*
Drepanidini, 274
Dromadidae, 59
Dromaius novaehollandiae, 93
Dromiceidae, 58
Drymodes brunneopygia, 268
Duck, Black-headed, see *Heteronetta*
Duck, Steamer, see *Tachyeres*
Duica, 283
Dumetella carolinensis, 335, 339, 348
Dunlin, see *Calidris alpina*

Eagle, Bald, see *Haliaeetus leucocephalus*

Emberiza
    citrinella, 336
    sp., 265
Empidonax virescens, 342
Emu, see *Dromaius novaehollandiae*
Eopsaltria
    albigularis, 180
    capito, 180
    leucops, 180
    nana, 180
Ephthianura,
    albifrons, 268
    sp., 275
Eudromias morinellus, 347
Eudyptes spp., 47, 52
Euphagus cyanocephalus, 344
Euplectes
    orix, 258–59
    sp., 265
Eurypigidae, 59

Fairy-Wren, Variegated, see *Malurus lamberti*
Falco
    columbarius, 345
    peregrinus, 345
    tinnunculus, 345
Falcon, Peregrine, see *Falco peregrinus*
Falconiformes, 35
Ficedula
    hypoleuca, 336
    sp., 265
Finch
    Cassin's, see *Carpodacus cassinii*
    House, see *Carpodacus mexicanus*
    Society, see *Lonchura domestica*
Flicker, see *Colaptes*
Flycatcher,
    Acadian, see *Empidonax virescens*
    Ash-throated, see *Myiarchus cinerascens*
    Great Crested, see *Myiarchus crinitus*
    Pied, see *Ficedula hypoleuca*
Fowl, domestic, see *Gallus gallus*
    Mallee, see *Leipoa ocellata*
Fringilla
    coelebs, 336
    montifringilla, 339
Fulica atra, 42, 85
Furnariidae, 259
Furnarius leucopus, 259

Galliformes, 35, 40, 43, 46, 47, 58
Gallirex porphyreolophus, 199
Gallus
　domesticus, see *Gallus gallus gallus*, 69, 195, 254, 270, 295ff
Gavia immer, 85
Gaviidae, 58
Gaviiformes, 35, 58
Geothlypis trichas, 343
Glareolidae, 59
Gnateater, Chestnut-crowned, see *Conopophaga castaneiceps*
Godwit, Black-tailed, see *Limosa limosa*
Goldfinch, American, see *Carduelis tristis*
Goose, Cape Barren, see *Cereopsis novaehollandiae*
　Lesser Snow, see *Chen caerulescens*
　Magpie, see *Anseranas semipalmata*
Grackle
　Boat-tailed, see *Quiscalus major*
　Common, see *Quiscalus quiscula*
Grallina cyanoleuca, 268
Grebe
　Clark's, see *Aechmophorus clarkii*
　Western, see *Aechmophorus occidentalis*
Greenshank, Common, see *Tringa nebularia*
Grosbeak, see *Pheucticus*
Grouse, Red, see *Lagopus lagopus*
Gruidae, 59, 47,
Gruiformes, 35
Gull
　California, see *Larus californicus*
　Herring, see *Larus argentatus*
Gymnorhina tibicen, 336

Haematopodidae, 59
Haematopus
　ostralegus, 41, 345
　unicolor, 196
Halcyon
　sancta, 259, 284
　senegalensis, 284
Haliaeetus leucocephalus, 345
Haplospiza, 283
Hawk
　Harris's, see *Parabuteo unicinctus*
　Sharp-shinned, see *Accipiter striatus*
　Sparrow, see *Accipiter nisus*
Heliornithidae, 59
Heron, Great Blue, see *Ardea herodias*

Heteronetta sp., 48
Himatione sanguinea, 274
Hirundo
　fulva, 332
　pyrrhonota, 341
　rustica, 86, 333, 342
　rustica X *fulva*, 333
Honeycreeper, Hawaiian, see Drepanidini
Hornero, Pale-legged, see *Furnarius leucopus*
Hydrobatidae, 59
Hylocistes subulatus, 259

Ibidorhynchidae, 58
Icteria virens, 343
Icterus
　bullockii, 132, 140–145
　galbula, 132, 140–145, 212

Jacanidae, 58
Jay
　Blue, see *Cyanocitta cristata*
　Scrub, see *Aphelocoma ceerulescens*
Junco, Dark-eyed, see *Junco hyemalis*
Junco hyemalis, 191–196, 198, 199, 357ff

Kestrel, European, see *Falco tinnunculus*
Kingfishers, see Alcedinidae
Kingfisher
　Belted, see *Ceryle alcyon*
　Common Paradise, see *Tanysiptera galatea*
　Green, see *Chloroceryle americana*
　Hook-billed, see *Melidora macrorhina*
　Sacred, see *Halcyon sancta*
Kiwi, Brown, see *Apteryx australis*

Lagopus
　lagopus, 43, 345
　leucurus, 345
　mutus, 345
Laniarius barbarus, 271, 273
Lanius
　collurio, 339
　ludovicianus, 342
Lapwings, see *Vanellus* spp.
Lari, 34, 35, 41
Laridae, 59
Larus
　argentatus, 44, 253, 266, 333, 345, 348

Larus (cont.)
 californicus, 46, 266, 333
 delawarensis, 44, 333, 345
 glaucescens, 266
 glaucoides, 266
 hyperboreus, 266
 occidentalis, 266
 ridibundus, 43
Leafgleaner
 Chestnut-crowned, see Automolus rufipileatus
 Striped, see Hylocistes subulatus
Leafscraper, Black-tailed, see Sclerurus caudacutus
Leipoa ocellata, 85
Lichmera, 265, 275
Limosa limosa, 345
Lonchura domestica, 96
Longspur, Lapland, see Calcarius lapponicus
Loon, Common, see Gavia immer
Luscinia svecica, 339, 342

Macrosphenus, 284
Magpie
 Australian, see Gymnorhina tibicen
 Black-billed, see Pica pica
 Yellow-billed, see Pica nuttallii
Mallard
 see Anas platyrhynchos
Maluridae, 274–276
Malurus
 alboscapulatus, 274
 amabilis, 180
 dulcis, 180
 lamberti, 268, 274, 275
 rogersi, 180
 splendens, 274
Martin, House, see Delichon urbica
 Purple, see Progne subis
Meadowlark
 Eastern, see Sturnella magna
 Western, see Sturnella neglecta
Megapodiidae, 58
Meleagrididae, 58
Melaenornis, 265
Melidora macrorhina, 259, 284
Meliphaga, 265
Meliphagidae, 265
Melopsittacus undulatus, 195

Melospiza
 georgiana, 343
 melodia, 266, 343, 372
Merganettini, 60
Mergini, 60
Merlin, see Falco columbarius
Mesitornithidae, 59
Microbates, 289
Mimus polyglottos, 342
Mockingbird, Northern, see Mimus polyglottos
Molothrus
 ater, 18, 363
 badius, 346
Motacilla
 alba, 342
 flava, 335, 342
 sp., 265
Motacillinae, 283
Myadestes townsendi, 273
Myiarchus
 cinerascens, 137
 crinitus, 137
Myna, Golden-crested, see Ampeliceps coronatus
Myzomela, 265

Nectarinia famosa, 335
Numenius arquata, 345
Numididae, 58
Nyctea scandiaca, 205

Oceanodroma leucorhoa, 277, 280
 spp., 277
Oenanthe oenanthe, 342
Onychognathus morio, 259
Opisthocomidae, 35
Oporornis
 philadelphia, 133
 tolmei, 133
Oriole
 Baltimore, see Icterus galbula
 Bullock's, see Icterus bullockii and Icterus galbula
Otidae, 59
Osprey, see Pandion haliaetus
Otus asio, 7, 8, 10–12, 18, 134
Ovenbird, see Seiurus aurocapillus
Owl
 Great Horned, see Bubo virginianus

Owl (cont.)
  Long-eared, see Asio otus
  Screech, see Otus asio
  Short-eared, see Asio flammeus
  Snowy, see Nyctea scandiaca
Oystercatcher
  European, see Haematopus ostralegus
  Variable, see Haematopus unicolor
Oxyurini, 60

Pachycephala pectoralis, 268, 277
Pachyptila vittata, 279, 280
Palaeognathiformes, 35
Pandion haliaetus, 345
Parabuteo unicinctus, 332
Paradisaea minor, 268
Parakeet, Shell, see Melopsittacus undulatus
Partridge, Red-legged, see Alectoris rufa
Parula americana, 342
Parus
  atricapillus, 138, 342
  atricristatus, 139
  bicolor, 132, 139
  carolinensis, 138
  inornatus, 139
  spp., 265
Passer, 265
Passerculus sandwichensis, 336, 343
Passerella iliaca, 199
Passeriformes, 35
Passerina
  amoena, 132, 146, 148
  cyanea, 132, 146, 148, 334, 342
Pedionomidae, 59
Pelecanoididae, 59
Pelecaniformes, 35
Penguin, Adelie, see Pygoscelis adeliae
Petrel
  Antarctic, see Thalassoica antarctica
  Bonin, see Pterodroma hypoleuca
  Herald, see Pterodroma arminjoniana
  Leach's, see Oceanodroma leucorhoa
Pewee, Western Wood, see Contopus sordidulus
Phainopepla, see Phainopepla nitens
Phainopepla nitens, 81
Phalarope, Wilson's, see Phalaropus tricolor
Phalaropus tricolor, 347
Phoebe, Eastern, see Sayornis phoebe
Phasianidae, 58
Phasianus colchicus, 195

Pheasant, Ring-necked, see Phasianus colchicus
Phalacrocorax
  carbo, 334
  niger, 195
Pheucticus
  ludovicianus, 132, 145, 146
  melanocephalus, 132, 145, 146
Phoenicopteridae, 35, 59
Phylidonyris, 275
Pica
  nuttallii, 336
  pica, 259, 335, 338, 339
Piciformes, 35
Pigeon, Rock, see Columba livia
Pipilo erythopthalmus, 132, 148–150, 343
Plectrophenax nivalis, 335, 343
Plectrorhyncha, 265
Ploceidae, 259
Ploceus
  capensis, 258–259
  cucullatus, 258–259
Plover
  Egyptian, see Pluvianus aegyptius
  Greater Golden, see Pluvialis apricaria
  Little Ringed, see Charadrius dubius
  Gray, see Pluvialis squatarola
  Mountain, see Charadrius montanus
  Ringed, see Charadrius hiaticula
  Snowy, see Charadrius alexandrinus
Pluvianus aegyptius, 85
Pluvialis
  apricaria, 345
  squatarola, 280
Pluvianellus, 59
Podiceps cristatus, 85
Podicipedidae, 59
Podicipediformes, 35, 41, 59
Podilymbus podiceps, 85
Poephila
  acuticauda, 180
  atropygialis, 180
  cincta, 180
  guttata, 71
  hecki, 180
  leucotis, 180
  personata, 180
Polioptila, 284
Prion, Broad-billed, see Pachyptila vittata
"Proavis", 108
Procellariidae, 59

BIRD NAME INDEX

Procellariiformes, 34, 35, 41, 52, 59, 212, 279, 280
Progne
   sp., 265
   subis, 342
Prosthemadura, 265
Protonotaria citrea, 342
Prunella, 265
Prunellinae, 283
Psittaciformes, 35
Psophidae, 59
Ptarmigan
   Rock, see *Lagopus mutus*
   White-tailed, see *Lagopus leucurus*
   Willow, see *Lagopus lagopus*
Pterodroma
   arminjoniana, 279, 280
   hypoleuca, 86
Ptilonorhynchus violaceus, 268
Ptiloris
   alberti, 180
   intercedens, 180
   magnificens, 180
   paradiseus, 180, 268
   victoriae, 180
Puffinus
   griseus, 279, 280
   pacificus, 77
Pycnonotus, 265
Pygoscelis adeliae, 18

Quail
   Bobwhite, see *Colinus virginianus*
   Coturnix, see *Coturnix coturnix*
Quelea quelea, 258–259
Quiscalus
   major, 344
   quiscula, 72, 344, 363

Rallidae, 40, 41, 59
Ramphocoenus, 284
Ramsayornis, 265
Ratites, 40, 58
Recurvirostra americana, 345
Recurvirostridae, 58
Redshank, Spotted, see *Tringa erythropus*
Redstart
   American, see *Setophaga ruticilla*
   Painted, see *Setophaga picta*
Rheidae, 58
Rhynchopidae, 59

Rhynochetidae, 59
Rifleman, New Zealand, see *Acanthisitta chloris*
Riparia riparia, 86, 342
Rissa tridactyla, 54
Robin, American, see *Turdus migratorius*
Rostratulidae, 59

Sanderling, see *Calidris alba*
Sandpiper
   Semipalmated, see *Calidris pusilla*
   Spotted, see *Tringa macularia*
Sapsucker, see *Sphyrapicus*
Sayornis phoebe, 342
Scolopacidae, 40, 58
Sclerurus caudacutus, 259
Seabirds, 21, 22
Seedeater, Hicks', see *Sporophila aurita*
Seiurus
   aurocapillus, 336, 343, 346
   noveboracensis, 343
Sericornis
   frontalis, 268
   sp., 275
Setophaga
   picta, 343
   ruticilla, 336, 341
Shearwater
   Sooty, see *Puffinus griseus*
   Wedge-tailed, see *Puffinus pacificus*
Shikra, see *Accipiter badius*
Shrike
   Loggerhead, see *Lanius ludovicianus*
   Red-backed, see *Lanius collurio*
Sialia
   currucoides, 130, 140, 335
   sialis, 130, 140, 333, 342, 346
Sicalis, 283
Silvereye, see *Zosterops lateralis*
Sitta spp., 277
Skua, Brown, see *Catharacta lonnbergi*
Solitaire, Townsend's, see *Myadestes townsendi*
Sparrow
   Chipping, see *Spizella passerina*
   Fox, see *Passerella iliaca*
   Lark, see *Chondestes grammacus*
   Savannah, see *Passerculus sandwichensis*
   Seaside, see *Ammodramus maritimus*
   Sharp-tailed, see *Ammodramus caudacutus*

Sparrow (cont.)
    Song, see Melospiza melodia
    Swamp, see Melospiza georgiana
    Tree, see Spizella arborea
    White-crowned, see Zonotrichia leucophrys
    White-throated, see Zonotrichia albicollis
Sphenisciformes, 35, 41, 58, 89
Sphyrapicus varius, 133
Spiza americana, 343
Sporophila aurita, 346
Spizella
    arborea, 206, 207, 350, 363
    passerina, 334, 343
Spreo bicolor, 259
Stachyris, 265
Starling, Common, see Sturnus vulgaris
    Pied, see Spreo bicolor
    Red-winged, see Onychognathus morio
    Singing, see Aplonis cantoroides
Stercorariidae, 59
Sterna
    hirundo, 44, 46
    paradisaea, 44
Stint, Little, see Calidris minuta
Stipiturus malachurus, 274
Streptopelia risoria, 195, 355
Strigiformes, 35
Sturnella
    magna, 131, 140, 343
    neglecta, 131, 140, 343
Sturnidae, 259
Sturnus
    sp., 265
    vulgaris, 259, 335, 363
Struthionidae, 58
Struthioniformes, 58
Sunbird, Malachite, see Nectarinia famosa
Swallow, Cliff, see Hirundo pyrrhonota
    Bank, see Riparia riparia
    Barn, see Hirundo rustica
    Cave, see Hirundo fulva
    hybrids, see Hirundo rustica X fulva
    Tree, see Tachycineta bicolor
    Violet-green, see Tachycineta thalassina
Swan, Mute, see Cygnus olor
Sylviidae, 284

Tachycineta
    bicolor, 342
    thalassina, 342

Tachyeres, 215
Tachyerini, 60
Tadornini, 60
Tanager, spp., 283
Tanysiptera galatea, 259, 284
Teratopius ecaudatus, 345
Tersina, 283
Tetraonidae, 58
Thalassoica antarctica, 279, 280
Thinocoridae, 58
Thryothorus, 265
Titmouse, see Parus
Tinamidae, 58
Tinamiformes, 58
Towhee, see Pipilo
Toxostoma, 265
Thornbill, Yellow-rumped, see Acanthiza chrysorrhoa
Treepie, Malaysian, see Dendrocitta occipitalis
Tringa
    erythropus, 347
    flavipes, 345
    macularia, 332, 371
    nebularia, 334, 345
Troglodytes
    aedon, 336, 342
    troglodytes, 331, 342
Troglodytidae, 284
Trogoniformes, 35
Turdoides, 265
Turdus
    merula, 339
    migratorius, 271, 273, 342
    sp., 265
Turkey, Brush, see Alectura lathami
Turnicidae, 59
Twite, see Acanthis flavirostris
Tympanuchus pallidicintus, 215

Vanellus
    armatus, 49
    chilensis, 49, 345
    crassirostris, 49
    vanellus, 345
Vireoninae, 277
Volatinia, 283

Wagtail
    Yellow, see Motacilla flava
    White, see Motacilla alba

BIRD NAME INDEX

Warbler
    Blackpoll, see *Dendroica striata*
    Kirtland's, see *Dendroica kirtlandii*
    MacGillivray's, see *Oporornis tolmei*
    Mourning, see *Oporornis philadelphia*
    Palm, see *Dendroica palmarum*
    Parula, see *Parula americana*
    Prairie, see *Dendroica discolor*
    Prothonotary, see *Protonotaria citrea*
    Wilson's, see *Wilsonia pusilla*
    Yellow, see *Dendroica petechia*
    Yellow-rumped, see *Dendroica coronata*
Waterthrush, Northern, see *Seiurus noveboracensis*
Weaver
    Cape, see *Ploceus capensis*
    Red-billed, see *Quelea quelea*
    Thick-billed, see *Amblyospiza albifrons*
    Village, see *Ploceus cucullatus*
Wheatear, see *Oenanthe oenanthe*
Wilsonia pusilla, 339, 342
Wren
    Fairy, see Maluridae

Wren (cont.)
    House, see *Troglodytes aedon*
    Marsh, see *Cistothorus palustris*
    New Zealand, see Acanthisittidae
    Sedge, see *Cistothorus platensis*
    Winter, see *Troglodytes troglodytes*
Wrens, see Troglodytidae
Woodpecker
    Golden-fronted, see *Centurus aurifrons*
    Lesser Spotted, see *Dendrocopos minor*
    Red-bellied, see *Centurus carolinus*

Xanthocephalus xanthocephalus, 344

Yellowlegs, Lesser, see *Tringa flavipes*
Yellowthroat, see *Geothlypis trichas*

Zonotrichia
    albicollis, 198
    capensis, 212
    leucophrys, 198, 212, 240, 343, 363, 366, 371
Zosterops lateralis, 367

SUBJECT INDEX

Air sacs, 124
Allometry, 71, 97
Avian development
   classified, 35
   in gaseous environments, 85
   and ecologic factors, 36
   and nest microclimate, 80
   phylogenetic factors, 36
   rates, 36
Avian egg
   allometry of, 71, 97
   contents of, 71, 73–74
   gas exchange of, 75–6, 81–90
   mass, 71–73
   nest microclimate, 80
   review of, 69–97
   shell formation, 90
   shell pores, 93–5
   shell structure, 87–90
   vapor pressure, 80–81
   water exchange, 76
   water loss, 77–80

Base sequences, 247ff
Bet-hedging, 25, 39
Biochemical systematics
   data acquisition, 212–24
   and DNA annealing, 246ff
   electrophoresis and, 213–14
   and genetic distance, 220–221
   and genetic heterozygosity, 214

Biochemical systematics (cont.)
   and genotypic frequencies, 214
   and F statistics, 223
Biological species concept
   definition, 161
   and classificatory convenience, 162
   and differentiated taxonomic units, 165
   and generic limits, 229–237
   and genetic structure, 237
   and hybridization, 155
   and units of evolution, 162
Biogeography
   dispersalist, 182, 183
   vicariance, 183
Body size
   and clutch size, 45
   and migration, 362, 365, 367
Brood size
   and clutch size, 40, 47, 54, 57
   and parental behavior, 49–51

Calcium carbonate
   in eggshell, 88
   of medullary bone, 91
   transport by shell gland, 91
Calcium phosphate, 255
Carbon dioxide
   diffusion through eggshell, 81
   production during incubation, 82
   variation in embryonic production, 82–83

Carbonic anhydrase, 91
Chromosomes
    review of, 189–208
    C-banding, 193–197
    culturing techniques, 190–191
    and differential banding procedures, 193–208
    diploid number problem, 190
    G-banding, 197–205
    and meiotic tissues, 193
    NOR-banding, 206
    R-banding, 205–206
    sequential banding, 206–207
Cladogram, general area, 181ff
Classification
    and DNA hybridization data, 288
    and evolutionary differentiation, 286
    and genetic similarity, 287
    and overall similarity, 287
    and synapomorphies, 286
Clock, molecular, 277, 279, 280
Clutch size
    and adult survival, 3, 38
    and body weight, 45
    and brood patch area, 38
    and competition, 38
    and courtship feeding, 46
    and daylength
    in determinate layers, 46
    and egg formation, 42
    and environmental variation, 4, 38
    and feeding young, 40
    and food, 3, 4, 38
    and incubation, 50
    in indeterminate layers, 46
    and latitudinal variation, 4, 44, 53
    and life histories, 3, 4, 38
    limits, 41
    models, 3
    multiple factor control, 54, 55
    and parental effort, 47, 57
    and polygyny, 56
    and population density, 4
    of precocial birds, 58–59
    and predation, 3, 4, 38, 52
    and reproductive rate, 4
    review of, 33–61
    and seasonal variation, 53, 54
    variation in general, 3, 33, 55
    of waterfowl, 60

Demography
    comparative, 5, 15
    defined, 1
    experimental, 5
    properties of, 4, 26–28
    and reproductive effort, 37
Density dependence
    and latitudinal gradients, 21
    and optimization, 20
Developmental regulation, 298, 303
    and polar coordinate theory, 310–313
    of somites, 313–315
    and zone of polarizing activity, 303–310
Diffusion
    Fick's first law of, 76
DNA annealing, see DNA–DNA hybridization
DNA haploid genome content, 255
DNA delta values,
    and relative time, 275
DNA–DNA hybridization, 246
    and adaptive radiation, 282
    and experimental error, 273
    and genetic distance, 260
    and heterologous hybrids, 256
    and hierarchic clustering, 201
    and homologous hybrids, 256
    and monophyly, 201
    and morphologic congruence, 278
    and rate of DNA change, 275
    reassociation rate, 257
    and salt concentration, 232
    and sequence organization, 247
    and thermal environment, 247, 251
DNA evolution
    and calibration, 281
    and relative rate test, 272
    uniform average rate of, 270
Differentiation
    clinal, 132
    by geographic variation, 132
    and secondary contact, 132
    and taxonomic status, 163

Ectoderm, 295
Egg contents
    and developmental mode, 74
    and evolution of altricial young, 75
    of pipped eggs, 74–75

SUBJECT INDEX

Egg contents (*cont.*)
  plumping of, 91
  variation in, 73, 77–80
Egg mass
  and calories, 72
  and embryonic supplies, 72
  and fledging success, 72, 73
  as proportion of adult size, 72
Eggshell, *see* Avian egg
Egg size, *see* Egg mass
Evolution
  and age of origin, 285
  convergent, 283, 284
  developmental basis of, 297
  and differentiation, 286
  of DNA, 270, 272, 281
  and DNA delta values, 275
  mosaic, 106
  and origin of birds, 105–126
  and origin of flight, 121–126
  and taxonomy, 168
  units of, 163
Extrapair copulation, 232–236

Fecundity
  and delayed reproduction, 3
  and specialization, 20
  tradeoff with survival, 3, 12, 14, 16, 38
Fitness
  criterion for, 6
  evolutionary increase in, 8
  geometric mean of, 7
  and life histories, 5, 6
  and overlapping generations, 7
  and population density, 19
  and resource allocation, 4
Flight, 121–125
Flight feathers
  asymmetry in *Archaeopteryx*, 109

Genetic differentiation, 269
Genetic distance
  estimating, 220–221
  and F_{st}, 226
  and genetic identity, 222
  and genetic revolutions, 225–227
Genetic heterozygosity, 214–220
Genetic structure,
  and avian systematics, 211–242

Genetic structure (*cont.*)
  and F statistics, 223
  and higher taxa, 227–237
  and objectivity of genera, 229
  at the species level, 237–239
  within species, 239–241
Great Plains, delimited, 133

Homology, 267
Hybridization
  of DNA
    see DNA–DNA hybridization
  and Hybrid Index, 136
  vs. intergradation, 133
  and neutral clines, 151
  stability of zones, 151
  and step-clines, 132
  and suture-zones, 132, 153–155
Hydroxyapatite, *see* Calcium phosphate
Hypothesis
  Arrival-Time, 370–375
  Body-Size, 363–365
  Dominance, 365–370
  Migration-Threshold, 375–387
  Red Queen's, 277
  Saunders–Zwilling, 295, 320

Introns, 251

K-selection, 4, 23, 24, 39

Life histories
  as correlated character sets, 25, 39
  and density dependence, 18
  differences in, 15
  and fitness, 5, 6
  optima, 9
Life table, 2
Limb bud development, 293–321
  and embryonic innervation, 316–320

Mate fidelity, 329–351
Mating systems, 329ff
Mesoderm, 295
Migration, 357–399
  and age ratio, 367–369
  and arrival time, 370–375
  and body size, 362–365, 367
  and dominance, 365–370
  evolution of, 362

Migration (*cont.*)
   and migration thresholds, 375–387
   and selective factors, 389
   and social behavior, 365, 367
   and winter distribution, 360–362
Monogamy, 329–331
   and contrast with polygamy, 330
Morphogenesis
   and apical ectoderm, 295
   and zone of polarizing activity, 295

Nucleotide sequence, 247ff

Origin of birds
   and coelurosaurians, 115–121, 125
   crocodilian relationships, 111–115, 125
   and origin of flight, 121
   and ornithischians, 110
   and pseudosuchians, 110
Ornithischians, *see* Origin of birds
Oxygen
   consumption in embryos, 84
   diffusion through eggshell, 81

Parental behavior
   and brood size, 49, 50, 51
   and clutch size, 47, 50
Phylogeny, 245
   and classification, 245–288
Polyandry, 330–331, 346–347
Polygamy, 330–331
Polygyny, 331
   facultative, 340
   occurrence in non-passerines, 345–346
   occurrence in passerines, 342–344
   opportunistic, 340
Population growth
   and age structure, 24
   parameters, 381–387
Precocial birds, 34–35
Pseudosuchians, 110

Reproductive effort
   definition, 36
   and demography and environment, 37
   and parental investment, 36
r-selection, 4, 23, 39
Resource allocation, 4

Sex ratio, 358, 359, 367–369
Shell gland, 91
Skeleton
   forelimb, 293ff
   innervation of, 319
   pneumaticity of, 124
   prospective map of, 298
Speciation analysis
   and area of endemism, 177–178
   and the biological species concept, 175
   and genetic distance, 225
   and reproductive isolation, 175
Species concepts, 159–184
   classificatory vs. evolutionary, 160
   morphological, 162
   ornithological, 165
   polytypic-biological, 161
   phylogenetic, 169–174
   and reproductive isolation, 162, 171–172
   taxonomic, 163
Stylopod, 300
Subspecies
   and allopatric populations, 168
   and degree of differentiation, 168
   as demes, 239
   and evolutionary rates, 168
   and geographic variation, 166
   as irrelevant, 171
   as subjective partitions, 167
   as taxonomic units, 166, 168
   as units of evolution, 164–165, 168
   and zones of intergradation, 241
Successive monogamy, *see* Monogamy
Survival, 2
   and fecundity, 12, 14, 16, 38
   and specialization, 20
Suture-zones, *see* Hybridization
Synapomorphy, 286
Systematics
   definition, 211
   and genetic structure, 211

Taxonomy
   and evolutionary rate, 168
   equivalence of categories, 285
   and geographic variation, 166
   and ontology of subspecies, 167
   traditional, 166

SUBJECT INDEX

Territoriality, 336–340
 and migration patterns, 370–375, 389
Transposons, 251

Vapor pressure, 80–81

Zeugopod, 300
Zone of polarizing activity,
 definition, 308
 and limb bud development, 303–310
 vs. polar coordinate model, 310
 in vitro identification, 305